Mechanisms of Soil Stabilization

Mechanisms of Soil Stabilization

Contributors

Ali Ateş et al.

www.aurisreference.com

Mechanisms of Soil Stabilization

Contributors: Ali Ateş et al.

Published by Auris Reference Limited
www.aurisreference.com

United Kingdom

Copyright 2016
Printed in 2017 for Sale in the Indian Subcontinent

The information in this book has been obtained from highly regarded resources. The copyrights for individual articles remain with the authors, as indicated. All chapters are distributed under the terms of the Creative Commons Attribution License, which permit unrestricted use, distribution, and reproduction in any medium, provided the original author and source are credited.

Notice

Contributors, whose names have been given on the book cover, are not associated with the Publisher. The editors and the Publisher have attempted to trace the copyright holders of all material reproduced in this publication and apologise to copyright holders if permission has not been obtained. If any copyright holder has not been acknowledged, please write to us so we may rectify.

Reasonable efforts have been made to publish reliable data. The views articulated in the chapters are those of the individual contributors, and not necessarily those of the editors or the Publisher. Editors and/or the Publisher are not responsible for the accuracy of the information in the published chapters or consequences from their use. The Publisher accepts no responsibility for any damage or grievance to individual(s) or property arising out of the use of any material(s), instruction(s), methods or thoughts in the book.

Mechanisms of Soil Stabilization

ISBN: 978-1-78154-974-2

British Library Cataloguing in Publication Data
A CIP record for this book is available from the British Library

Printed in the United Kingdom

Exclusively distributed by CBS Publishers & Distributors Pvt. Ltd.

Sales & Distribution Rights only for India, Pakistan, Bangladesh, Sri Lanka, Nepal and Bhutan. This book is not to be sold outside these territories.

Contents

	List of Abbreviations vii
	List of Contributors.. ix
	Preface.. xiii
Chapter 1	The Effect of Polymer-Cement Stabilization on the Unconfined Compressive Strength of Liquefiable Soils... 1
Chapter 2	Stabilization of Organic Matter by Biochar Application in Compost-amended Soils with Contrasting pH Values and Textures...... 19
Chapter 3	Behavior of Clayey Soil Stabilized with Rice Husk Ash & Lime 39
Chapter 4	Cement Kiln Dust Chemical Stabilization of Expansive Soil Exposed at El-Kawther Quarter, Sohag Region, Egypt....................... 51
Chapter 5	Phosphorus: Chemism and Interactions 69
Chapter 6	Poultry Litter Fertilization Impacts on Soil, Plant, and Water Characteristics in Loblolly Pine (Pinus taeda L.) Plantations and Silvopastures in the Mid-South USA.. 103
Chapter 7	Physiological and Biochemical Mechanisms of Plant Adaptation to Low-Fertility Acid Soils of the Tropics: The Case of Brachiariagrasses.. 141
Chapter 8	Soil Indicators of Hillslope Hydrology 179
Chapter 9	Updated Brazilian's Georeferenced Soil Database – An Improvement for International Scientific Information Exchanging.. 217
Chapter 10	An Application Approach to Kalman Filter and CT Scanners for Soil Science... 247
	Citations.. 287
	Index... 289

List of Abbreviations

BMPs	Best management practices
CASH	Calcium aluminosilicate hydrates
CAH	Calcium aluminum hydrate
CSH	Calcium silicate hydrate
CBR	California Bearing Ratio
CEC	Cation exchange capacity
CKD	Cement kiln dust
DUL	Drained Upper Limit
EC	Electrical conductivity
HOST	Hydrology of Soil Types
LSD	Least significant difference
OMC	Optimum moisture content
PWP	Permanent Wilting Point
PEPC	Phosphoenolpyruvate carboxylase
PPT	Phosphoenolpyruvate/phosphate translocator
PAW	Plant Available Water
PI	Plasticity index
PK	Pyruvate kinase
ROC	Readily-oxidizable carbon
RHA	Rice Husk Ash
SEM	Scanning electron microscope
SDS	Sodium dodecyl sulfate
SOM	Soil organic matter
SPS	Sucrose phosphate synthase
SQDG	Sulfoquinovosyl diacylglycerol
TPT	Triose phosphate translocater
TPT	Triose-phosphate translocator
UPV	Ultrasonic pulse velocity
UCS	Unconfined compressive strength
VM	Volatile matter

List of Contributors

Ali Ateş
Geotechnical Division, Department of Civil Engineering, Technology Faculty, Düzce University, 81620 Düzce, Turkey

Shih-Hao Jien
Department of Soil and Water Conservation, National Pingtung University of Science and Technology, Pingtung 91201, Taiwan

Chung-Chi Wang
Department of Soil and Water Conservation, National Pingtung University of Science and Technology, Pingtung 91201, Taiwan

Chia-Hsing Lee
Department of Agricultural Chemistry, National Taiwan University, Taipei 10617, Taiwan

Tsung-Yu Lee
Department of Geography, National Taiwan Normal University, Taipei 10610, Taiwan

B.Suneel Kumar
Graduate Student, Department of Civil Engineering, Geotechnical Engineering, SRM University, Kattankulathur-60320, Tamil Nadu, India

T.V.Preethi
Assistant professor, Department of Civil Engineering, SRM University, Kattankulathur-603203, Tamil Nadu, India

Hesham A. H. Ismaiel
Geology Department, Faculty of Science, South Valley University, Qena, Egypt

E. Saljnikov
Institute of Soil Science Serbia

D. Cakmak
Institute of Soil Science Serbia

Michael A. Blazier
Louisiana State University Agricultural Center USA

Hal O. Liechty
University of Arkansas Monticello USA

Lewis A. Gaston
Louisiana State University Agricultural Center USA

Keith Ellum
University of Arkansas Monticello USA

T. Watanabe
Graduate School of Agriculture, Hokkaido University, Kita-ku, Sapporo, Japan

M. S. H. Khan
Department of Soil Science, HMD Science and Technology University, Dinajipur, Bangladesh

I. M. Rao
Centro Internacional de Agricultura Tropical (CIAT), A.A.6713, Cali, Colombia

J. Wasaki
Graduate School of Biosphere Science, Hiroshima University, Higashi-Hiroshima, Japan

T. Shinano
National Agricultural Research Center for Hokkaido Region, Sapporo, Japan

M. Ishitani
Centro Internacional de Agricultura Tropical (CIAT), A.A.6713, Cali, Colombia

H. Koyama
Faculty of Applied Biological Sciences, Gifu University, Gifu, Japan

S. Ishikawa
National Institute for Agro-Environmental Science, Tsukuba, Japan

K. Tawaraya
Faculty of Agriculture, Yamagata University, Tsuruoka, Japan

M. Nanamori
Graduate School of Agriculture, Hokkaido University, Kita-ku, Sapporo, Japan

N. Ueki
Faculty of Agriculture, Yamagata University, Tsuruoka, Japan

T. Wagatsuma
Faculty of Agriculture, Yamagata University, Tsuruoka, Japan

Johan van Tol, Pieter Le Roux
Department Soil, Crop and Climate Sciences, University of the Free State South Africa

Malcolm Hensley
Department Soil, Crop and Climate Sciences, University of the Free State South Africa

Marcelo Muniz Benedetti
Centro de Tecnologia Canavieira

Nilton Curi
Universidade Federal de Lavras

Gerd Sparovek
Escola Superior de Agricultura "Luiz de Queiroz" /USP

Amaury de Carvalho Filho
Empresa Brasileira de Pesquisa Agropecuária Brazil

Sérgio Henrique Godinho Silva
Universidade Federal de Lavras

Marcos A. M. Laia
Embrapa Instrumentation
Physics Institute of São Carlos, University of São Paulo Brazil

Paulo E. Cruvinel
Embrapa Instrumentation
Physics Institute of São Carlos, University of São Paulo Brazil

J. Shamshuddin
Department of Land Management, Faculty of Agriculture, Universiti Putra Malaysia, 43400 Serdang, Selangor

Noordin Wan Daud
Department of Crop Science, Faculty of Agriculture, Universiti Putra Malaysia, 43400 Serdang, Selangor Malaysia

Ahmed Amdihun
Addis Ababa University, EiABC, Chair of CAD and Geo-informatics

Ephrem Gebremariam
Addis Ababa University, EiABC, Chair of CAD and Geo-informatics

Lisa-Maria Rebelo
International Water Management Institute (IWMI)

Gete Zeleke
Land and Water Resource Center

Milad Mirzaei Aminiyan
Soil Science Department, College of Agriculture, Bu-Ali Sina University

Ali Akbar Safari Sinegani
Soil Science Department, College of Agriculture, Bu-Ali Sina University

Mohsen Sheklabadi
Soil Science Department, College of Agriculture, Bu-Ali Sina University

Preface

Soil stabilization is a group of earthwork technologies to improve soil characteristics during the construction process and to improve mechanical and load bearing properties. The text *Mechanisms of Soil Stabilization* summarizes current knowledge on soil organic matter (SOM) dynamics and stabilization and to synthesize a conceptual SOM model based on physicochemically defined SOM pools. In first chapter, a laboratory experiment has been conducted to evaluate the effects of waterborne polymer on unconfined compression strength and to study the effect of cement grout on pre-venting of liquefiable sandy soils. Second chapter conducts a short-term incubation experiment to assess the effects of biochar application on the decomposition of added bagasse compost in three rural soils with different pH values and textures. The behavior of clayey soil stabilized with rice husk ash and lime has been discussed in third chapter. Fourth chapter deals with a chemical stabilization of an expansive high plastic soil of Pliocene deposits exposed at El-Kawther quarter using cement kiln dust (CKD) and cement kiln dust with lime (L) to reduce their swelling and improve their geotechnical properties. Fifth chapter focuses on mechanisms and distribution of different forms of phosphorus, its transformation and dynamics in the soil. Sixth chapter discusses on changes in soil nutrition, physical properties and microbes, tree nutrition and growth, and water nutrient contents in loblolly pine plantations and silvopastures in response to fertilization with conventional fertilizer and poultry litter. Seventh chapter reviews the progress made in defining the physiological and biochemical mechanisms of adaptation of brachiariagrasses to low fertility acid soils. Eighth chapter deals with soil indicators of hillslope hydrology. The database used in ninth chapter reveals an adequate representativeness of the soil classes distributed over the Brazilian territory. An application approach to Kalman filter and computerized tomography (CT) scanners for soil science has been presented in last chapter.

Chapter 1

THE EFFECT OF POLYMER-CEMENT STABILIZATION ON THE UNCONFINED COMPRESSIVE STRENGTH OF LIQUEFIABLE SOILS

Ali Ateş

Geotechnical Division, Department of Civil Engineering, Technology Faculty, Düzce University, 81620 Düzce, Turkey

ABSTRACT

Soil stabilization has been widely used as an alternative to substitute the lack of suitable material on site. The use of nontraditional chemical stabilizers in soil improvement is growing daily. In this study a laboratory experiment was conducted to evaluate the effects of waterborne polymer on unconfined compression strength and to study the effect of cement grout on pre-venting of liquefiable sandy soils. The laboratory tests were performed including grain size of sandy soil, unit weight, ultrasonic pulse velocity, and unconfined compressive strength test. The sand and various amounts of polymer (1%, 2%, 3%, and 4%) and cement (10%, 20%, 30%, and 40%) were mixed with all of them into dough using mechanical kneader in laboratory conditions. Grouting experiment is performed with a cylindrical mould of 150 × 300 mm. The samples were subjected to unconfined compression tests to determine their strength after 7 and 14 days of curing. The results of the tests indicated that the waterborne polymer significantly improved the unconfined compression strength of sandy soils which have susceptibility of liquefaction.

INTRODUCTION

Soil stabilization refers to produce in which a special soil, a cementing material, or other chemical materials are added to the liquefiable sandy soils to improve one or more of their properties. There are two methods to enhance the properties of sandy soils, one of them is the mechan-ical stabilization which is mixed the natural soil and stabilizing material together for ob-taining a homogeneous mixture and the second one is adding stabilizing material into un-disturbed soils to obtain interaction by letting it permeate through soil voids [1, 2]. Chemical stabilization is the modification of properties of a locally available soil to improve its engineering performance. The two most commonly used chemical stabilization methods are lime stabilization and cement stabilization. Additives such as fly ash and phosphogypsum can be added to the lime soil and cement soil mixtures to enhance the properties of the stabilized soil. The use of polymeric materials in setting soil improvement is growing daily. Unfortunately, little research has been completed to distinguish between products that deliver enhanced performance and those that do not. The nature of soil stabilization dictates that products may provide soil-specific properties and/or provide compatibility with environment. In other words, some products may work well in specific soil types in a given environment but perform poorly when applied to dissimilar materials in a different environment. Application of stabilizing agent on soils has a long history [3]. Cement was first used as stabilizing agent at the beginning of the 20th century to mix with the soils and form road material in the United States. Since then many other materials such as lime [4, 5], organic polymers [6], and their mixtures [7] have been used as stabilizing agents. Also several researchers [8–10] have discussed aqueous polymer applications while others [11–13] have provided useful data on polymer-soil interactions that determine the effectiveness of polymer solution in various applications. The objective of this research is to evaluate the effect of waterborne polymer and cement on unconfined compressive strength of liquefiable sandy soils.

MATERIALS

The Properties of Cement

The cement used for the study is Portland cement labelled CEMI 42.5 R and the properties of which are given in Table 1.

Table 1: Properties of the cement

Properties	Values
Grade	42.5 R
Specific gravity	3.17
Fitness (%)	3750
Consistency (%)	1.3
Initial and final setting time (min)	200–260

Epoxy Resin

A commercial product of epoxy resin was used, which is an emulsion synthetic elastic chemical substance that increases the bound with the substrate as additive in optimum moisture, as well as the cohesion and the strength. A multi component resin grout usually provides very high-tensile compression and bond strengths. Some important properties were given in Table 2.

Table 2: Important physicochemical properties of as-received emulsion

Name	Vinyl Acrylic-Copolymer watered solution
Physical state	Liquid-white colour
Solvability in water	Solution
Boiling point	100°
Viscosity	280
Sparkling point	Over 600
Non-self-burning	Nonexplosive
Temperature	Over 3200
Density (g/cm^3)	1.11 (20°)
PH	4–7.5

The Properties of Soil Sample

Locally available sandy soil was utilized for this study and was obtained from source of sand in Duzce City in Turkey. It is necessary for the samples to be stabilized to lie in the interval of upper and lower borders on the sieve analysis graph determined by Union of Japanese Civil Engineers. For this purpose, the gradation of samples was prepared on the upper and lower borders of sieve analysis graph to be applied the emulsion according to the Union of Japanese Civil Engineers. The samples were sieved by the number of the 8 mm, 4.76 mm, 2 mm, 1 mm, and 0.425 mm and the gradation was supplied mixing the grains remained on those sieves [14, 15].

The grain-size distribution of the soils was presented in Figure 1. The properties of the tested soils in term of particle size and sandy soil parameters were given in Table 3.

Table 3: Engineering properties of sandy soils

Property	Sample
Specific gravity	2.73
Grain size	
Gravel (>20 mm) (%)	0
Sand (75–20 mm) (%)	100
Max. void ratio (e_{max})	0.77
Min. void ratio (e_{min})	0.45
Relative density, D_r (%)	45
Void ratio, e (%)	1.024
Optimum moisture content (%)	13
Maximum dry unit weight (g/cm³)	1.92
Soil classification (USCS)	S

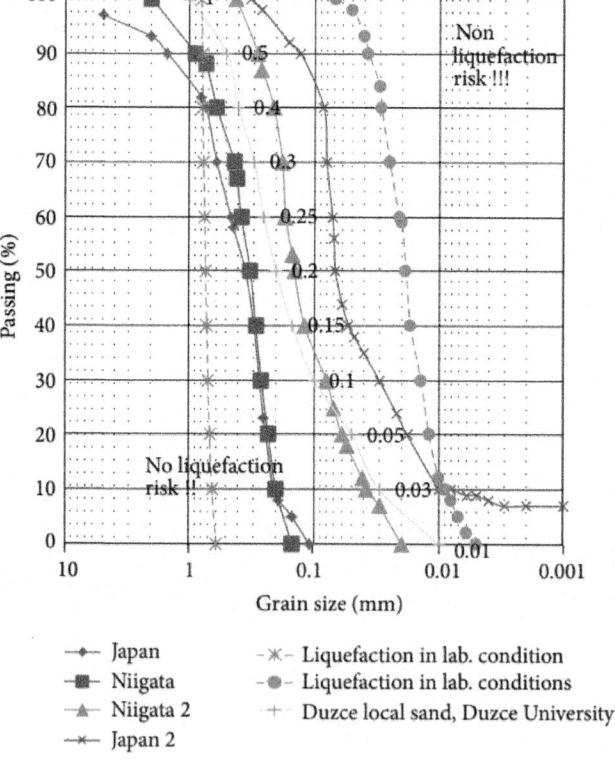

Figure 1: Grain size distribution of sandy soils.

Laboratory Studies

This experimental work has been conducted to investigate the influence of curing time and percentage of waterborne polymer on the unconfined compression strength of sandy soils in Department of Civil Engineering, Faculty of Technology in Duzce University. Use of unconfined compressive strength for soil stabilizers is a quick and simple test and provides a convenient basis for comparison between stabilizer types.

Sandy soils with a different rate of cement mixing were used and various percentages of waterborne polymers were added to soils to investigate the ultrasonic pulse velocity, unit weight, and compressive strength of stabilized samples. The soils were dried before using in the mixtures. First, the required amounts of polymer as a percentage of dry weight of sample and cement were blended and then added to dry soils. The amount of aqueous polymer was chosen as 1, 2, 3, and 4% by total weight of dry sample and the amount of cement was chosen as 10, 20, 30, 40% by weight of dry sample, respectively. The mixing sample was placed into the mould. After 24 hours later, the specimens were taken out of the moulds and specimens were stored in the curing room at the temperature ranging from 21 to 25 centigrade and then tested at 7 and 14 days.

Preparation of the Samples

The preparation of the soil sample is of great importance for laboratory research. Because of using polymer which was insoluble in water, in all experiments, sands were owned dried for 24 hrs to eliminate sand's moisture and then specified amount of polymer added to the dry cement-sand. After completing this treatment, the specimens were prepared by mixing the polymer with sand in the loose dry density of 19 kN/m^3. The polymer mixture was developed in to dough using mechanical kneader. The uniformly mixed dough was subsequently placed into a steel mold measuring 150 mm in height and 300 mm in diameter (Figure 2). Finally, the molded specimens were left to cure at room's temperature. Specimens containing 1, 2, 3, and 4 wt.% (% by dry weight) polymer and 10, 20, 30, and 40 wt.% (% by dry weight) cement were prepared using this method. Other sets of samples were prepared using the same method and submerged in water to be cured for 7 and 14 days. In the following stage, the moisture content of the sandy soil was determined.

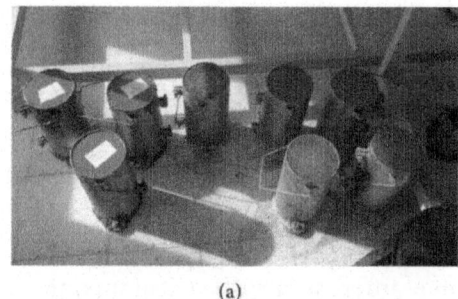

(a)

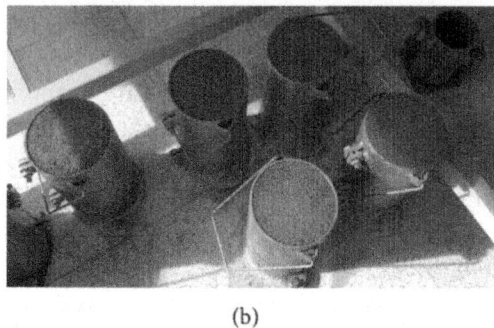

(b)

Figure 2: Viev of preparing the specimens in laboratory.

Curing Time

Curing times of 7 and 14 days were used in this research. Three samples for each curing time were prepared in order to provide an indication of reproducibility as well as to provide sufficient data for accurate interpolation of the results.

Submerge of Specimens in Water

In this study, each specimen was arranged vertically in a steel mould and submerged in a water tank after the specimen in each steel mould was allowed to be cured under room temperature. The specimens submerged in water for 7 and 14 days and then taken out from the water and tested within 24 hrs of removal from the soaking reservoir, and their unconfined compressive strength was recorded. Thus, the effect of curing period on compressive strength, unit weight, and ultrasonic pulse velocity of stabilized samples was evaluated.

Unconfined Compressive Strength Testing

Unconfined compressive strength testing was performed on all extracted specimens with a constant stress rate by manually controlled test machine

(Figure 3). A data acquisition system was used to record the applied load. Each specimen was loaded until peak load was obtained.

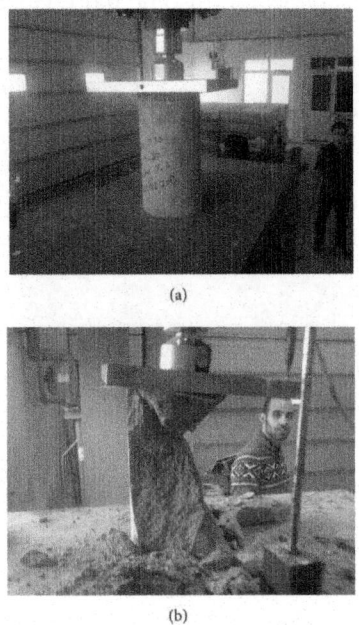

Figure 3: Arrangement of specimen for compressive test in lab.

Ultrasonic Pulse Velocity Testing

The ultrasonic pulse velocity (UPV) measuring devices can be applied for the complex investigation of the soil properties (density, porosity, and strength, as well as its integrity). However, in many investigations up to now, there have been frequent references on the effect of soil stabilization close to the tested product soil cement mixing as it increases the UPV in soil. To investigate this issue more comprehensively, the specimens were made, where soil and cement with polymer of same diameter and length were assigned. The acoustic velocity wave speed of a given soil-concrete (that mixing soil is strengthened like concrete) specimen can easily be obtained with the travel time of a stress wave and the length of the specimen. The pulse is sent from the sending transducer to the receiving transducer through the soil concrete specimen as seen in Figure 4. The relationship of a specimen's acoustic velocity is simply calculated from a time and a length measurement. It should be noted that cracks, flaws, voids, and other anomalies within a material specimen could increase time of travel therefore decreasing the materials acoustic velocity. However, assuming the specimen in Figure 4 is free of anomalies, its acoustic velocity can be calculated simply by sending the pulse. The length of the specimen is 300 mm. Figure 4 shows typical ultrasonic pulse velocity test procedure.

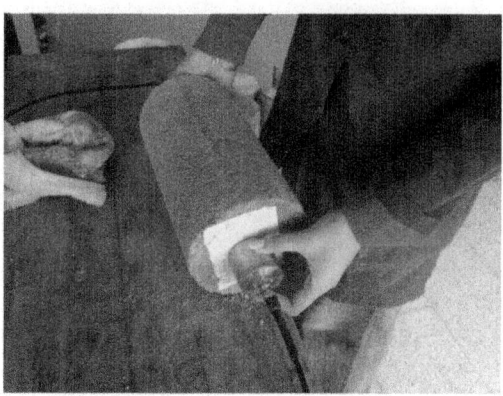

Figure 4: Ultrasonic pulse velocity measurement.

TEST RESULTS

For this research, it was desired to evaluate the effectiveness of the soils mixed with stabilizer at dry and wet conditions being cured after 7 and 14 days. Low to high dosage rates were used for soils at dry and wet conditions. The stabilizer dosage rates used for this study were 1, 2, 3, and 4% as polymers and 10, 20, 30, and 40% as cement.

Effect of Curing Time on Strength

The results of curing times on unconfined compression strength results were shown in Figures 5 and 6. The unconfined compression strength of stabilized samples increases with curing time. Both specimens containing polymer content of 1–4% by wt.% and cement content of 10–40 wt.% were cured in air and water during 7 and 14 days. So, by increasing the polymer contents, cross-linking between polymer network increased and the strength of soil increased. The strength of the specimens containing 3% polymer and 30% cement content at 7 days of curing time achieved 90% of the 14 days compressive strength. It is clear from Figures 5 and 6 that compressive strength of the stabilized soils was increased while increasing the curing time in both water and air curing conditions. As it is expected the effect of water curing on strength is more effective than air curing.

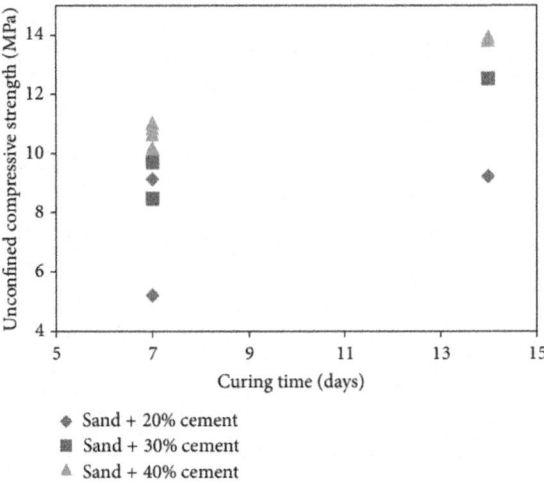

Figure 5: The unconfined compressive strength for specimens in water.

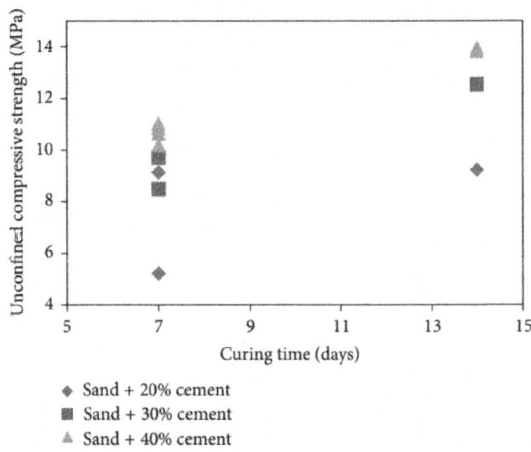

Figure 6: The unconfined compressive strength for specimens in air.

Effect of Curing Condition on Strength

Effects of wet conditions on unconfined compression strength stabilized soils containing polymer and cement were presented in Figure 5. As discussed earlier in this paper, the treated samples were tested using dry and wet curing conditions to provide an indication of the material's moisture content. After 7-day curing period, placing the specimens in water tank provided an excellent indicator of the material's strength under wet conditions. Strength properties of the specimens that consist of 20–40 wt.% cement and 2–4% polymer content increased after submerging in water relatively. The reason of increasing the unconfined compressive strength of specimens is based on the role of water as catalyst.

Effect of Cement and Polymer Content on Strength

The results of unconfined compressive strength on stabilized soils with 10–40 wt.% cement and 1–4 wt.% polymer were presented in Figure 7. Stabilized soils with the 1% polymer have lower strength than 2, 3, and 4% polymer and could not record the value of unconfined compressive strength in this study. For each specimen, the unconfined compression strength increases with increment cement and polymer content. This means that the sandy soil becomes hardened with an increase in the cement content. This phenomenon is explained by the fact that the fine grains of cement are positioned around and among the sand grains; the polymer covered all of sample's area.

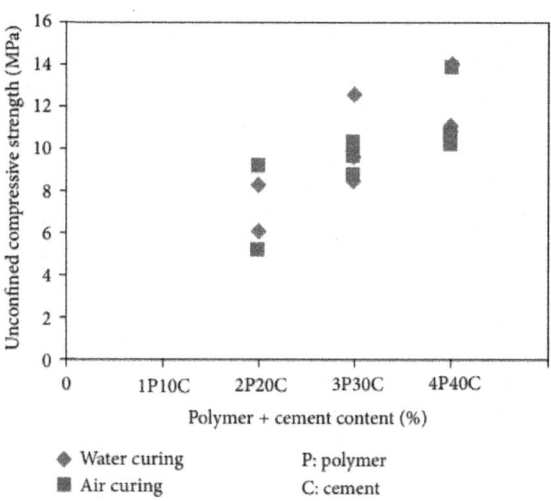

Figure 7: Unconfined compressive strength of soils modified with cement and polymer.

The unconfined compressive strength increased linearly with a polymer concentration up to 4 wt.% for all of the sandy soil used. Comparing the performance of the emulsion in improving the unconfined compressive strength, it could be seen that the unconfined compression strength increases with increment polymer content. This phenomenon is explained by the fact that an increment in polymer content resulted in enhancing bond mechanisms of the sand-emulsion interface. It could be seen that specimens consist of

4 wt.% polymer gave the highest value for all specimens and specimens consist of 1 wt.% polymer gave the lowest values that furthermore the values could not be recorded for polymer at a rate of 1% successively. The unconfined compressive strength value of sand at 2 and 4 wt.% polymer content was 5.21 and 13.83 MPa, respectively, which are more than 100% increased.

The results obtained in the simple compression tests at 7 days have shown an increase in the strength of the stabilized material against the natural material whose strength was 10.65 MPa. The increases in strength obtained were proportional to the amount of polymer added, except for combination of 1% which reached 5.12 MPa, only achieved by the combination of 2–4%. This has been interpreted as a result of the combination of the effect of the polymer matrix, not modified by low addition of polymer (at a rate of polymer 1%), and, on the other hand, the binding effect of the polymer. For content of 2% and upwards may modify the cement matrix. Therefore the increased strength was due only to the polymer. The results after 14 days showed values between 10.19 and 10.25 MPa in all combinations except that of 1%, which approximately reached 0.35 MPa. Even more, it was so weak that the result could not be recorded. The value of 10.65 Mpa was obtained in the combination of 3% and 4% polymer and the highest was obtained in the whole experiment. The values after 14 days, the results were very similar at those of 7 days, reaching between 5.21–6.12 MPa except for the combination of 1%. For this curing time the greatest strength was obtained for the combination of 4%, with 10.65 MPa, a value very close to those obtained with combination of 3%, with 9.74 MPa. The analysis of test results of unconfined compression shows that the addition of 1% of polymer is insufficient since, while it changes the polymer nature of the soil, it is not enough for the formation of chemical reactions over the curing time. On the other hand, the combination of 4% clearly shows the excess values of polymer in the soil. The combination which improved optimal point from the standpoint of improving the resistant properties of soil tested against the addition of polymer was 4% and a compressive strength of 10.65 MPa.

Effect of Polymer Content on Maximum Dry Density

The effect of polymer content on maximum dry and specific density was presented in Figures 8, 9, 10, and 11. For any particular amount of polymer, an increase in polymer content causes an increment in dry density. As already explained, this is due to increment of average unit weight of solids in the sandy soil polymer mixture. It is observed from the Figures 8, 9, 10, and 11 that the plot of variation of dry density with respect to polymer content is more or less proportional in shape.

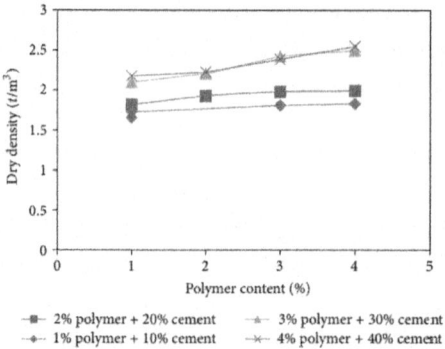

Figure 8: Effect of polymer content on maximum dry density of soil in dry conditions for 7 days.

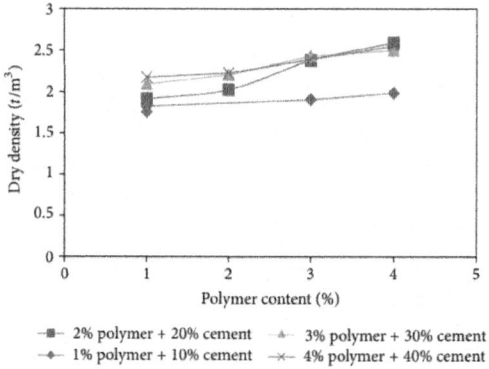

Figure 9: Effect of polymer content on maximum dry density of soil in wet conditions for 7 days.

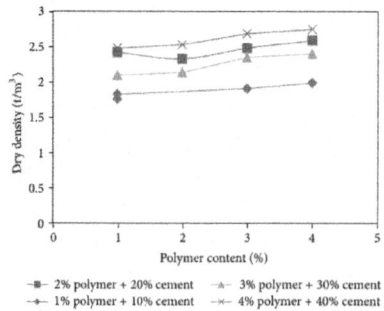

Figure 10: Effect of polymer content on maximum dry density of soil in dry conditions for 15 days.

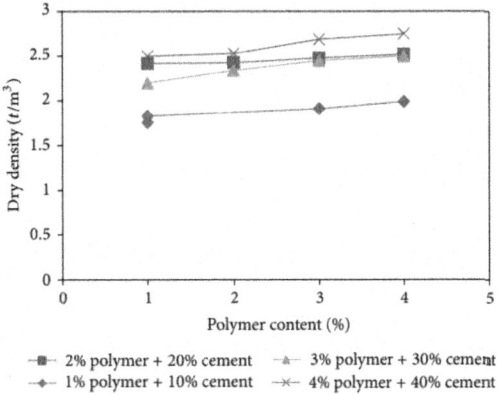

Figure 11: Effect of polymer content on maximum dry density of soil in wet conditions for 15 days.

Effect of Polymer Content on Optimum Moisture Content

Optimum moisture content (OMC) is increased with increase in polymer content. The maximum OMC is recorded 13.8% for 4% polymer content, and as low as 5.4 for soil with 1% of polymer content. The initial inclusion of content of 1% caused a more sudden hike in the OMC than that of the ordinary sandy soil, and a further increase in polymer content increased the OMC. But in all cases, the OMC is greater than that of raw sandy soil. The effect of polymer content on OMC was presented in Figures 12, 13, 14, and 15.

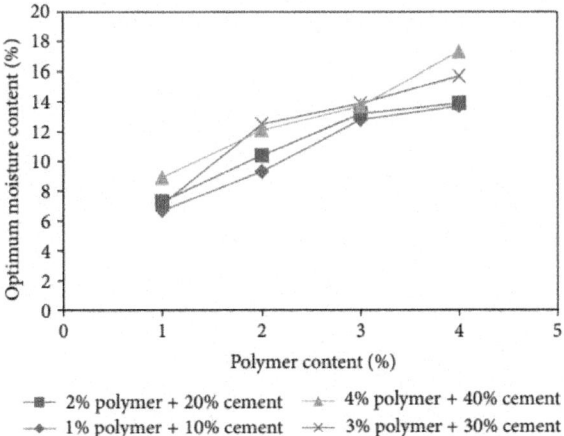

Figure 12: Effect of polymer content on OMC in dry conditions for 7 days.

14   Mechanisms of Soil Stabilization

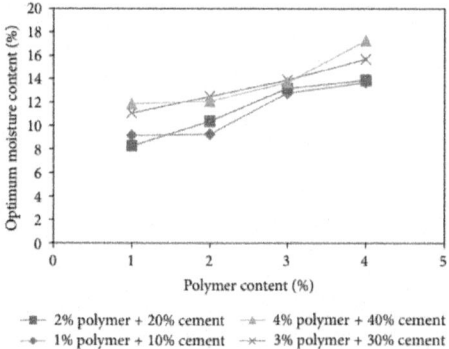

Figure 13: Effect of polymer content on OMC in wet conditions for 7 days.

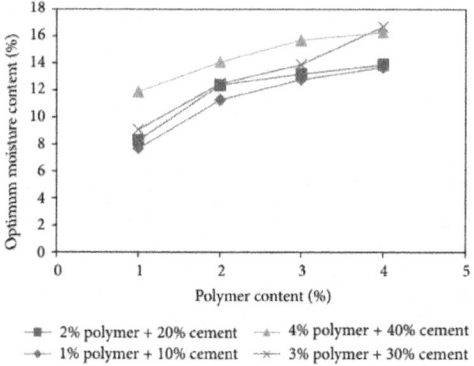

Figure 14: Effect of polymer content on OMC in dry conditions for 15 days.

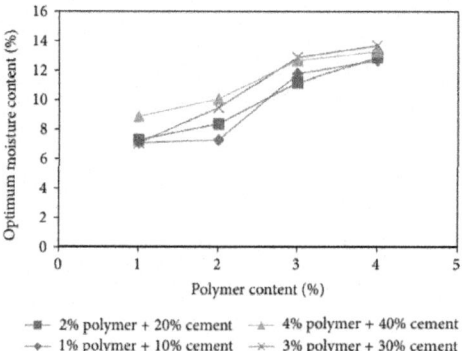

Figure 15: Effect of polymer content on OMC in wet conditions for 15 days.

Ultrasonic Pulse Velocity Test Results

The velocity versus density that results in pulse passing through in soil cement with polymer mixing was presented in Figures 16, 17, 18, and 19.

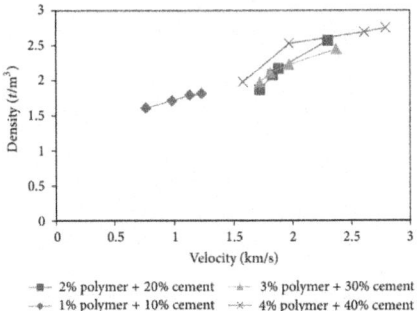

Figure 16: Ultrasonic pulse velocity test procedure in dry conditions for 7 days.

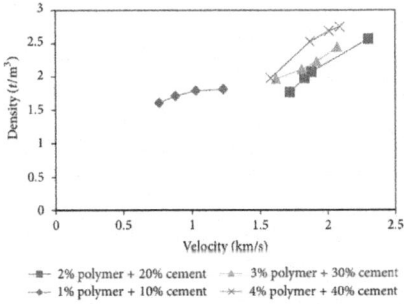

Figure 17: Ultrasonic pulse velocity test procedure in wet conditions for 7 days.

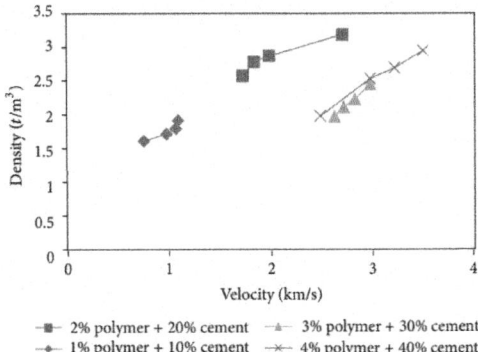

Figure 18: Ultrasonic pulse velocity test procedure in dry conditions for 14 days.

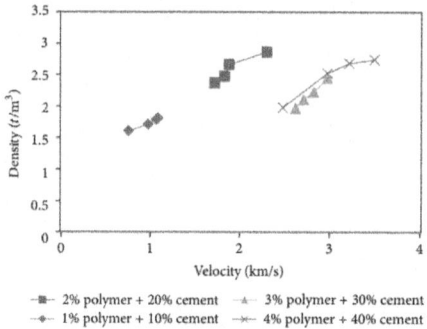

Figure 19: Ultrasonic pulse velocity test procedure in wet conditions for 14 days.

An increase in polymer content causes increment in velocity of soundings. This is due to increment of average unit weight of solids in the soil cement polymer mixture. It was observed from Figures 16, 17, 18, and 19 that the plot of variation of velocity versus density with respect to polymer content is more or less proportional.

CONCLUSIONS

This study was undertaken to investigate the influence of polymer percentage, curing time, and dry and wet conditions on the unconfined compressive strength of stabilized cement with sandy liquefiable soils. The results of the study were presented in following conclusions.(i)When the soil has stabilized with polymer, it increases the dry density of the soil due to specific gravity and unit weight of polymer. The variation is linear for both cases. Initial introduction of polymer content in to the soil causes an increase in OMC, so further increase in polymer content may cause to enhance in OMC.(ii)The addition of polymer to the natural soil produced an improvement in its mechanical capacities that were determined by unconfined compression tests, from the first period of curing examination. From the strength aspect of liquefiable sandy soils, the optimum polymer content estimated polymer at 3%, since in all tests it yielded the best result with the rate of 3% dosage of polymer.(iii)The strength of sandy soil mixtures has increased with increment of cement contents up to about 30% and above 30% cement content; the strength of the soil almost becomes constant. This phenomenon is explained by the fact that the fine grains of cement were covered and positioned around and among the sand grains.(iv)The unconfined compressive strength of specimens has increased with increment of polymer contents; this phenomenon is explained by the fact that increment of polymer and the polymer cover all of sample's area and increases cross-links.(v)For specimens submerging in water, there are two types of results.

First one is about specimensens that consist of 30 and 40 wt.% cement content at 4% polymer and 20 and 30 wt.% cement content at 3% polymer; the strength properties of these specimens were reduced after submerging in water. Second is about the rest of specimens. After 7 days of submerging, the unconfined compressive strength of the specimens increased while it was less than the dry condition. The reason of increasing the unconfined compressive strength of specimens is the role of water as catalyst. When specimens submerged in water, the H^+ ions of water react with three-member epoxies' rings and the epoxies' ring was opened, then the hardener easily can react with resin. So, the strength of specimens increased.(vi)The unconfined compressive strength significantly increased with curing time. The unconfined compressive strength of the mixtures rapidly increased at 7 days.(vii)The maximum unconfined compressive strength values have increased with increase of the polymer content. Sandy soils stabilized with 3, 4% polymer at 30, 40% cement content have higher values on unconfined compressive strength than other percentages.

REFERENCES

1. W. H. Perloff, Soil Mechanics, Principal and Applications, John Wiley & Sons, New York, NY, USA, 1976.
2. B. K. G. Theng, "Clay-polymer interactions: summary and perspectives," Clay and Clay Minerals, vol. 30, no. 1, pp. 1–10, 1982.
3. K. E. Clare and A. E. Cruchly, "Laboratory experiments in the stabilization of clays with hydrated lime,"Geotechnique, vol. 7, pp. 97–110, 1957.
4. B. K. G. Theng, "Clay-polymer interactions: summary and perspectives," Clay and Clay Minerals, vol. 30, no. 1, pp. 1–10, 1982.
5. D. Dermatas and X. G. Meng, "Utilization of fly ash for stabilization/solidification of heavy metal contaminated soils," Engineering Geology, vol. 70, no. 3-4, pp. 377–394, 2003.
6. S. M. Lahalih and N. Ahmed, "Effect of new soil stabilizers on the compressive strength of dune sand,"Construction and Building Materials, vol. 12, no. 6-7, pp. 321–328, 1998.
7. B. Indraratna, "Utilization of lime, slag and fly ash for improvement of a colluvial soil in New South Wales, Australia," Geotechnical and Geological Engineering, vol. 14, no. 3, pp. 169–191, 1996.
8. V. S. Green, D. E. Stott, L. D. Norton, and J. G. Graveel, "Polyacrylamide molecular weight and charge effects on infiltration under simulated rainfall," Soil Science Society of America Journal, vol. 64, no. 5, pp. 1786–1791, 2000.

9. A. B. Moustafa, A. R. Bazara, and A. R. Nour El Din, "Soil stabilization by polymeric materials,"Angenandte MaKromoleKular Chemie, vol. 97, no. 1, pp. 1–12, 2003.

10. S. A. Naeini and A. Mahdavi, Effact of polymer on shear strength of silty sand [M.S. thesis], Civil Engineering Department, Imam Khomeini International University, Qazvin, Iran, 2009.

11. J. L. Daniels, H. I. Inyang, and I. K. Iskandar, "Durability of Boston Blue Clay in waste containment applications," Journal of Materials in Civil Engineering, vol. 15, no. 2, pp. 144–152, 2003.

12. J. L. Daniels and H. I. Inyang, "Contaminant barrier material textural response to interaction with aqueous polymers," Journal of Materials in Civil Engineering, vol. 16, no. 3, pp. 265–275, 2004.

13. S. A. Naeini and M. Ghorbanalizadeh, "Effect of wet and dry conditions on strength of silty sand soils stabilized with epoxy resin polymer," Journal of Applied Sciences, vol. 10, no. 22, pp. 2839–2846, 2010.

14. ASTM, "Standard practice for classification of soils for engineering purposes (Unified Soil Classification System)," ASTM D2487, ASTM, West Conshohocken, Pa, USA, 2000,http://www.astm.org/Standards/D2487.htm.

15. H. Tsuchida, "Prediction and countermeasure against liquefaction in sand deposits," in Proceedings of the Seminar of the Port and Harbour Research Institute, vol. 3, pp. 1–3, Ministry of Trans-port, Yokosuka, Japan, 1970.

Chapter 2

STABILIZATION OF ORGANIC MATTER BY BIOCHAR APPLICATION IN COMPOST-AMENDED SOILS WITH CONTRASTING PH VALUES AND TEXTURES

Shih-Hao Jien[1], Chung-Chi Wang[1], Chia-Hsing Lee[2] and Tsung-Yu Lee[3]

[1]Department of Soil and Water Conservation, National Pingtung University of Science and Technology, Pingtung 91201, Taiwan

[2]Department of Agricultural Chemistry, National Taiwan University, Taipei 10617, Taiwan

[3]Department of Geography, National Taiwan Normal University, Taipei 10610, Taiwan

ABSTRACT

Food demand and soil sustainability have become urgent concerns because of the impacts of global climate change. In subtropical and tropical regions, practical management that stabilizes and prevents organic fertilizers from rapid decomposition in soils is necessary. This study conducted a short-term (70 days) incubation experiment to assess the effects of biochar application on the decomposition of added bagasse compost in three rural soils with different pH values and textures. Two rice hull biochars, produced through slow pyrolization at 400 °C (RHB-400) and 700 °C (RHB-700), with application rates of 1%, 2%, and 4% (w/w), were separately incorporated into soils with and without compost (1% (w/w) application rate). Experimental results indicated that C mineralization rapidly increased at the beginning in all treatments, particularly in those involving 2% and 4% biochar. The biochar addition increased C mineralization by 7.9%–48% in the compost-amended soils after 70 days incubation while the fractions of mineralized C to applied C significantly decreased. Moreover, the estimated maximum of C mineralization amount in soils treated with both compost and biochar were obviously lower than expectation calculated by a double exponential model (two pool model). Based on the micromorphological observation, added compost was wrapped in the soil aggregates formed after biochar application and then may be protected from decomposing by microbes. Co-application of compost with biochar may be more efficient to stabilize and sequester C than individual application into the studied soils, especially for the biochar produced at high pyrolization temperature.

INTRODUCTION

Climate change and food demand are currently the two most crucial concerns for agricultural scientists throughout the world. How to maintain the soil organic matter (SOM) levels in soils is a key consideration in agricultural productivity and carbon sequestration, particularly in agricultural lands in subtropical and tropical regions. Mekuria *et al.* (2014) [1] mentioned that mulches, compost, or manure can be effective in enhancing soil organic carbon pool and agricultural productivity in the tropic regions, but these amendments were often short-lived. The added organic matters were usually mineralized to CO_2 rapidly leading to large-scale leakage in subtropical/tropical regions. Therefore, developing strategies for reducing the mineralization of added OM and increasing carbon sequestration in subtropical/tropical rural soils is necessary to facilitate land sustainability [2,3].

Biochar is the by-product of the pyrolysis of organic wastes and is regarded as a chemically- and biologically-stable C pool [4]. Applying biochar to agricultural soils is considered to improve soil quality effectively [5,6,7,8] while sequestering carbon and reducing greenhouse gas emission from soils [7,8]. Recently, co-application of biochars and other organic amendments has been determined to be an effective management practice for compensating for the limitations of applying biochar or an organic amendment alone [9,10,11,12]. Nevertheless, most of those studies were conducted at arid, Mediterranean, and temperate regions, and the biochars were incorporated into the soils for stimulating microbial activities to facilitate decomposition of added organic matters. Awad *et al.* (2013) [11] considered that rapid decomposition of plant residue was desired in double-cropping systems in temperate climates in order to maintain a proper supply of nutrients for crop growth and substrates for soil microorganisms. Several inconsistent results including facilitation and inhibition of SOM decomposition in soils co-amended with biochar and other organic amendments have been reported [2,12,13,14], which resulted from the original characteristics of biomass, differences in biochar properties (pyrolization processes and temperatures), and soil environments, such as soil pH value, native SOC contents, and soil texture. Van Veen and Kuikman (1990) [15] and Qayyum *et al.* (2012) [10] indicated that fine soil texture can physically protect SOM from decomposition caused by microbes through the adsorption of organic matter onto clay surfaces. Regarding C mineralization in biochar-amended soils, Hamer *et al.* (2004) [16] and Kuzyakov *et al.* (2009) [14] reported that co-metabolism occurs in soils incorporated with a biochar and a fresh C source, thus facilitating cumulative CO_2 emission from the amended soils. Zimmerman *et al.* (2011) [17] observed both positive (< 90 days) and negative priming (>250 days) effects of incubating grass and wood biochars in sandy soils. This negative effect could be due to the adsorption of native SOM onto the surfaces and pore spaces of biochars, thus protecting the SOM from decomposition. Furthermore, Keith *et al.* (2011) [18] indicated that adding biochar can lead to a positive priming effect on native SOM but not on added organic matter (OM); however, the added OM can cause a positive priming effect on the biochar.

Many of the aforementioned studies clearly demonstrated the enhancing effects of biochar application on the physiochemical properties of soil. However, to further clarify the interaction between biochar and added organic fertilizers, we aimed to (1) determine the effects of rice husk biochar application on the C mineralization of compost-amended soils with various soil pH values and textures, and (2) examine the micro-structure by using a polarized microscope to determine the possible processes of interaction among the biochar, added organic fertilizer, and clay particles.

MATERIALS AND METHODS

Soil Collection and Biochar Preparation

Surface soil samples (0–15 cm) were collected from three agricultural slopelands in Taiwan. Laopi (Lo) soil is generated from quaternary-aged materials and is widely distributed in the terrace landscapes of Southern Taiwan. Shanhuipu (Sp) and Choutseunlun (Ct) soils are slate alluvial sediments along streams in Southern Taiwan. These three soils were selected because of their wide range of physical and chemical properties (Table 1). The soil samples were air-dried, sieved through a 2-mm screen, and stored in covered plastic containers at 25 °C.

Rice hull was used as a feedstock to separately produce two types of biochar through slow pyrolysis in a furnace equipped with an N_2-purged retort referred to Streubel *et al.* (2011) [19] at 400 °C (RHB-400) and 700 °C (RHB-700), respectively. The furnace was initially heated to 100 °C, and the temperature was then increased to 400 °C and 700 °C, respectively at a rate of 5 °C min^{-1} with a resident time of 30 min. The biochars were subsequently cooled overnight while the N_2 flush was maintained. The biochars were then gently crushed and ground to pass a 2-mm sieve before use and analysis.

Analytical Methods

The pH values of the soil samples and biochars were mixed with deionized water and determined using a glass electrode (1:1 *w/v* for soil; 1:10 *w/v* for biochar) [20]. The electrical conductivity (EC) of the saturated paste extracts of soils was measured using a conductivity meter [21]. The soil particle-size distribution was determined using a pipette method [22]. The cation exchange capacity (CEC) was determined using an ammonium acetate method (pH 7.0) [23]. The organic C content of the tested soils was determined using a wet oxidation method [24]. The total N was measured using the semi-micro-Kjeldahl procedure [25]. The inorganic N was extracted using 2 M KCl (1:10 *w/v*); concentrations of NH_4^+-N and NO_3^--N were estimated using steam distillation involving MgO and the Devarda alloy [26]. The calcium carbonate contents were determined by simple titrimetric method [27], which finely-ground soil and biochar samples (2.0 g) were reacted with 2 M HCl for 16 h. The emitted CO_2 in the reacted bottle was captured by NaOH, and then the base solution was titrated with 0.1 M HCl to calculate carbonate contents. All chemical analyses were conducted intriplicate. Table 1 and Table 2 present summaries of the relevant properties of the soils and biochars.

The readily-oxidizable carbon (ROC) was determined using a method proposed by Blair *et al.* (1995) [28]. Air-dried soil samples containing 15 mg of C were weighed into centrifuge tubes and reacted with 333 mM $KMnO_4$ for 1 h at 25 °C. After centrifugation, the supernatants were diluted at a ratio of 1:250 with deionized water. The absorbance of the diluted samples and standards was recorded using a split-beam spectrophotometer at 565 nm. The change in the $KMnO_4$ concentration was used to estimate the amount of C oxidized, assuming that 1 mM $KMnO_4$ wasconsumedin the oxidation of 0.75 mM or 9 mg of C. The $KMnO_4$-C fraction, suggested by Blair *et al.* (1995) [28], encompasses all the organic components that can be readily oxidized by $KMnO_4$, including labile humic material and polysaccharides [29], and accounts for 5%–30% of total organic carbon.

Incubation Experiment

A 70-days incubation experiment was conducted to investigate the effects of applying biochars and compost on CO_2 emission in three agricultural slopeland soils. Twenty grams of each air-dried soil sample was placed in small plastic cups. A commercial bagasse compost was added as a substrate to the soils for each treatment at a rate of 20 t ha^{-1}. The biochars were then thoroughly mixed with the soilsat 0%, 2%, and 4% (w/w) (approximately 0, 40, and 80 t ha^{-1}, respectively). The experimental design consisted of 10 treatments for each soil in triplicate: (1) O (control; soil only), (2) O + C (soil with 1% compost), (3) O

+ C + 2% RHB-400, (4) O + C + 4% RHB-400, (5) O + C + 2% RHB-700, (6) O + C + 4% RHB-700, (7) O + 2% RHB-400, (8) O + 4% RHB-400, (9) O + 2% RHB-700, and (10) O + 4% RHB-700. Deionized water was added to the soils to reach 60% water-holding capacity. Each cup with the treated soil and a plastic vessel containing 10 mL of 1 N NaOH solution was placed in a wide-mouth plastic jar, which was subsequently sealed. Jars without treated soils were used as blanks. After 3, 7, 14, 21, 28, 42, and 70 days, the emitted CO_2 was measured in nondestructive determination by titrating the NaOH solution with 0.5 N HCl following addition of $BaCl_2$. The jars were then sealed again for incubation until the next measurement. The incubation experiment was performed in the dark at 25 ± 2 °C [30]. The 70 days of incubation duration was conducted based on Novak et al. (2010) [31] and Streubel et al. (2011) [19] who denoted that the CO_2 evolution rate may approach to a minimum value after 67–75 days.

Calculations and Statistical Analysis

The percentage of applied C mineralized (ACM) in the treatments involving compost and/or biochar was calculated according to Ribeiro et al. (2010) [32]:

$$ACM, \% = \frac{CMC_{treatment(70\ day)} - CMC_{control(70\ day)}}{organic\ C\ applied} \times 100 \qquad (1)$$

where CMC is the cumulative C mineralized in the form of CO_2-C emitted during incubation.

The measured carbon emission results under the treatment of co-application of the biochar and the compost (O + C + RHB) indicate the final results of an overall interaction between the organic amendments. Expected results were calculated from the values of related treatments as follows: (O + C) + (O + RHB) − (O). The expected values indicate no interactions between the compost and biochars. Although the priming effects between compost and biochars were difficult to be determined, the difference between the expected and measured values could be explained as the interaction effect. We used the double exponential model to fit the expected and measured carbon emission of the incubation experiment. The two-pool model involved a labile fraction and a resistant fraction, which can well describe the decomposition of soil N and C [31]:

$$C_{min} = C_l \times \left(1 - \exp^{(-k_l \times t)}\right) + C_r \times \left(1 - \exp^{(-k_r \times t)}\right) \qquad (2)$$

where C_{min} is the mineralized C amount at time t (day), C_l and C_r mean the amounts of potentially mineralizeable C (mg C/g C applied) of the labile and resistant fractions, respectively, and k_l and k_r are the respective mineralization rate constants (day^{-1}). The model fitting was carried out using the statistical program of Sigmaplot 8. The maximum values of the unstable C pool were calculated by C_l plus C_r which were derived from the model fitting.

Soil Micromorphology

Kubiena boxes were used to collect undisturbed soil blocks from the experimental pots (Stoops, 2003) [33]. The same solid mixture of each treatment was placed into a pot with a size larger than the Kubiena box followed by the same incubation process as described. After the 70-days incubation, the soil blocks were taken using Kubiena boxes. Thin sections of 30-μm thickness were then prepared following air drying using a microtome by Spectrum Petrographics Inc. (Washington, USA). The thin sections were then used for observing distribution and structure of organic matters among soil particles under a polarized microscope (Leica DM EP, TX, USA).

Statistical Methods

The effects of soil type, biochar type, and application rate of the biochar on the total CO_2-C evolution and their interactions were tested using a multivariate analysis of variance (MANOVA). Significant effects were identified when $p < 0.05$. Multiple mean comparisons were performed using Fisher's protected least significant difference (LSD) procedures at $p < 0.05$. All the statistical analyses were performed with IBM SPSS Statistics, Version 22 (Somers, NY, USA).

RESULTS

Characteristics of Soils, Compost, and Biochars

Table 1 shows the characteristics of the studied soils. Lo soil is acidic and enriched with silt. Sp soil exhibits a texture similar to that of Lo soil but is pH neutral. The soil textures are silty clay loam and silt loam for Lo and Sp soils, respectively, based on Soil Taxonomy (Soil Survey Staff, 2010) [34]. By contrast, Ct soil exhibits high pH and EC levels, and it has a considerably higher proportion of sand than the other two soils. The Ct soil was classified as a sandy soil [34]. Low total C content was determined in all studied soils; it was below 0.2% in the Ct soil. The CECs of the Lo and Sp soils were approximately 15 cmol (+) kg^{-1}, which was considerably higher than that of the Ct soil. The CECs were consistent with the contents of clay and organic carbon in these soils. Regarding the total nitrogen content (TN) and inorganic nitrogen content (IN; NH_4^+-N + NO_3^--N), the Ct soil exhibited a lower TN and IN than the other two soils did. Table 2 lists the properties of the bagasse compost and biochars produced at various temperatures. The pH value of the compost was 5.5, and the pH values of the two biochars were approximately 8.0. The total carbon (TC) of the compost and biochars was in the range of 30%–33%, and the compost contented more TN than the biochars did. The TN content in RHB-400 and RHB-700 were <0.5%. The C/N ratios of the two biochars were >70, which was considerably higher than that of the compost. The exchangeable K content of the compost and biochars were considerably higher than other exchangeable cations. In addition, the compost exhibited more exchangeable Ca and Mg than the biochars did.

Table 2 also reveals that the ROC contents were 3.2 g kg^{-1} and 2.1 g kg^{-1} in the RHB-400 and RHB-700, respectively. The biochars could contribute 64 mg and 42 mg ROC kg^{-1} soil at the application rate of 2% and 128 mg and 84 mg kg^{-1} ROC at the rate of 4%, respectively.

Table 1: Selected properties of the studied soils

Properties	Soils		
	Laopi (Lo)	Shashuipu (Sp)	Choutseulun (Ct)
pH	4.43	6.90	8.25
EC * (dS m^{-1})	0.08	0.19	1.96
Sand (%)	14	17	95
Silt (%)	57	66	4.0
Clay (%)	29	17	1.0
Texture $^{#}$	SiCL	SiL	S
OC (%)	1.78	1.38	0.12
CaCO$_3$ (g kg^{-1})	0.12	0.27	5.08
CEC (cmol(+) kg^{-1}) ¶	15.0	14.6	4.00
TN § (%)	0.16	0.19	0.01
NH$_4^+$-N (mg kg^{-1})	2.84	11.2	ND ^{e}
NO$_3^-$-N (mg kg^{-1})	14.2	47.8	8.33
ROC (g kg^{-1}) $^{¢}$	0.42	0.75	ND

*: Electric conductivity; #: SiCL: silty clay loam; SiL: silt loam; S: sand; ¶: Cation exchange capacity: CEC (cmol(+) kg^{-1}); §: Total nitrogen content; ¢: Readily oxidizable carbon; e: data not detected.

Table 2: Characteristics of the bagasse compost and the rice hull biochars in this study

	Compost	RHB-400	RHB-700
pH (1:10 w/v)	5.50	7.99	8.03
TC (%)	30.2	31.0	32.9
TN (%)	1.08	0.41	0.35
C/N ratio	28	76	94
CEC (cmol(+) kg^{-1})	82.5	26.1	35.6
Exchangeable K (g kg^{-1})	6.94	7.01	7.02
Exchangeable Na (g kg^{-1})	0.44	0.28	0.24
Exchangeable Ca (g kg^{-1})	4.61	0.47	0.54
Exchangeable Mg (g kg^{-1})	2.12	0.22	0.23
Carbonate (g kg^{-1})	ND	1.52	1.88
ROC (g kg^{-1})	36.0	3.22	2.10

RHB-400 and RHB-700 are the rice hull biochars produced at 400 °C and 700 °C, respectively. Explanation of the abbreviations are the same of those in Table 1.

Carbon Dioxide Emissions from Soils

The CO_2 evolution rates and cumulative CO_2 emission of the treated soils during 70 days of incubation are shown in Figure 1. The CO_2 evolution rate was slightly lower in RHB-700 than in RHB-400 and exhibited a similar trend between the two biochars for a given treatment. As a representative, the results of RHB-400 treatments were shown in Figure 1.

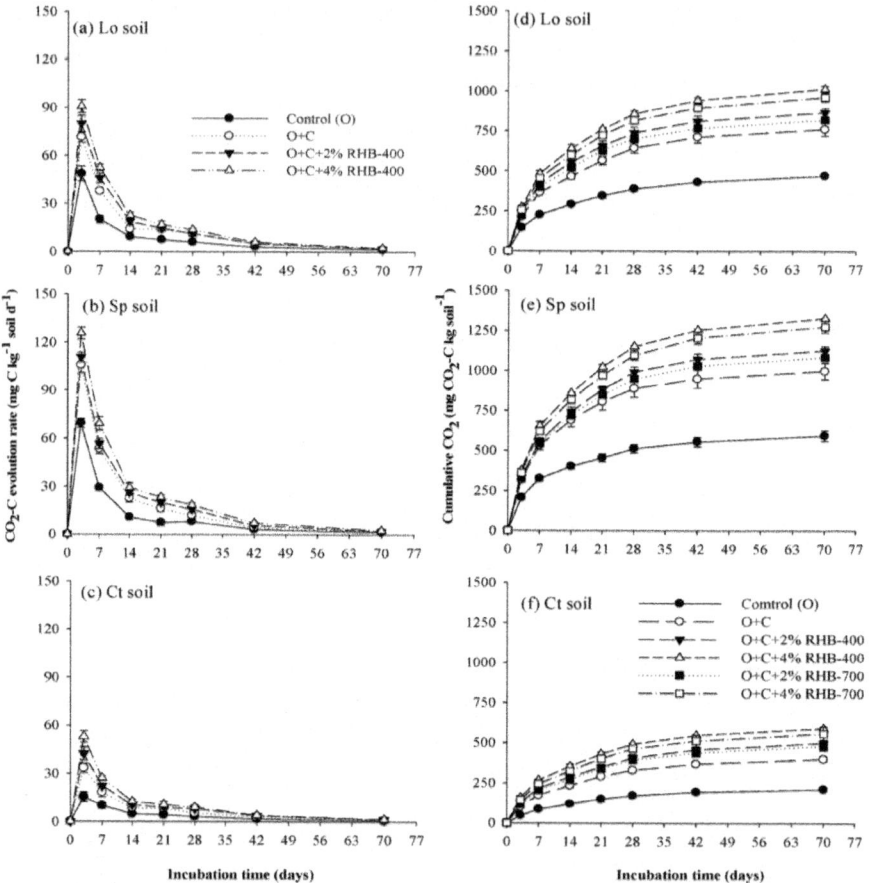

Figure 1: CO_2 evolution rate ((a), (b), and (c)) and cumulative CO_2 evolution ((d), (e), and (f)) for the Lo, Sp, and Ct soils amended with compost and biochars. The vertical error bars indicate the standard deviation. O (control): without compost and biochars; O + C: only compost (1%).

The CO_2 evolution rates were obviously higher in the first two weeks for all tested soils than in the following period. The control (O) maintained a low CO_2 evolution rate and consequent cumulative CO_2 emission throughout the incubation period compared with the other treatments (Figure 1), which exhibited final accumulative CO_2 emission of 470, 594, and 213 mg CO_2-C kg soil^{-1} for the Lo, Sp, and Ct soils, respectively. Compared with control (O), application of the compost (O + C) considerably increased the cumulative emission of CO_2 in all studied soils, while co-application of biochars with the compost resulted in even higher values. Maximum amounts of CO_2 emission was observed in the treatment involving co-application of compost and 4% biochar, particularly for the RHB-400 and for the Sp soil. The amounts of the cumulative CO_2 emission of each treatment followed the order of Sp soil > Lo soil > Ct soil ($p < 0.05$) (Table 3).

Table 3: Characteristics of the bagasse compost and the rice hull biochars in this study

	Control	O + C	O + C + 2% RHB-400	O + C + 4% RHB-400	O + C + 2% RHB-700	O + C + 4% RHB-700
Lo	470 ± 4.00 [a]	761 ± 40.6 [a]	865 ± 27.5 [a]	1014 ± 20.2 [a]	821 ± 19.9 [a]	963 ± 25.7 [a]
Sp	594 ± 33.1 [b]	999 ± 51.7 [b]	1127 ± 26.4 [b]	1327 ± 2.12 [b]	1085 ± 11.8 [b]	1274 ± 34.9 [b]
Ct	212 ± 11.3 [c]	400 ± 15.0 [c]	499 ± 21.2 [c]	591 ± 8.09 [c]	482 ± 27.9 [c]	558 ± 43.9 [c]

Different letters along the column (different soil types) mean significant difference ($p < 0.05$) between each soil.

The three soils revealed the similar trends of differences in the ACM (%) among treatments (Figure 2). Application of the compost (O + C treatment) exhibited the highest ACM (%), namely 10.2%, 14.2%, and 6.6% for the Lo, Sp, and Ct soils, respectively, while the values were clearly lower in the treatments of co-application of the compost and biochars. Compared with the O+C treatment, the ACM (%) for the soils amended with compost and biochars significantly decreased by 66%–76%, 65%–76%, and 61%–72% for the Lo, Sp, and Ct soils, respectively. However, for a given biochar application rate, the pyrolization temperature of the biochar did not result in significant differences in the proportion of the ACM. For all treatments, the ACM was apparently higher in the Sp soil than in the other soils, and the Ct soil exhibited the lowest ACM. According to the MANOVA results (Table 4), the soil type and application rate significantly affected the cumulative CO_2 emission ($p < 0.001$), while biochar type had no significant effect. Moreover, a significant interaction between the soil type and the application rate was found ($p < 0.001$).

Table 4: Multivariate Analysis of Variance (MANOVA) of the total amounts of CO_2-C evolved after 70 day incubation for each treatment

Parameter	mg C kg^{-1} soil		
	Freedom degree	F-value	Significance
Soil type (S)	2	2626.6	< 0.001
Biochar type (B)	1	0.26	0.6094
Application rate (R)	2	243.2	< 0.001
S × R	4	9.02	< 0.001
S × B	2	0.75	0.4802
B × R	2	0.20	0.8166
S × B × R	4	0.04	0.9960

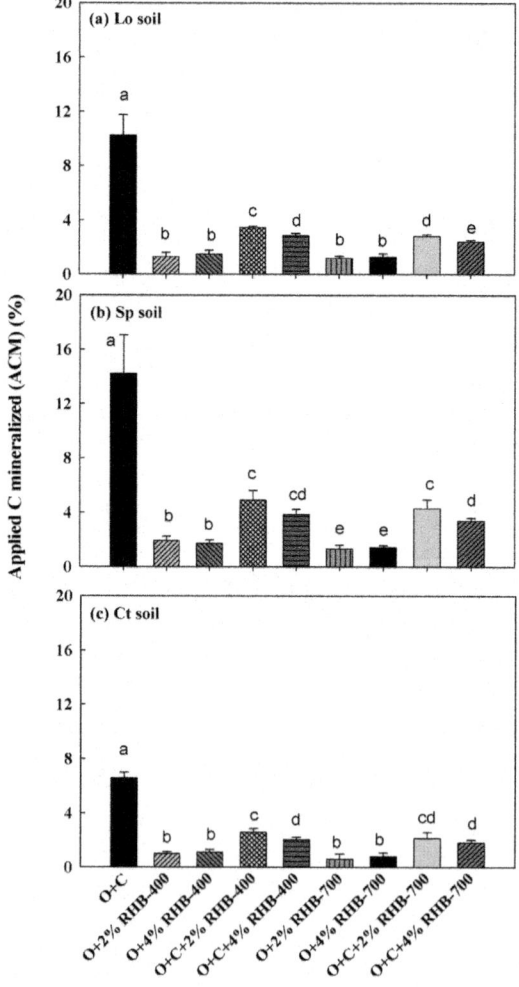

Figure 2: Percentage of the applied carbon mineralized (ACM %) of the Lo (a), Sp (b),

and Ct (c) soils amended with compost (1%) and biochars at the end of the incubation. The vertical error bars indicate the standard deviation.

Kinetics of Carbon Mineralization

To clarify the interaction between compost and biochars, a comparison was conducted between the expected values and measured values of cumulative CO_2 emission in the treatment of co-applications (O + C + RHB) (Figure 3a,b). The expected values were calculated with the values from the individual applications of the compost and biochars as follows: (O + C) + (O + RHB) – O. Therefore, the differences between expected and measured values could be attributed to the effect of co-applications. In this study, the three soils exhibited a similar trend in the comparison. The trend found in the Sp soil was illustrated in Figure 3 as a representative of the three soils. The cumulative CO_2 emission curves of the measured values approximately reached a plateau while those of the expected values kept increasing, which indicates that the unstable C pool might decline by co-application.

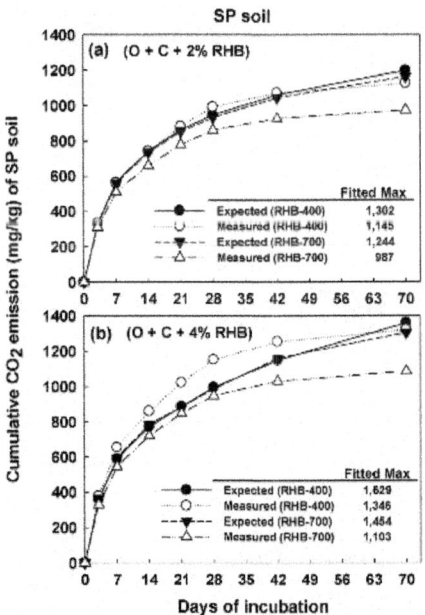

Figure 3: The expected and measured cumulative CO_2 emission in the Sp soil amended with the compost at the rate of 1% and biochars at the rates of 2% (a) and 4% (b). Expected values were calculated from the treatments of control (O), compost only (O + C) and biochar only (O + RHB). Fitted Max: the maximum value estimated via two-pool kinetic model fitting.

Double-exponential model (two-pool kinetic model) was used to describe C mineralization of soils amended with biochar and compost. Molina *et al.* (1980) [35] has proposed a two-pool kinetic model of nitrogen mineralization, which was also successfully used to predict carbon mineralization [36]. We fitted the results with the model and estimated the maximum of unstable carbon pool as given in Figure 3. The results of Sp soil were revealed only because the similar trends were found among three studied soils. Co-application of biochar with compost obviously decreased the maximum of unstable carbon pool by 12.1%–17.4% and 20.7%–24.1% for RHB-400 and RHB-700 (Figure 3a,b), respectively.

Soil Micromorphology

To determine the interactions among the compost, biochar, and soil particles, micro-structures were observed using a polarized microscope. As representatives, the microscope images of the thin section of the Lo soil and Sp soil treated with 1% compost and 2% biochars are shown in Figure 4.

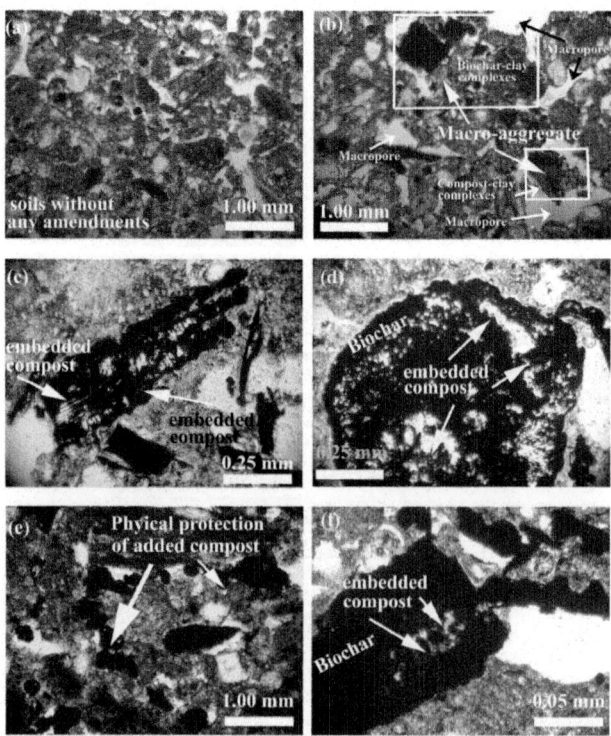

Figure 4: Microstructural observations in the biochar- and compost-amended soils (Lo soil and Sp soil) by using a polarized microscope: (a) un-amended

Lo soil; (b) (c)the treatment of 2% RHB-400 + 1% compost in Lo soil, plain polarized light (PPL); (d)–(f) the treatment of 2% RHB-700 + 1% compost in Sp soil with PPL.

After 70 days, macroaggregates formed during the mutual interaction among the soil particles, biochar, and compost (Figure 4a,b). Microstructure changed from single spaced porphyric (unamended treatment) to single spaced equal enaulic (biochar treatment) based on the micromorpholigical description guidelines [33]. Figure 4c–f further indicated that added compost (brown color) was obviously embedded or adsorbed into the micropores and surface of the biochar.

DISCUSSIONS

CO_2 Emissions from Soils Amended with the Compost and Biochars

Some researchers have suggested that biochar application in soils might facilitate the decomposition rate of organic matter to maintain nutrients for crops in temperate regions [9,10,11,18,35]. On contrary, in subtropical or tropical regions, it is better to stabilize OM from rapidly decomposing inducing financial lose while reducing compost application for land sustainability. Therefore, a new management strategy to slow down decomposition of organic amendments is necessary, particularly in subtropical and tropical regions.

The current results show that the treatments involving biochar exhibited higher cumulative CO_2 emission than biochar-free treatments (Figure 1). This could be due largely to the significantly higher emission of CO_2 in the first two weeks. Similar results have been reported by some studies [3,9,18], which indicated that labile C in biochars could effectively lead to an increase in CO_2 emission because of priming effects. Deenik et al. (2011) [37] demonstrated that biochars with high volatile matter (VM) contents also provide a source of bioavailable C, which stimulates microbial growth and increases C mineralization in soils. Therefore, we deduced that the increased CO_2 emission occurred after biochar addition because of (1) the mineralization of labile C (including VM) in the biochars, (2) interactive priming effects among the biochars, compost, and native SOM, and (3) the facilitation of soil aeration by biochar addition, which could be demonstrated by microstructure observation (Figure 4).

With incubation time, this study verified that added OM could be gradually stabilized through biochar addition. Figure 3indicates that the cumulative CO_2 emission curves of the measured values approximately reached a plateau while those of the expected values kept increasing, which expressed that the unstable C pool might decline by co-application.

Furthermore, in this study, the biochar produced at a lower pyrolization temperature seemed to induce more cumulative CO_2 emissions in the biochar-amended soils than the biochar produced at a higher temperature did (Figure 1). This may be attributable to a greater proportion of recalcitrant C [12] and lower ROC content in RHB-700 than in RHB-400 (Table 2) [17,18]. We supposed that co-application of biochars with composts may be a better way to stabilized SOM and sequestrate carbon in the soils than individual application, especially for a biochar produced with higher temperatures.

To clarify the interaction among biochar, compost, and soil component, a micro-scale observation was carried out by polarized microscope. From our microstructure observation (Figure 4), a mechanism of SOM stabilization by biochar addition could be deduced as follows: soil structure was changed and some macro-aggregate were formed after biochar incorporation (Figure 4a;4b), which was also provided by our previous studies [38,39]. The formation of the new aggregates wrapped the biochar and compost in the aggregates, and therefore might prevent from rapidly decomposing by microbes (Figure 4c–f). Accordingly, the decreases in unstable carbon pool (Figure 3) may also result from the sorption of compost-derived carbon onto the biochar, either within the biochar pores (Figure 4c,d,f) or onto the external biochar surfaces (Figure 4e). Cornelissen *et al.* (2005) [40] and Sobek *et al.* (2009) [41] reported that biochars exhibit extremely high adsorption affinity for organic matter and might suppress organic C mineralization. In addition, Kasozi *et al.* (2010) [42] reported that the organic matter sorption onto biochar surfaces is kinetically limited by slow diffusion into the subnanometer-sized pores dominating biochar surfaces. The various organomineral interactions lead to aggregations of clay particles and organic materials, which stabilizes both soil structure and the carbon compounds within the aggregates.

Effect of Soil Type on Carbon Mineralization with Compost and Biochar Amendment

According to the results shown in Figure 1, the differences in cumulative CO_2 emission between the treatments with and without biochar were approximately 59.9–252, 85.8–327, and 81.0–190 mg/kg for the Lo, Sp, and Ct soils, respectively. Sigua*et al.* (2014) [43] has incorporated several biochars into loamy and sandy soils, and the loamy soil exhibited a cumulative CO_2-C emission that was two to three-fold higher than that of the sandy soil, which was explained by the higher content of labile SOC in the loamy soil. In this study, the highest cumulative CO_2 emission was observed in the Sp soil for all treatments (Figure 1), which may be attributable to the higher ROC content of the Sp soil compared with other soils (Table 1).

Except for the effects of the initial labile C pool on the cumulative CO_2 emission, soil texture, and soil pH are also suggested to be critical control factors in carbon decomposition or CO_2 emission. Sissoko and Kpomblekou-A (2010) [44] indicated that the stabilizing effect of organic matter in soils contributed to the encapsulation between clay particles and entrapment of organic matter in small pores of aggregates, which are inaccessible to microbes. Qauuym *et al.* (2012) [11] revealed that charcoal is considerably more stable in Oxisol than in Alfisol, because Oxisol generally contains higher clay and Fe/Al oxide content. Furthermore, fine soil texture can physically protect SOM against decomposition by microbes through the adsorption of organic matter onto the inorganic clay surface and the entrapment of organic matter in small aggregates [15,45]. However, the lowest rate and cumulative amount of CO_2 emission for all treatments were found in the Ct soil, which had much lower clay than the other two soils (Figure 1). Moreover, the ACM (%) was higher in the Sp soil for all treatments than the Lo and Ct soils (Figure 2), which indicates that carbon derived either from the compost or biochars decomposed more rapidly in the Sp soil despite its clay content not being the lowest.

The highest pH value (pH 8) of the Ct soil might alter the microbial population and therefore lead to a lower cumulative CO_2 emission than that of the other two soils. Soil pH value might influence the microbial activity and consequently limit the decomposition of applied organic amendments [46,47]. Therefore, the neutral Sp soil may be more suitable for microbial activity and consequently exhibited higher CO_2 emission than the other two soils. Our results implied that the efficiency of carbon stabilization caused by biochar application may be more sensitive to the soil pH than the clay content.

CONCLUSIONS

One of the best management practices (BMPs) for land sustainability in subtropics and tropics is long-term stabilization of SOM. Based on our results, the potential benefits of biochar application could reduce the C mineralization of the added compost through mutual interaction of biochars and compost and, thus, extend the efficiency of the compost application. Co-application of compost with biochar may be more efficient to sequester C than individual application into soils, especially for the biochar produced at high pyrolization temperature. In this study, applying 4% of both husk biochars produced at 400 and 700 °C to soils with 1% compost provided the highest efficiency in reducing the C loss in soils.

ACKNOWLEDGMENTS

The authors thank the Ministry of Science and Technology, Republic of China, for financially supporting this research under contract number NSC-101-2313-B-020-013-MY2. The authors are also grateful to Chuan-Chi Chien from the Industrial Technology Research Institute, Tainan, Taiwan, for providing the rice hull biochar.

AUTHOR CONTRIBUTIONS

Dr. Shih-Hao Jien designed all research, made all tables and figures and finished this paper writing; Chung-Chi Wang, who was the master student graduated from Dr. Jien's lab performed this research and analyzed the data; Dr. Chia-Hsing Lee and Tsung-Yu Lee provided their valuable opinions during the manuscript writing. All authors read and approved the final manuscript.

REFERENCES

1. Mekuria, W.; Noble, A.; Sengtaheuanghoung, O.; Hoanh, C.T.; Bossio, D.; Sipaseuth, N.; McCartney, M.; Langan, S. Organic and Clay-Based Soil Amendments Increase Maize Yield, Total Nutrient Uptake, and Soil Properties in Lao PDR. *Agroecol. Sustain. Food Syt.* 2014, *38*, 936–961.
2. Bolan, N.S.; Kunhikrishnan, A.; Choppala, G.K.; Thangarajan, R.; Chung, J.W. Stabilization of carbon in composts and biochars in relation to carbon sequestration and soil fertility. *Sci. Total Environ.* 2012, *424*, 264–270.
3. Troy, S.M.; Lawlo, P.G.; O'Flynn, C.J.; Healy, M.G. Impact of biochar addition to soil on greenhouse gas emissions following pigmanure application. *Soil Biol. Biochem.* 2013, *60*, 173–181.
4. Lehmann, J.; Czimczik, C.; Laird, D.; Sohi, S. Stability of biochar in the soil. In *Biochar for Environmental Management: Science and Technology*; Lehmann, J., Joseph, S., Eds.; Earthscan: London, UK, 2009; pp. 183–205. [Google Scholar]
5. Yuan, J.H.; Xu, R.K.; Zhang, H. The forms of alkalis in the biochar produced from crop residues at different temperatures. *Bioresour. Technol.* 2011, *102*, 3488–3497.
6. Zhao, X.; Wang, J.W.; Xu, H.J.; Zhou, C.J.; Wang, S.Q.; Xin, G.X. Effects of crop-straw biochar on crop growth and soil fertility over a wheat-millet rotation in soils of China. *Soil Use Manag.* 2014, *30*, 311–319.
7. Gaunt, J.; Lehmann, J. Energy balance and emissions associated with biochar sequestration and pyrolysis bioenergy production. *Environ. Sci.*

Technol. 2008, *42*, 4152–4158.
8. Laird, D.A. The charcoal vision: A win-win-win scenario for simultaneously producing bioenergy, permanently sequestering carbon, while improving soil and water quality. *Agron. J.* 2008, *100*, 178–181.
9. Rogovska, N.; Laird, D.; Cruse, R.; Fleming, P.; Parkin, T.; Meek, D. Impact of Biochar on Manure Carbon Stabilization and Greenhouse Gas Emissions. *Soil Sci. Soc. Am. J.* 2011, *75*, 871–879.
10. Awad, Y.M.; Blagodatskaya, E.; Ok, Y.S.; Kuzyakov, Y. Effects of polyacrylamide, biopolymer, and biochar on decomposition of soil organic matter and plant residues as determined by ^{14}C and enzyme activities. *Eur. J. Soil Biol.* 2012, *48*, 1–10.
11. Awad, Y.M.; Blagodatskaya, E.; Ok, Y.S.; Kuzyakov, Y. Effects of polyacrylamide, biopolymer and biochar on the decomposition of ^{14}C-labelled maize residues and on their stabilization in soil aggregates. *Eur. J. Soil Sci.* 2013, *64*, 488–499.
12. Qayyum, M.F.; Steffens, D.; Reisenauer, H.P.; Schubert, S. Biochars influence differential distribution and chemical composition of soil organic matter. *Plant Soil Environ.* 2014, *60*, 337–343. [Google Scholar]
13. Fernández, J.M.; Nieto, M.A.; López-de-sá, E.G.; Gascó, G.; Méndez, A.; Plaza, C. Carbon dioxide emmisions from semi-arid soils amended with biochar alone or combined with mineral and organic fertilizers. *Sci. Total Environ.* 2014, *482–483*, 1–7.
14. Kuzyakov, Y.; Subbotina, I.; Chen, H.; Bogomolova, I.; Xu, X. Black carbon decomposition and incorporation into soil microbeal biomass estimated by ^{14}C labeling. *Soil Biol. Biochem.* 2009, *41*, 210–219.
15. Van Veen, J.A.; Kuikman, P.J. Soil structure aspects of decomposition of organic matter by micro-organisms. *Biogeochemistry* 1990, *11*, 213–233.
16. Hamer, U.; Marschner, B.; Brodowski, S.; Amelung, W. Interactive priming of black carbon and glucose mineralisation. *Org. Geochem.* 2004, *35*, 823–830.
17. Zimmerman, A.R.; Gao, B.; Ahn, M.Y. Positive and negative carbon mineralization priming effects among a variety of biochar-amended soils. *Soil Biol. Biochem.* 2011, *43*, 1169–1179.
18. Keith, A.; Singh, B.; Singh, B.P. Interactive Priming of Biochar and Labile Organic Matter Mineralization in a Smectite-Rich Soil. *Environ. Sci. Technol.* 2011, *45*, 9611–9618.
19. Streubel, J.D.; Collins, H.P.; Garcia-Perez, M.; Tarara, J.; Granatstein,

D.; Kruger, C.E. Influence of contrasting biochar types on five soils at increasing rates of application. *Soil Sci. Soc. Am. J.* 2011, *75*, 1402–1413.
20. Thomas, G.W. Soil pH and soil acidity. In *Methods of Soil Analysis: Soil Science Society of America Book Series 5 Part 3—Chemical Methods*; Sparks, D.L., Ed.; ASA and SSSA: Madison, WI, USA, 1996; pp. 487–488.
21. Rhoades, J.D. Soluble salts. In *Methods of Soil Analysis Part 2—Chemical and Microbiological Properties*; Page, A.L., Ed.; ASA and SSSA: Madison, WI, USA, 1982; pp. 167–179.
22. Gee, G.W.; Bauder, J.W. Particle-size analysis. In *Methods of Soil Analysis Part 1—Physical and Mineralogical Methods*; Klute, A., Ed.; ASA and SSSA: Madison, WI, USA, 1986; pp. 383–411.
23. Sumner, M.E.; Miller, W.P. Cation exchange capacity and exchange coefficients. In *Methods of Soil Analysis: Soil Science Society of America Book Series 5 Part 3—Chemical Methods*; Sparks, D.L., Ed.; ASA and SSSA: Madison, WI, USA, 1996; pp. 1218–1220.
24. Nelson, D.W.; Sommers, L.E. Total carbon, organic carbon, and organic matter. In *Methods of Soil Analysis: Soil Science Society of America Book Series 5 Part 3—Chemical Methods*; Sparks, D.L., Ed.; ASA and SSSA: Madison, WI, USA, 1996; pp. 961–1010.
25. Bremner, J.M.; Mulvaney, C.S. Nitrogen-total. In *Methods of Soil Analysis, Part 2—Chemical and Microbiological Properties*; Page, A.L., Ed.; ASA and SSSA: Madison, WI, USA, 1982; pp. 595–624.
26. Mulvaney, R.L. Nitrogen-Inorganic forms. In *Methods of Soil Analysis: Soil Science Society of America Book Series 5 Part 3—Chemical Methods*; Sparks, D.L., Ed.; ASA and SSSA: Madison, WI, USA, 1996; pp. 1123–1184.
27. Loeppert, R.H.; Suarez, D.L. Carbonate and gypsum. In *Methods of Soil Analysis Part 2—Chemical and Microbiological Methods*, 2nd ed.; Agronomy Monograph 9; America Society of Agronomy and Soil Science Society of America: Madison, WI, USA, 1982; pp. 437–451.
28. Blair, G.J.; Lefroy, R.D.B.; Lisle, L. Soil carbon fractions based on their degree of oxidation, and the development of a carbon management index for agricultural systems. *Aust. J. Agric. Res.* 1995, *46*, 1459–1466.
29. Conteh, A.; Lefroy, R.D.B.; Blair, G.J. Dynamics of organic matter in soils as determined by variations in $^{13}C/^{12}C$ isotopic ratios and fractionation by ease of oxidation. *Aust. J. Soil Res.* 1997, *35*, 881–890.
30. Zibilske, L.M. Carbon mineralization. In *Methods of Soil Analysis:*

Soil Science Society of America Book Series 5 Part 2—Microbial and Biochemical Properties; Weaver, R.W., Ed.; ASA and SSSA: Madison, WI, USA, 1994; pp. 835–863.

31. Novak, J.M.; Busscher, D.W.; Watts, D.W.; Laird, D.A.; Ahmedna, M.A.; Niandou, M.A.S. Short-term CO_2 mineralization after additions of biochar and switchgrass to a TypicKandiudult. *Geoderma* 2010, *154*, 281–288.

32. Riberio, H.M.; Fanqueiro, D.; Alves, F.; Vasconcelos, E.; Coutinho, J.; Bol, R.; Cabral, F. Carbon-mineralization kinetics in an organically managed Cambic Arenosol amended with organic fertilizers. *J. Plant Nutr. Soil Sci.* 2010, *173*, 39–45.

33. Stoops, G. *Guidelines for Analysis and Description of Soil and Regolith Thin Sections*; Soil Science Society of Amenrica, Inc.: Madison, WI, USA, 2003.

34. Soil Survey Staff. *Keys to Soil Taxonomy, 11th edn USDA-NRCS, Agricultural Handbook No. 436*; US Government Printing Office: Washington, DC, USA, 2010.

35. Molina, J.A.E.; Clapp, C.E.; Larson, W.E. Potentially mineralizable nitrogen in soil: the simple exponential model does not apply to the first 12 weeks of incubation. *Soil Sci. Soc. Am. J.* 1980, *44*, 442–443.

36. Liang, B.Q.; Lehmann, J.; Solomon, D.; Kinyangi, J.; Grossman, J.; O'Neill, B.; Skjemstad, J.O.; Thies, J.; Luizão, F.J.; Petersen, J.; et al. Black carbon increases cation exchange capacity in soils. *Soil Sci. Soc. Am. J.* 2006, *70*, 1719–1730.

37. Deenik, J.L.; Diarra, A.; Uehara, G.; Campell, S.; Sumiyoshi, Y.; Antal, M.J., Jr. Charcoal ash and volatile matter effects on soil properties and plant growth in an acid Ultisol. *Soil Sci.* 2011, *176*, 336–345.

38. Jien, S.H.; Wang, C.S. Effects of biochar on soil properties and erosion potential in a highly weathered soil. *Catena* 2013, *110*, 225–233.

39. Hseu, Z.Y.; Jien, S.H.; Chien, W.S.; Liou, R.C. Impacts of biochar on physical properties and erosion potential of a mudstone slopeland soil. *Sci. World J.* 2014.

40. Cornelissen, G.; Gustafsson, O.; Bucheli, T.D.; Jonker, M.T.O.; Koelmans, A.A.; VanNoort, P.C.M. Extensive sorption of organic compounds to black carbon, coal, and kerogen in sediments and soils: Mechanisms and consequences for distribution, bioaccumulation, and biodegradation. *Environ. Sci. Technol.* 2005, *39*, 6881–6895.

41. Sobek, A.; Stamm, N.; Bucheli, T.D. Sorption of phenyl urea herbicides to black carbon. *Environ. Sci. Technol.* 2009, *43*, 8147–8152.

42. Kasozi, G.N.; Zimmerman, A.R.; Nkedi-Kizza, P.; Gao, B. Catechol and humic acid sorption onto a range of laboratory-produced black carbons (biochars). *Environ. Sci. Technol.* 2010, *44*, 6189–6195.
43. Sigua, G.C.; Novak, J.M.; Watts, D.W.; Cantrell, K.B.; Shumaker, P.D.; Szogi, A.A.; Johnson, M.G. Carbon mineralization in two ultisols amended with different sources and particle sizes of pyrolyzed biochar. *Chemosphere* 2014, *103*, 313–321.
44. Sissoko, A.; Kpomblekou-A, K. Carbon decomposition in broiler litter-amended soils. *Soil Biol. Biochem.* 2010, *42*, 543–550.
45. Sørensen, L.H. Size and persistence of the microbial biomass formed during the humification of glucose, hemicellulose, cellulose, and straw in soils containing different amounts of clay. *Plant Soil* 1983, *75*, 121–130.
46. Motavalli, P.P.; Palm, C.A.; Parton, W.J.; Elliott, E.T.; Frey, S.D. Soil pH and organic C dynamics in tropical forest soils: evidence from laboratory and simulation studies. *Soil Biol. Biochem.* 1995, *27*, 1589–1599.
47. Huang, C.C.; Chen, Z.S. Carbon and nitrogen mineralization of sewage sludge compost in soils with different initial pH. *Soil Sci. Plant Nutr.* 2009, *55*, 715–724.

Chapter 3

BEHAVIOR OF CLAYEY SOIL STABILIZED WITH RICE HUSK ASH & LIME

B.Suneel Kumar[1] & T.V.Preethi[2]

[1]Graduate Student, Department of Civil Engineering, Geotechnical Engineering, SRM University, Kattankulathur-60320, Tamil Nadu, India

[2]Assistant professor, Department of Civil Engineering, SRM University, Kattankulathur-603203, Tamil Nadu, India

ABSTRACT

In India the soil mostly present is Clay, in which the construction of sub grade is problematic. In recent times the demands for sub grade materials has increased due to increased constructional activities in the road sector and due to paucity of available nearby lands to allow excavate fill materials for making sub grade. In this situation, a means to overcome this problem is to utilize the different alternative generated waste materials, which cause not only environmental hazards and also the depositional problems. Keeping this in view stabilization of weak soil in situ may be done with suitable admixtures to save the construction cost considerably. The present investigation has therefore been carried out with agricultural waste materials like Rice Husk Ash (RHA) which was mixed with soil to study improvement of weak sub grade in terms of compaction and strength characteristics. Silica produced from rice husk ashes have investigated successfully as a pozzolanic material in soil stabilization. However, rice husk ash cannot be used solely since the materials lack in calcium element. As a result, rice husk ash shall be mixed with other cementitious materials such as lime and cement to have a solid chemical reaction in stabilization process. Lime is calcium oxide or calcium hydroxide. It is the name of the natural mineral (native lime) CaO occurs as a product of coal seam fires and in altered lime stone xenoliths in volcanic ejection. In this study RHA and Lime is mixed in different percentage like (RHA as 5%, 10%, and 15%) and (Lime as 3%, 6%, 9%) and laboratory test CBR is done with a curing period of 4, 7 and 14 days with different percentages of RHA & Lime and Lime+ RHA.

INTRODUCTION

Soil improvement could either be by modification or stabilization or both. Soil modification is the addition of a modifier (cement, lime etc.) to a soil to change its index properties, while soil stabilization is the treatment of soils to enable their strength and durability to be improved such that they become totally suitable for construction beyond their original classification. Over the times, cement and lime are the two main materials used for stabilizing soils. These materials have rapidly increased in price due to the sharp increase in the cost of energy since 1970s (Neville, 2000). The over dependence on the utilization of industrially manufactured soil improving additives (cement, lime etc), have kept the cost of construction of stabilized road financially high. This hitherto, has continued to deter the underdeveloped and poor nations of the world from providing accessible roads to their rural dwellers who constitute the higher percentage of their population and are mostly, agriculturally dependent. Thus the use of agricultural waste (such as rice husk ash) will considerably reduce the cost of construction and as well reducing the environmental hazards they causes. Therefore, replacing proportions of the Portland cement in soil stabilization with a secondary cementitious material like RHA will reduce the overall environmental impact of the stabilization process. Silica produced from rice husk ashes have investigated successfully as a pozzolanic material in soil stabilization. However, rice husk ash cannot be used solely since the materials lack in calcium element. As a result, rice husk ash shall be mixed with other cementitious materials such as lime and cement to have a solid chemical reaction in stabilization process.

Lime is a general term for calcium-containing inorganic materials in which carbonates, oxides and hydroxides predominate. Strictly speaking, lime is calcium oxide or calcium hydroxide. It is the name of the natural mineral (native lime) CaO occurs as a product of coal seam fires and in altered lime stone xenoliths in volcanic ejection. The word "lime" originates with its earliest use as building mortar and has a sense of "sticking and or adhering". "Burning" converts them into the highly caustic material quicklime (calcium oxide, Cao) and through subsequent addition of water, into less caustic (but still strongly alkaline) slaked lime or hydrated lime (calcium hydroxide, $CA(OH)_2$ =74.10), the process of which is called slaking of lime.

Rice husk is an agricultural waste obtained from milling of rice. About 108 tonnes of rice husk is generated annually in the world. Meanwhile, the ash has been categorized under pozzolana, with about 67-70% silica and about 4.9% and 0.95%, Alumina and iron oxides, respectively (Oyetola and Abdullahi, 2006). The silica is substantially contained in amorphous form, which can react with the CaOH librated during the hardening of cement to further form cementations compounds. Light compaction energy, the effect of Rice Husk Ash on the soil was investigated with respect to compaction characteristics, California Bearing Ratio (CBR) and unconfined compressive strength (UCS) tests. Results obtained, there was also a tremendous improvement in the CBR and UCS with increase in the RHA and lime at specified contents to their peak values at 6% lime and 10% RHA. The UCS values also improved with curing age and in the combination of lime + RHA, 6% lime+10% RHA showed good improvement in UCS and CBR value with increase in curing period. Ario Muhammad, (2007) Silica produced from rice husk ashes has investigated successfully as a pozzolanic material in soil stabilization. However, rice husk ash cannot be used solely since the materials lack in calcium element. As a result, rice husk ash shall be mixed with other cementitious materials such as lime and cement to have a solid chemical reaction in stabilization process. For the stabilized soils, the admixture materials, i.e. lime, rice husk ash, were mixed in 12 % and 24 % of the dry weight of soil matrix respectively.

Biswas, (2007) The present investigation has therefore been carried out with agricultural waste materials like Rice Husk Ash (RHA) which was mixed with soil or lime-soil mixture to study improvement of weak sub grade in terms of compaction and strength characteristics. The laboratory test results show marked improvement of strength of soil on addition of admixtures in terms of California Bearing Ratio (CBR). In this study RHA is mixed with lime stabilized soil in which lime is 3%, 6% & 9%. Musa Alhassan, (2008) Soil sample collected from Maikunkele area of Minna, was stabilized with 2-12% rice husk ash (RHA) by weight of the dry soil. Performance of the soil-RHA was investigated with respect to compaction characteristics, California bearing ratio (CBR) and unconfined compressive strength (UCS) tests. The results obtained, indicates a general decrease in the maximum dry density (MDD) and increase in optimum moisture content (OMC) with increase in RHA content. There was also slight improvement in the CBR with increase in the RHA content. Chakraborty & Saibal, (2010) in recent times the need for suitable road materials has increased due to demand of construction activities in the road sector. Keeping this in view stabilization of weak soil in situ may be done with suitable admixtures. Investigation has therefore been carried out with agricultural waste materials like Rice Husk Ash (RHA) which was mixed with soil or lime-soil mixture to study improvement of weak sub grade in terms of compaction and strength characteristics. Abu Siddique (2011), the effects of lime stabilisation on plasticity, shrinkage, swelling, moisture-density relations and strength characteristics of an expansive soil have been investigated. The soil was stabilised with lime contents of 3%, 6%, 9%, 12% and 15%. With the increase in lime content, maximum dry density decreased while the optimum moisture content increased. California Bearing Raito (CBR) of the stabilised samples at all levels of compaction increased significantly with increasing lime content.

MATERIALS USED

Soil Used

Soil sample is collected from a proposed for the construction of road alignment in guduvanchery area, Chennai. Standard tests were conducted to determine the physical properties of the soil and the results are given in Table 1.

Table 1: Physical properties of soil

SL. no	Test Conducted	Result	
1	Wet sieve analysis	% passing 75 microns sieve is 56%, so it is fine grained soil	
2	Liquid limit	36.5 %	
3	Plastic limit	12.62 %	
4	Plasticity index on A-line	11.61 %	
5	Shrinkage limit	21.68 %	
6	Specific gravity	2.23	
7	Free swell index	18.18 %	
8	Standard proctor	OMC 15.15 %	MDD 1.797 kg/cm^3
9	UCS (kg/sq cm)	0.624	
10	CBR	5.48 %	

Classification of Soil Sample

Based upon the tests performed in laboratory for soil sample and according to the results obtained, the soil sample is classified as follows,

- 56% of soil is passing through 75 microns sieve so it is fine grained soil.
- According to A-line Chart, the soil can be classified as clay with Intermediate Compressibility –CI
- According to free swell index value, the soil is classified as low compressible.

Rice Husk Ash (RHA)

Husk Ash (RHA) Rice husk ash, basically a waste material, is produce by rice - mill industry while processing rice from paddy. Rice husk ash is a pozzolanic material that could be potentially used in soil stabilization, though it is moderately produced and readily available. About 20 – 22% rice husk is generated from paddy and about 25% of this total husk become ash when burn. It is non – plastic in nature. RHA has a good pozzolanic property. The chemical properties of RHA are shown in Table 2.

Table 2: Chemical properties of rice husk ash

Chemical	Percentage (%)
Silica(SIO_2)	83.60
Aluminium(Al_2O_3)	3.5
Iron(FEO_3)	1.10
Calcium (CAO)	1.80
Magnesium(MGO)	1.28
Sodium(NA_2O)	0.17
Potassium(K_2O)	0.29

Silica produced from rice husk ashes have investigated successfully as a pozzolanic material in soil stabilization. However, rice husk ash cannot be used solely since the materials lack in calcium element. Rice husk ash shall be mixed with other cementitious materials such as lime and cement to have a solid chemical reaction in stabilization process.

Lime

Lime is a general term for calcium-containing inorganic materials in which carbonates, oxides and hydroxides predominate. Strictly speaking, lime is calcium oxide or calcium hydroxide. It is the name of the natural mineral (native lime) CaO occurs as a product of coal seam fires and in altered lime stone xenoliths in volcanic ejects. The word "lime" originates with its earliest use as building mortar and has a sense of "sticking and or adhering".

These materials are still used in large quantities as building and engineering materials (including limestone products, concrete and mortar) and as chemical feedstock's, and sugar refining, among other uses. Lime industries and the use of many of the resulting products date from prehistoric periods in both the old & new worlds.

The rocks and minerals from which these materials are derived, typically limestone or chalk, are composed primarily of calcium carbonate. They may be cut, crush or pulverized and chemically altered. "Burning" converts them into the highly caustic material quicklime (calcium oxide, Cao) and through subsequent addition of water, into less caustic (but still strongly alkaline) slaked lime or hydrated lime (calcium hydroxide, $CA(OH)_2 = 74.10$), the process of which is called slaking of lime.

LABORATORY STUDIES

The testing program conducted on the clayey soil samples included determination of the physical and chemical properties of soils at their natural state. On the other hand, the testing program conducted on the clayey soil samples mixed with different percentages of rice husk ash and lime materials, included unconfined compression test and CBR test.

A. Unconfined compression test UCS test is performed in accordance with IS:2720 part 10 (1973). The sample sizes were of 38 mm diameter and 76 mm length. At the optimum moisture content (OMC) and maximum dry unit weight, the tests were performed. B. California bearing ratio (CBR) is a penetration test for evaluation of the mechanical strength of road sub grades and base courses. It was developed by the California Department of Transportation before World War II.

The test is performed by measuring the pressure required to penetrate a soil sample with a plunger of standard area. The measured pressure is then divided by the pressure required to achieve an equal penetration on a standard crushed rock material. The CBR test is fully described in IS: 2720 part 16 (1987).

RESULTS AND DISCUSSIONS

UCC Test Results

UCC test was conducted in laboratory on soil sample with addition of different percentages of lime and RHA and the results obtained are shown in table no 3 and the figures 1 & 2 shows the graphs for UCC. Figure 1 shows the UCS value for different percentages of lime and figure 2 shows the UCS value for different percentages of RHA.

Table 3: UCC Test Results on soil sample

Additives	UCS Value (Kg/Sq Cm)		
	4 days	7 days	14 days
3% LIME	2.338	5.32	6.464
6% LIME	4.558	6.26	8.62
9% LIME	1.662	3.698	4.504
5% RHA	2.248	3.29	4.068
10% RHA	3.344	4.552	6.14
15% RHA	1.788	2.75	3.148

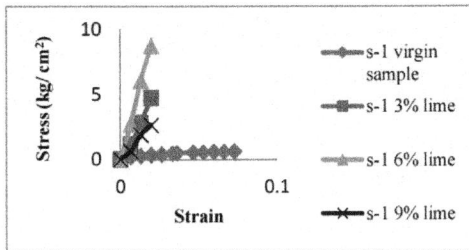

Figure 1: UCS value for lime.

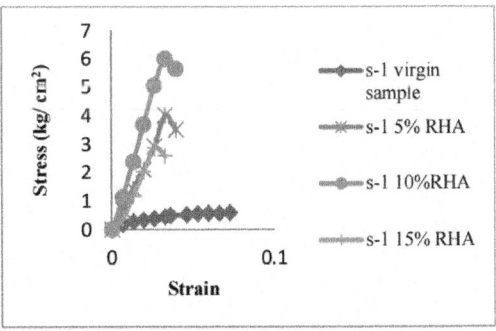

Figure 2: UCS value for RHA.

Discussion for Soil Sample

Based on the ultimate UCS value as shown in Figure 3 & 4, the CBR test was performed for 6% of lime and 10% of RHA individually and combination of lime + RHA is done for 6% lime with different percentages of RHA i.e. (5%, 10%, and 15%).

CBR Test Results

CBR test was conducted in laboratory on soil sample with addition of different percentages of lime and RHA and the results obtained are shown in Table 4.

Table 4: CBR Test Results for Soil Sample

Additives	CBR Value For Curing Period			
	0 day	4 days	7 days	14 days
6% LIME	13.95	33.11	50.8	66.75
10% RHA	7.9	8.21	9.8	13.77
6% LIME + 5% RHA	14.56	26.7	27.54	39.73
6% LIME + 10% RHA	17.21	29.66	45.8	56.68
6% LIME+ 15 % RHA	7.9	20.13	21.33	27.8

Figure 3 shows the graph drawn for CBR value for 6% of lime and 10 % of RHA which can be compared with virgin sample. Percentage of lime and RHA are taken from ultimate UCS value.

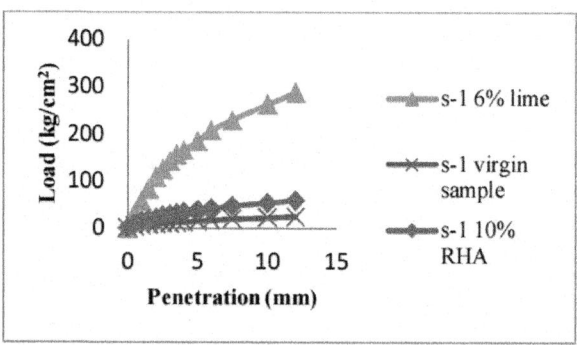

Figure 3: CBR value for lime and RHA.

Figure 4 shows the graphs drawn for combination of lime +RHA in which ultimate UCS value of 6% lime is taken and mixed with RHA in different percentages i.e. (5, 10, and 15%). For which 6%lime + 10% RHA is giving the ultimate value.

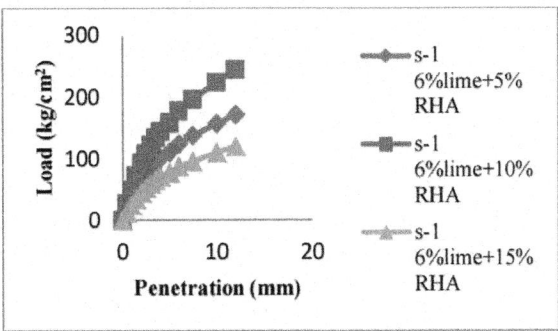

Figure 4: CBR value for lime+ and RHA.

CONCLUSION

- Based on the UCS value comparison, 6% addition of lime showed the good improvement of 92.74% in UCS value for 14 days curing compared to virgin soil and for 10% of RHA with 89.93% for a same period of curing.
- CBR test was performed based on addition of 6% of lime for curing period of 14 days which showed an improvement of 91.79% and for the 10% RHA, CBR value was increased up to 60.20% at 14th day. CBR test is also conducted for the combination of lime+RHA in which lime is taken as 6% and the combination of RHA is done in different percentages i.e (5%, 10%, & 15%) in which maximum improvement of CBR is observed in the combination of 6%lime+10%RHA for which % of CBR improvement are 90.65% for 14th day of curing comparing with virgin sample.
- Addition of industrial waste (RHA) alone gave an average improvement of 60% when compared with virgin sample. When the additive lime is added to it the CBR value increased to great extent which had been mentioned above (91%). So based on the respective results, quality of soil is increasing from poor condition to excellent condition based on CBR test values. As per the Pavement Design, when the California Bearing Ratio increases, the sub grade thickness can be reduced. So the RHA and Lime can be used to improve the CBR ratio respectively.

REFERENCE

1. Agus Setyo Muntohar, "Uses of Lime -Rice Husk Ash And Plastic Fibers as Mixtures-material in High-plasticity Clayey Sub Grade", Journal Ilmiah Semesta Teknika, Vol. 10, 146 No. 2, 2007: 145 – 154.

2. Chakraborty & Saibal, "Stabilization of Sub grades of Flexible Pavements with Admixtures", Indian Geotechnical Conference – 2010, GEOtrendz December 16–18, 2010 IGS Mumbai Chapter & IIT Bombay.
3. IS: 2720- Part 5-1985 ,"Determination of liquid limit and Plastic limit".
4. IS: 2720 - Part 40- 1977 ," Determination of free swell".
5. IS: 2720- Part 3 –sect. 1-1980, "Determination of specific gravity".
6. IS: 2720- Part 6-1972, "Determination of Shrinkage limit".
7. Sudhira rath, "Lime Stabilization of Weak Sub-Grade for Construction of Rural Roads", International journal of earth sciences AND engineering Issn 0974-5904, vol. 05, no. 03 (01), june 2012, PP. 554-561.
8. Koteswara Rao, D., "Stabilization Of Expansive Soil With Rice Husk Ash, Lime And Gypsum", International Journal of Engineering Science and Technology (IJEST) ISSN : 0975-5462 Vol. 3, No. 11 November 2011.
9. " S.K.Khanna And C.E.G.Justo, "Highway engineering" khanna publications ninth edition (2011)
10. Brooks, R. M., (2009), "Soil Stabilization with Fly ash and Rice Husk Ash", International Journal of Research and Reviews in Applied Sciences, Volume 1, Issue 3, pp. 209- 217.
11. Gidigasu, M.D., (1976), "Laterite Soil Engineering: Pedogenesis and Engineering Principles", Elsevier, Amsterdam, the Netherlands.
12. Ito, K. K, Senge, M., Adomako, J. T., and Afandi, (2008), "Amendment of Soil Physical and Biological Properties Using Rice Husk and Tapioca Wastes", Journal of Jpnanese Society of Soil Physics, No. 108, pp. 81-90.
13. Experimental Study", International Journal of Engineering Science and Technology, Vol. 3 No. 11, pp. 8076 – 8085.
14. Mtallib, M. O. A., and Bankole, G. M., (2011), "The Improvement of the Index Properties and Compaction Characteristics of Lime Stabilized Tropical Lateritic Clays with Rice Husk Ash (RHA) Admixtures", Electronic Journal of Geotechnical Engineering, Vol. 16, Bund. I, pp. 984-996.
15. Muntohar, S., and Hantoro, G., (2000), "Influence of Rice Husk Ash and Lime on Engineering Properties of a Clayey Sub-grade", Electronic Journal of Geotechnical Engineering, Vol. 5.
16. Neville A. M., (2000), "Properties of Concrete", 4th edition. Pearson Education Asia Ltd, Malaysia.
17. Ola, S.A., (1975), "Stabilization of Nigeria Lateritic Soils with Cement, Bitumen and Lime", Proc. 6th Reg. Conf. Africa on Soil Mechanics and

Foundation Engineering. Durban, South Africa.

18. Osinubi K.J., (1999), "Evaluation of Admixture Stabilization of Nigeria Black Cotton Soil", Nigeria Soc. Engin. Tech. Trans., Vol. 34, No. 3, pp. 88-96.

19. Osinubi, K.J. and Katte, V.Y., (1997), "Effect of Elapsed Time after Mixing on Grain Size and Plasticity Characteristic, I: Soil-Lime Mixes", NSE Technical Transactions Vol. 32, No. 4.

20. Osula D. O. A., (1991), "Lime Modification of Problem Laterite", Engineering Geology, Vol. 30, pp. 141-149.

Chapter 4

CEMENT KILN DUST CHEMICAL STABILIZATION OF EXPANSIVE SOIL EXPOSED AT EL-KAWTHER QUARTER, SOHAG REGION, EGYPT

Hesham A. H. Ismaiel

Geology Department, Faculty of Science, South Valley University, Qena, Egypt

ABSTRACT

This work dealt with a chemical stabilization of an expansive high plastic soil of Pliocene deposits exposed at El-Kawther quarter using cement kiln dust (CKD) and cement kiln dust with lime (L) to reduce their swelling and improve their geotechnical properties. Several specimens of the studied expansive soil were collected from El-Kawther quarter. Chemical analysis of the used cement kiln dust and the lime was conducted. Microstructural changes were examined using scanning electron microscope (SEM) before and after chemical treatment of the studied soil. Geotechnical properties including plasticity, compaction parameters, unconfined compressive strength (qu), ultrasonic velocities and free swelling of the studied soil were measured before and after the treatment. An optimum content of the cement kiln dust was 16% (CKD). The optimum content of the cement kiln dust with the lime was 14% (CKD) with 3% (L) according to pH-test. The results showed that the addition of cement kiln dust and cement kiln dust with lime led to a decrease in maximum dry density and an increase in optimum water content. Unconfined compressive strength values were increased using cement kiln dust and cement kiln dust with lime at 7 days curing time. Ultrasonic longitudinal (Vp) and shear (Vs) velocities values were also increased by addition of the cement kiln dust and the cement kiln dust with lime at 7 days curing time. Increment of the curing time from 7 to 28 days led to an increase in both unconfined compressive strength and ultrasonic velocities values. Free swelling percent of the studied soil was reduced from 80.0% to 0.0% after the treatment.

INTRODUCTION

Environmental conditions (arid or humid regions) are major factor affecting swelling behavior of the expansive soils. The desert area as arid region (at El-Kawther quarter east Sohag city, Figure 1) is the most promising area for development of territories outside the Nile valley. The sediments of the arid region are characterized by an occurrence of expansive clay minerals; these clays may lead to a heave of the roads construction. For this reason, there is urgency to treat the expansive clayey soils to avoid the heave [1]. The chemical stabilization of the problematic soils (expansive soils and soft fine-grained soils) is very important for many of the geotechnical engineering applications such as pavement structures, roadways, building foundations, channel and reservoir linings, irrigation systems, water lines and sewer lines to avoid the damage due to the swelling action (heave) of the expansive soils or the settlement of the soft soils [2].

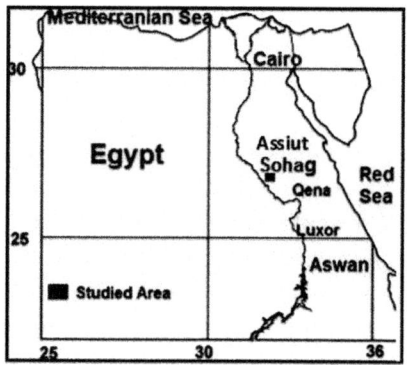

Figure 1. Location map of the studied area.

In Egypt, lime was used in-situ chemical stabilization of expansive clay soil at New Valley Governorate [3]. An addition of lime slurry showed an improvement of physical and chemical soils properties with respect to the heave [3].

Cement kiln dust is a by-product in the production of Portland cement clinker. Disposal of cement kiln dust is an environmental problem. The utilization of this waste material has received increasing attention because it not only solves a potential solid waste problem but also provides an alternative stabilizing agent using in chemical stabilization of problematic soils and provides an alternative construction material. The use of the cement kiln dust for chemical stabilization applications may be an environmental solution of the problems associated with its disposal process where a very huge amount of the cement kiln dust as by-product is daily produced from the cement factories in Egypt. The composition of the cement kiln dust is similar to raw materials of cement but the amount of alkalis, chlorides and sulfates is usually considerably higher in the cement kiln dust [4]. Cement kiln dust is not a hazardous waste material under united state environmental protection agency guidelines [5].

A roadway section consists of a complete pavement system [6] shown in Figure 2. The sub-grade refers to the in situ soils on which the stresses from the overlying roadway will be distributed. The sub-base or sub-base course and the base or base course materials are stress distributing layer overlying sub-grade layer and underlying of the pavement layer. The pavement structure consists of a relatively thin wearing surface constructed over a base course and a sub-base course, which rests upon an in situ sub-grade. The wearing surface is primarily asphalt layer.

The quality of the sub-grade soils used in pavement application is classified into 5 types (soft, medium, stiff, very stiff and hard sub-grade) depending on unconfined compressive strength values [7]. The sub-grade soils, which are classified as A7-5 and A7-6, have general rating as fair to poor according to American association of state highway and transportation officials (AASHTO). These types are considered as unstable sub-grades and need to be improved and stabilized, especially in terms of pavement applications [8]. Sub-grade soil of the studied area is an expansive and leads to the heave.

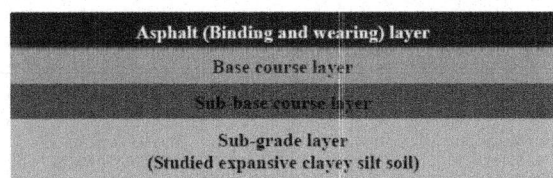

Figure 2. A typical flexible pavement structure.

Location of Study Area

The study area (El-Kawther quarter, east Sohag city) lies between old flood plain and El-Maaza limestone plateau. It lies between latitudes 26°30'N and 26°45'N and longitudes 31°45'E and 32°00'E (Figure 1). Two sections at El-Kawther sector were studied. The location of the first section is at 1 km + 400 m and the second one at 3 km + 100 m in the northwest direction.

Previous Studies

Many geological investigations were carried out on the study area like [9-13] and others. Few geotechnical investigations for lime chemical stabilization in the studied area and in Egypt were achieved like [3] (at Idku city), [14] (at Toshka area), [15] (at New Valley Governorate) and [1] (at El-Kawther area).

Scope of Present Study

This work dealt with an investigation of the geotechnical properties of an expansive high plastic clayey silt soil at El-Kawther quarter (Sohag region). The studied soil occurred as sub-grade of upper Egypt-Red Sea (SohagSafaga) road. The main goal of this study was a chemical stabilization program of the studied sub-grade soil using cement kiln dust and cement kiln dust with hydrated lime to reduce its swelling (Figure 3) and improve its geotechnical properties. Finally, the increasing trend towards cement production, in Egypt, through limestone/shale and/or marl combustion has aggravated the problems associated with the disposal of the cement kiln dust (byproduct), so that the use of the cement kiln dust as a stabilizing agent play an environmental and economical role. Economically, use of cement kiln dust as a chemical additive in chemical soil stabilization and for geotechnical applications is cheaper than Portland cement and other (expansive) chemical stabilization. Environmentally, reuse of the cement kiln dust for soil chemical stabilization may be a solution of the environmental problems associated with its disposal process where a very huge amount of the cement kiln dust as by-product are daily produced from the cement factories in Egypt.

Figure 3. A heave in asphalt layer of the studied road due to swelling of the expansive soil sub-grade at El-Kawther sector.

GEOLOGICAL SETTING OF STUDY AREA

Geology of the studied area is characterized by a large carbonate plateau called El-Maaza plateau and several terraces of variable levels intercalated occasionally by marl and carbonate sediments and covered with Quaternary sediments. The distribution of the sediments is mapped in some details in a geological map as shown in Figure 4. El-Maaza plateau which belongs to Eocene age is characterized by low relief topography with general inclination towards the west direction. The high of plateau reaches to 560 m above sea level [1]. The carbonate plateau is dissected by a number of Wadis running in a general NW-SE direction. Some of these Wadis are run in the direction of the general topographic slope and parallel to the Nile valley and other streams run in the direction opposite to both dips of the formations and the consequent streams. The major wadis dissecting the limestone plateau (East Sohag) are summarized as Wadi Abu Nafukh, Wadi Kiman, and Wadi Dir El-Hadied which connect with river Nile [10]. The studied soil (expansive high plastic clayey silt) is belonging to Issawia Formation having Pliocene age. Issawia Formation is composed of chocolate brown clay and silt. It overlies older strata mostly Eocene and underlies alluvial deposits of Pleistocene age [16].

MATERIALS AND METHODS

Materials

Fourty two specimens of the studied expansive soils were collected from the two sections exposed at El-Kawther sector as a part of the upper Egypt-Red Sea (SohagSafaga) road, east of Sohag city. The studied soil represented the sub-grade of Sohag-Safaga road at El-Kawther sector. It was composed of fine sand (20.00%), silt (55.00%) and clay (25.00%). This clayey silt is classified as (MH) according to Unified Soil Classification System (USCS) and as A7-5 according to AASHTO and it is poor to fair as sub-grade. The studied expansive soil is mainly composed of silica, iron, aluminum and calcium. The mineralogical composition of studied the expansive soil is clay (37.70%), calcite (27.80%), quartz (23.5%), gypsum (7.90%) and Hematite (3.10%). The soluble salts of the studied soil including sulfates and chlorides are 0.01% and 0.04% respectively [1]. In the present study, cement kiln dust and cement kiln dust with hydrated lime were used for chemical stabilizing of the studied expansive clayey silt soil. The chemical composition of the used cement kiln dust and the hydrated lime was illustrated in Table 1. Cement kiln dust is by-product from Qena cement plant which lies at industrial Qift city. Huge amount of the cement kiln dust (180 Tons) is daily produced. The cement kiln dust is non-plastic fine grained materials having silt size grains.

The chemical analysis showed that it was mainly composed of the oxides of calcium, aluminium, iron and silica. Hydrated lime is calcium hydroxide, Ca(OH)$_2$. It is produced by reacting quicklime (CaO) with sufficient water to form a white powder.

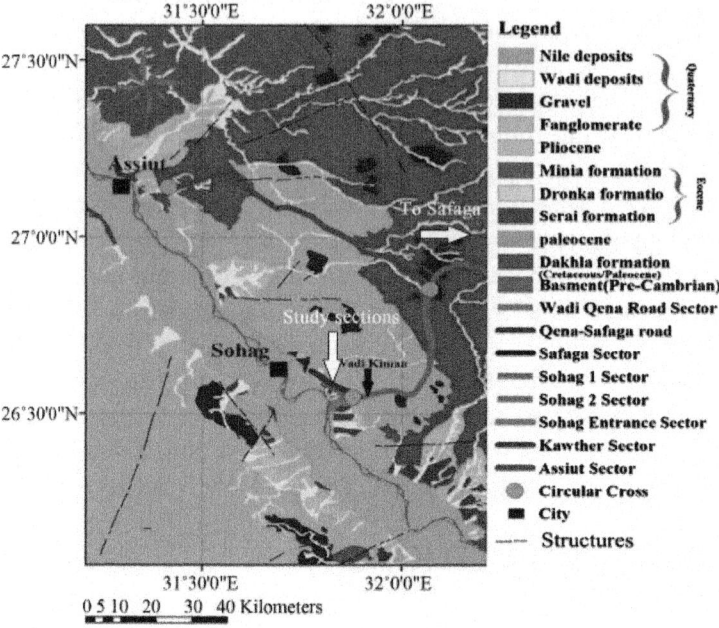

Figure 4: Geological map of the studied area modified after [1].

Table 1: Chemical composition of the used cement kiln dust and the hydrated lime

Chemical oxides (%) Type of additive	MgO	Al$_2$O$_3$	SiO$_2$	K$_2$O	CaO	TiO$_2$	Fe$_2$O$_3$	P$_2$O$_5$	Na$_2$O	SO$_3$	Cl	LOI
Cement kiln dust (CKD)	0.78	5.13	16.96	1.18	57.39	-	3.28	-	-	4.43	0.82	10.03
Hydrated lime (L)	1.90	0.10	1.16	0.27	62.6	0.01	1.40	0.01	0.55	-	-	32.0

LOI = Los of ignition.

Methods

Chemical analysis, X-ray fluorescence (XRF), of the used cement kiln dust and the hydrated lime was carried out. Chemical stabilization program (Laboratory) using cement kiln dust and cement kiln dust with lime was conducted including three main steps. The first step was a preparation of the soil sample; soil sample was dried in the air and then it was put into the oven at 50°C for 24 hours. The dried soil was crushed in crushing-machine. The second step was a determination of an optimum cement kiln dust and an optimum cement kiln dust with lime contents to stabilize the studied soil using pH-test [17]. The third step was a preparation of the cement kiln dust-stabilized samples compacted at a maximum dry density and an optimum water content and a preparation of the cement kiln dust with lime-stabilized samples compacted at a maximum dry density and an optimum water content. There are two methods to evaluate the geotechnical properties of the stabilized soils. The first method is destructive and including a measurement of several geotechnical parameters like unconfined compressive strength.

The second method is nondestructive and including a measurement of ultrasonic velocity of the studied soil before and after the chemical stabilization. Geotechnical properties including plasticity [18], compaction parameters [19], unconfined compressive strength (qu) [20], ultrasonic velocities including both longitudinal [21] and shear velocities using JAMES instrument (V-METER MK IV), free swelling percent [22] of the studied soil were measured before and after the treatment. Microstructural changes of the studied soil were examined using scanning electron microscope (SEM) before and after the treatment, where the changes of the microstructure and the microstructural development of the soils due to chemical stabilization play a significant role in the geotechnical properties and the mechanical behavior of the stabilized soils [2].

RESULTS

Chemical Analysis Results

Table 2 illustrated the relation between pH-values and chemical additives according to pH-test [17]. The results showed that the optimum content of the cement kiln dust was 16% (CKD) and the optimum content of the cement kiln dust with lime, to stabilize the studied soil, was 14% (CKD) with 3% (L).

Geotechnical Results

Plasticity Test Results

Table 3 showed the values of Atterberg limits including liquid limit (LL), plastic limit (PL) and plasticity index (PI) of the studied soil before and after the addition of the chemical additives. The addition of cement kiln dust led to a reduction of these values from 91.60%, 61.00% and 30.60% to 34.50%, 28.20% and 6.30% respectively. The addition of cement kiln dust with lime resulted also in a reduction of these values to 32.40%, 27.30% and 5.10% respectively.

Compaction Test Results

Figure 5 illustrated the moisture-density curves (compaction or Proctor curves). The results showed that the maximum dry density (Proctor density) and the optimum water content of the natural studied soil were 1.79 g/cm^3 and 15.10% respectively. In general, the addition of cement kiln dust resulted in an increase in the optimum moisture content from 15.10% to 25.00% and a decrease in the maximum dry density from 1.79 to 1.68 g/cm^3. The addition of cement kiln dust with lime led to an increase in the optimum moisture content to 27.00% and a decrease in the maximum dry density to 1.65 g/cm^3 (Table 3). The bell-shaped compaction curve of the studied soil was converted to flattened-shaped curves.

Unconfined Compressive Strength Test Results

Table 3 showed the results of unconfined compressive strength tests of both natural compacted and stabilized soil. The addition of the optimum cement kiln dust content to the study soil led to an increment of the strength value from 209.00 to 1528.80 KN/m^2 after 7 day curing. Increasing curing time from 7 to 28 days led to an increase in the qu-value to 2292.99 KN/m^2. The addition of optimum cement kiln dust with lime content resulted in an increment of the strength value to 1656.06 KN/m^2 after 7 day curing. Increment of the curing time from 7 to 28 days led to an increase in the qu-value to 2802.55 KN/m^2. Figure 6 illustrated the stabilized samples after the unconfined compressive strength test.

Table 2. Relation between the pH-values and the optimum contents of the chemical additives

Type of chemical additive	Percent (%)	pH-value	Temperature
Without additive	0	8.8	25
Cement kiln dust (CKD)	8 CKD	12.20	25
	10 CKD	12.26	25
	12 CKD	12.33	25
	14 CKD	12.39	25
	16 CKD*	12.40	25
	18 CKD	12.44	25
	20 CKD	12.46	25
	22 CKD	12.48	25
	24 CKD	12.49	25
Cement kiln dust with lime (CKD + L)	8 CKD + 3 L	12.31	25
	10 CKD + 3 L	12.34	25
	12 CKD + 3 L	12.37	25
	14 CKD + 3 L*	12.40	25
	16 CKD + 3 L	12.44	25
	18 CKD + 3 L	12.45	25
	20 CKD + 3 L	12.49	25
	22 CKD + 3 L	12.50	25
	24 CKD + 3 L	12.51	25

* = Optimum content.

Table 3: Geotechnical properties of the studied expansive soil

Sample type / Geotechnical Properties	MDD (g/cm³)	OWC (%)	Vp (m/sec)	Vp-gain	Vs (m/sec)	Vs-gain	qu-value (KN/m²)	qu-gain	Consistency limits			Free swelling (%)
									LL (%)	PL (%)	PI (%)	
Untreated compacted soil	1.79	15.10	699.52	0.00	1246.63	0.00	209.00	0.00	91.60	61.00	30.60	80.00
Cement kiln dust-treated soil (7 days curing)	1.68	25.00	1379.22	1.97	2481.68	1.99	1528.80	7.31	34.50	28.20	6.30	0.00
Cement kiln dust-treated soil (28 days curing)			1983.33	2.84	3514.65	2.82	2292.99	10.97				0.00
Cement kiln dust with lime-treated soil (7 days curing)	1.65	27.00	905.87	1.29	1607.52	1.29	1656.06	7.92	32.40	27.30	5.10	0.00
Cement kiln dust with lime-treated soil (28 days curing)			2154.94	3.10	3829.20	3.10	2802.55	13.41				0.00

MDD = Maximum Dry Density; OWC = Optimum Water Content; Vp = Longitudinal Velocity; Vs = Shear Velocity.

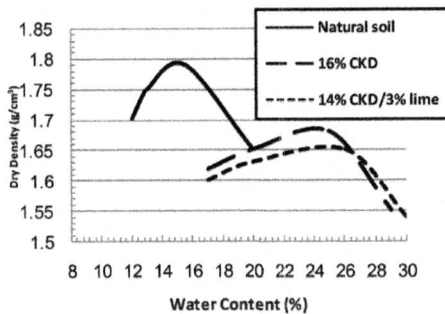

Figure 5: Compaction curves of the natural and the treated soil.

Ultrasonic Velocities Test Results

Table 3 illustrated the ultrasonic velocity values including both longitudinal (Vp) and shear (Vs) wave velocities of both natural and stabilized soil. The addition of the optimum cement kiln dust content to the soil resulted in an increase in both Vp and Vs-values from 699.52 to 1379.22 m/sec and from 1246.63 to 2481.68 m/sec after 7 day curing. Increment of the curing time from 7 to 28 days led to an increment of both Vp and Vs-values to 1983.33 and 3514.65 m/sec respectively. The addition of optimum cement kiln dust with lime content to the soil led to an increment of both Vp and Vs-values from 699.52 to 905.87 m/sec and from 1246.63 to 1607.52 m/sec after 7 day curing. With increasing curing time from 7 to 28 days, both Vp and Vs-values were increased to 2154.94 and 3829.20 m/sec respectively.

(a)

(b)

(c)

Figure 6: Treated samples after qu-tests; (a) stabilized sample after 7 days curing (CKD); (b) stabilized sample after 7 days curing (CKD with L); and (c) stabilized sample after 28 days curing (CKD with L).

Free Swelling Test Results

Table 3 showed the free swelling percent values of the studied expansive soil before and after the treatment. The results showed that the free swelling percent value of the natural studied soil were 80.00%. After 7 and 28 days curing, the addition of cement kiln dust and cement kiln dust with lime led to a reduction of the free swelling percent from 80.00% to 0.00%.

Microstructural Changes

Figure 7(a) illustrated the micrograph of the natural expansive clayey silt which showed flaky arrangements of clay particles as matrix between detrital fine grains (silt and fine sand) [1]. Figure 7(b) showed the micrograph of the stabilized soil with 16% CKD cured for 7 days. The micrograph showed crumbs of floccules with a porous nature and cementitous compounds, calcium aluminum hydrate (CAH) and calcium silicate hydrate (CSH), coating the relics of the silt particles and the flocs. The edges of the relics of the particles were attacked by cement kiln dust and their boundaries had a ragged-form.

Additionally, the reaction of cement kiln dust with clay led to a formation of an aggregate of various sizes and that was responsible for the increase in porosity of the soil system. Similar microfabric structure was observed by [1,2,23-27].

Figure 7(c) illustrated the micrograph of the treated soil with 14% CKD with 3% L cured for 7 days. The microstructure showed both the fibrous and the gel hydration reaction products. The hydration reaction products coated both the cement kiln dust and the soil-particles and filled the voids partially between the particles. The microstructure is highly porous due to the flocculation and the increase in the diameter of the flocs by production of the cementitous compounds surrounded these flocs.

Figure 7(d) illustrated the micrographs of the stabilized soil with 16% CKD cured for 28 days. The micrograph showed cementitious compounds (due to pozzolanic reaction) coated and joined the soil and the cement kiln dust particles. The pores were partially filled with the cementitious compound and were relatively reduced. Figure 7(e) showed the micrograph of the treated soil with 14% CKD with 3% L cured for 28 days. The microstructure had relatively small pores due to a formation of a large amount of cementitious compound resulted from the stronger reaction between cement kiln dust and lime together with the soil.

p class="E-Title1"> 5. Discussions and Conclusions

The optimum content of the cement kiln dust to stabilize the studied soil was

16% and the optimum content of the cement kiln dust with lime was 14% (CKD) with 3% (L), using pH-test [17]. The addition of the optimum cement kiln dust resulted in a reduction of the liquid limit (LL), the plastic limit (PL) and the plasticity index (PI) values from 91.60%, 61.00% and 30.60% to 34.50%, 28.20% and 6.30% respectively. The addition of the cement kiln dust with lime also led to a reduction of these values to 32.40%, 27.30% and 5.10% respectively. This reduction was due to a decrease in the thickness of the double layer of the clay particles. That is because of cation exchange reaction, which causes an increase in the attraction force leading to a flocculation of the particles [28]. The maximum dry density (Proctor density) and the optimum water content of the natural studied soil were 1.79 g/cm^3 and 15.10% respectively. Generally, the addition of the cement kiln dust to the studied soil led to an increment of the optimum moisture content from 15.10% to 25.00% and a decrement of the maximum dry density from 1.79 to 1.68 g/cm^3. Similar results were recorded by the addition of the cement kiln dust with lime where the optimum moisture content was increased to 27.00% and the maximum dry density was reduced to 1.65 g/cm^3. The bell-shaped compaction curve of the studied soil was converted to flattened-shaped curve. The typical flattening of the compaction curve of the studied stabilized soil makes it easier to achieve the required density over a wider range of possible moisture contents [2]. The change in the shape and characteristics of the peak of the compaction curves can allow for significant savings in time, effort and energy [29]. The addition of the cement kiln dust resulted in an increase in the strength value from 209.00 to 1528.80 KN/m^2 after 7-day curing.

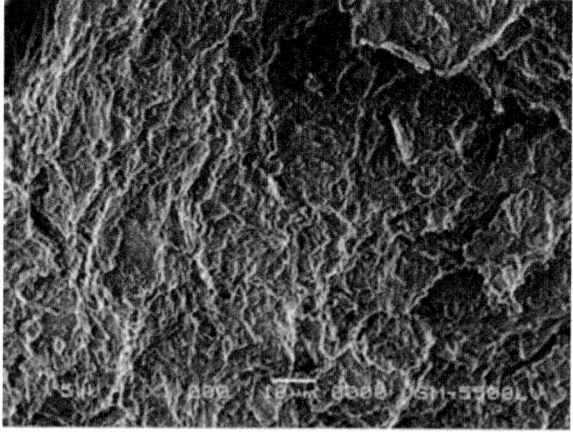

(a)

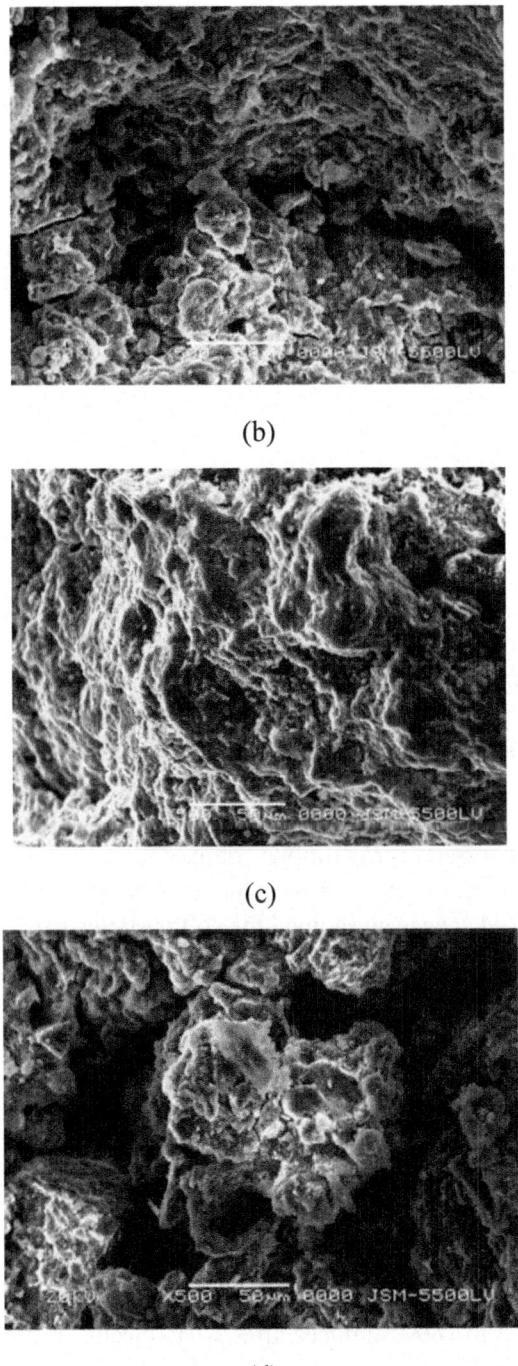

(b)

(c)

(d)

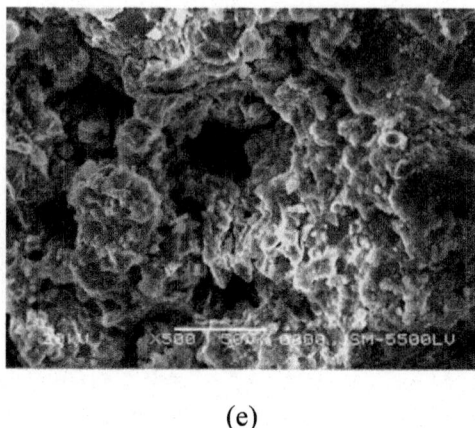

(e)

Figure 7: Micrographs of natural and stabilized soil using scanning electron microscope (SEM).

Increasing curing time from 7 to 28 days resulted in an increase in the strength value to 2292.99 KN/m². The strength gain (strength of stabilized soil/ strength of natural compacted soil) was also increased from 7.31 to 10.97. The addition of the cement kiln dust with lime led to an increment of the strength value to 1656.06 KN/m² after 7-day curing. Increment of the curing time from 7 to 28 days resulted in an increase in the strength value to 2802.55 KN/m². The strength gain was also increased from 7.92 to 13.41. The mechanical behavior of the stabilized soil had a brittle behavior due to a formation of cementitous compounds. The formed cementitious compounds (as a result of the chemical reactions between the silica and the alumina and the additives) reduced the volume of the void spaces and joined the soil particles [2].

$$Ca^{++} + 2(OH) + S_iO_2 \rightarrow CSH \text{ Calcium silicate hydrate}$$
$$\text{(Silica)} \qquad \text{(Gel)}$$
$$Ca^{++} + 2(OH) + Al_2O_3 \rightarrow CAH \text{ Calcium aluminumhydrate}$$
$$\text{(Alumina)} \qquad \text{(Fibrous)}$$

The addition of the cement kiln dust to the studied soil resulted in an increase in both Vp and Vs-values from 699.52 to 1379.22 m/sec and from 1246.63 to 2481.68 m/sec after 7-day curing. Increment of the curing time from 7 to 28 days led to an increment of both Vp and Vs-values to 1983.33 and 3514.65 m/sec, respectively. The addition of the cement kiln dust with lime to the soil led to an increment of both Vp and Vs-values from 699.52 to 905.87 m/sec and from 1246.63 to 1607.52 m/sec after 7-day curing. With increasing curing time from 7 to 28 days, both Vp and Vs-values were increased to 2154.94 and 3829.20 m/sec respectively. The increasing of the ultrasonic velocity due to the formation of new cementitous compounds and mineral crystals as a pozzolanic reaction produced through the curing [2].

The addition of cement kiln dust and cement kiln dust with lime led to a decrease in the free swelling value from 80.00% to zero% at curing times 7 and 28 days. The reduction of the free swelling due to consumption of the clay minerals during the hydration reaction formed the cementitious compounds. Additionally, the reaction of chemical additives with clay led to a formation of an aggregate of various sizes with low ability for swelling.

The improvements of the engineering properties of the studied soil due to use of the chemical additives can be explained by two basic reactions: short-term reactions consisting cation exchange and flocculation and the longterm reaction named pozzolanic activity. During the first stage of the reaction between the chemical additives and the clay, excesses of calcium ions in lime or in cement kiln dust replace all other monovalent cations in the clay and change the electrical charge density around the clay particles. This results in an increase in the interparticle attraction causing flocculation and aggregation and a consequent decrease in the plasticity of the soil [30]. The pozzolanic reaction is time-bound and tempreture dependent. During this process, the high pH causes silics and alumina to be dissolved out of the structure of the clay minerals and to combine with the calcium (occurred in the cement kiln dust or the lime) to produce the new cementitious compounds calcium silicate hydrates (CSH), calcium aluminate hydrates (CAH), calcium aluminosilicate hydrates (CASH) [30] and others.

Examination of the stabilized soil using scanning electron microscope (SEM) indicated that the microstructures of the tested soil were changed due to the cement kiln dust and the cement kiln dust with lime. The stabilization process caused a formation of a silt-fine sand like structure (open fabric) characterized by a highly porous system. The SEM-micrographs of the natural and the stabilized soil indicated the formation of new cementitous compounds and mineral crystals as a pozzolanic reaction product through the curing. These cementitious compounds have improved the geotechnical properties of the studied soil and reduced the swelling. The influence of the cement kiln dust with lime addition on the soil was greater than the influence of the cement kiln dust alone. Use of the cement kiln dust with small percentage of lime produced more dramatic results, especially after 28 days curing.

Finally, cement kiln dust (as by product of Qena cement plant, at Qift industrial city) contains relatively small percent of sulfates (4.43%) and can be utilized to treat and stabilize the expansive fine grained soil as economical (cheaper) alternative to Portland cement and other (expansive) chemical stabilizers. The use of cement kiln dust for chemical stabilization applications is an environmental solution of the problems associated with its disposal process.

ACKNOWLEDGEMENTS

Greatly thanks to staff members of Qena faculty of engineering, Al-Azhar University for offering the laboratory facility to conduct the geotechnical tests of this work. Special thanks for the staff members of Qena cement plant for valuable cooperation. Special thanks for Ms. Heba Taha, Ms. Mona Mohamed and Ms. Samar Yousef for laboratory help.

REFERENCES

1. H. A. H. Ismaiel and M. M. Badry, "Lime Chmical Stabilization of Expansive Deposits Exposed at El-Kawther Quarter, Sohag Region, Egypt," Geosciences, Vol. 3, No. 3, 2013, pp. 89-98.
2. H. A. H. Ismaiel, "Treatment and Improvement of the Geotechnical Properties of Different Soft Fine-Grained Soils Using Chemical Stabilization," Ph.D. Thesis, Mathematisch-Naturwissenschaftlich-Technischen Fakultät der Martin-Luther-Universität Halle-Wittenberg, 2006.
3. E. H. Ramadan, "In Situ Chemical Stabilization of Expansive Clay Soil by Lime Additive," 7th International Colloquium on Structural and Geotechnical Engineering, Ain Shams Uni., Cairo, 1996, pp. 403-419.
4. T. O. Al-Refeai, "Stabilization Characteristics of Cement Kiln Dust," Geoengineering in Arid Lands, 2000, pp. 133-137.
5. M. Zaman, I. G. Lagurous and A. Sayah, "Soil Stabilization Using Cement Kiln Dust," Proceedings of the 7th International Conference on Expansive Soil, Dallas, 1992.
6. Tensar Technical Note, "Chemical and Mechanical Stabilization of Sub-Grades and Flexible Pavement Sections," TTN, BR 10, 1998.
7. B. Das, "Principles of Geotechnical Engineering," 3rd Edition, PWS-Kent Publishing Company, Boston, 1994.
8. J. Bowles, "Engineering Properties of Soil and Their Measurements," 4th Edition, McGraw-Hill, Boston, 1992.
9. M. E. Mustafa, "Computerized Analysis of Geologic Structures in Central Eastern Desert, Egypt and Their Role in Distribution of Radioactive Mineralization," Ph.D. Thesis, Cairo Uni., Egypt, 1979.
10. A. A. Abdelmoneim, "Hydrogeology of the Nile Basin in Sohag Province," Master Thesis, Faculty of Sci., Assiut Uni., Egypt, 1988.

11. A. M. Youssef, "Mapping the Pliocene Clay Deposits Using Remote Sensing and Its Impact on the Urbanization Developments in Egypt, Case Study, East Sohag Area," Geotechnical and Geological Engineering, No. 26, 2008, pp. 579-591.
12. F. E. Salwa, B. B. Abdel Aziz and A. Z. El-Sayed, "Hazards Mitigation and Natural Resources Evaluation around Sohag-Safaga Highway, Eastern Desert, Egypt," Egyptian Journal of Remote Sensing and Space Sciences, Vol. 14, No. 1, 2011, pp. 15-28.http://dx.doi.org/10.1016/j.ejrs.2011.01.001
13. H. A. H. Ismaiel, M. M. Askalany and M. M. Badry, "Geotechnical Propertied and Classification of Both NonExpansive and Expansive Soils Exposed along the New Upper Egypt-Red Sea Road, Eastern Desert, Egypt," Scientific Journal of Banha University, No. 6, 2011, pp. 1-18.
14. A. Aly, "Assessment of Drying-Wetting Cycles for Mitigation the Potential of Expansive Soil in Upper Egypt," Journal of Applied Sciences Research, No. 12, 2009, pp. 2277-2284.
15. M. A. Sakr, M. A. Shahin and M. M. Yasser, "Utilization of Lime for Stabilizing Soft Clay Soil of High Organic Content," Geotechnical and Geological Engineering, No. 27, 2009, pp. 105-113.
16. Conoco Inc., "Stratigraphic Lexicon and Explanatory Notes to the Geological Map of Egypt 1: 500000," Coy H. Squyres General Chairman Map Project, Cairo, 1986.
17. J. L. Eades and R. E. Grim, "A Quick Test to Determine Lime Requirements for Lime Stabilization," Highway Research Board, National Research Council, Washington DC, No. 139, 1996, pp. 61-72.
18. AASHTO, T 90, "Standard Method of Test for Determining the Plastic Limit and Plasticity Index of Soils," Single User Digital Publication, American Association of State Highway and Transportation Officials, Washington DC, 2010.
19. AASHTO, T 99, "Standard Method of Test for MoistureDensity Relations of Soils Using a 2.5kg (5.5-lb) Rammer and a 305-mm (12-in.) Drop," Single User Digital Publication, American Association of State Highway and Transportation Officials, Washington DC, 2010.
20. AASHTO, T 208, "Standard Method of Test for Unconfined Compressive Strength of Cohesive Soil," Single User Digital Publication, American Association of State Highway and Transportation Officials, Washington DC, 2010.
21. N. Yesiller, J. L. Hanson, A. T. Rener and M. A. Usmen, "Ultrasonic Testing for Evaluation of Stabilized Mixtures," Geomaterials, Transportation

Research Record, No. 1757, 2001, pp. 32-39.
22. Egyptian Code, "Egyptian Code of Soil Mechanics," Foundations Carrying out and Designation, Part 2, Laboratory Tests, 6th Edition, 2001.
23. J. L. Eades and R. E. Grim, "Reaction of Hydrated Lime with Pure Clay Minerals in Soil Stabilization," Highway Research Board, No. 262, 1960, pp. 51-63.
24. J. B. Croft, "The Processes Involved in the Lime Stabilization of Clay Soils," Proceedings of Australian Road Research Board, Part 2, 1964, pp. 1169-1203.
25. R. S. Narasimha and G. Rajasekaran, "Strength and Deformation Behavior of Lime Treated Marine Clays," Proceedings of 3rd International Offshore and Polar Engineering, Singapore, 1993, pp. 185-196.
26. G. Rajasekaran, K. Murali and R. Srinivasarghavan, "Fabric and Mineralogical Studies on Lime Treated Marine Clays," Ocean Engineering, Vol. 24, No. 3, 1995, pp. 227-234. http://dx.doi.org/10.1016/S0029-8018(96)00010-8
27. G. Rajsekaran and R. S. Narasimha, "Particle Size Analysis of Lime-Treated Marine Clays," Geotechnical Testing Journal, Vol. 21, No. 2, 1998, pp. 109-119. http://dx.doi.org/10.1520/GTJ10749J
28. Z. Nalbantoglu and E. Gucbilmez, "Utilization of an Industrial Waste in Calcareous Expansive Clay Stabilization," Geotechnical Testing Journal, Vol. 25, No. 1, 2002, pp. 8-84.
29. P. Nicholson, V. Kashyap and C. Fuji, "Lime and Fly Ash Admixture Improvement of Tropical Hawaiian Soils," Transportation Research Record, Washington DC, No. 1440, 1994, pp. 71-78.
30. Z. Nalbantoglu and E. Gucbilmez, "Improvements of Calcareous Expansive Soils," Journal of Arid Environments, Vol. 47, No. 4, 2001, pp. 453-463. http://dx.doi.org/10.1006/jare.2000.0726

Chapter 5

PHOSPHORUS: CHEMISM AND INTERACTIONS

E. Saljnikov and D. Cakmak

Institute of Soil Science Serbia

INTRODUCTION

Phosphorus (P) is a limiting nutrient for terrestrial biological productivity. The availability of "new" P in ecosystems is restricted by the rate of release of this element during soil weathering. Soil P exists in inorganic and organic forms. Inorganic P forms are associated with amorphous and crystalline sesquioxides, and calcareous compounds. Organic P forms include the relatively labile phospholipids and fulvic acids and the more resistant humic acids. The intergrades and dynamic transformations between the forms occur continuously to maintain the equilibrium conditions (Hedley et al., 1982). Its low concentration and solubility (< 0.01 mg P kg^{-1}) in soils, however, make it a critical nutrient limiting plant growth.

In natural soil ecosystems the main source of inorganic phosphorus is rocks where the primary minerals are of the greatest importance, where in turn calcium phosphates are the most important (e.g. apatite) (Fig.1). In weathered soils, leaching of Ca ion results in formation of Al-phosphate (e.g. berilinite) and Fe-phosphate (e.g. stregnite); the complete list were given by Lindsay, (1979) and Lindsay, et al., (1989). By the definition, these minerals are characterized with three-dimensional atomic structure. As far as phosphorus concentration in the soils is concerned, it can be very low from 50 mg kg^{-1} and high up to 3500 mg kg^{-1} (Foth & Ellis, 1997; Frossard et al., 1995). Application of phosphorus from mineral phosphate results mainly in formation of amorphous compounds with soluble Al, Ca and Fe where the phosphorus is adsorbed on the surfaces of clay minerals, Fe and Al oxyhydroxides or carbonates and physically occluded by secondary minerals.

In natural ecosystems, P availability is controlled by sorption, desorption, and precipitation of P released during weathering and dissolution of rocks and minerals of low solubility (Sharpley, 2000). Due to high fixation and immobilization of phosphorus in the soil, the agriculturists apply high amounts of p-fertilizer, what results in greater input of P into soil that plant uptake. Application of phosphates can maintain or improve crop yields, but it can also cause changes in the chemical and physical properties of the soil, both directly and indirectly (Hera & Mihaila, 1981; Acton & Gregorich, 1995; Aref & Wander, 1998; Belay et al., 2002). The cumulative accumulation of available P in agricultural soils may partially saturate the capacity of a soil for P sorption, with resulting increase of P leaching into the subsoil layers (Ruban, 1999), or may sometimes reach depth more than 90 cm (Chang et al., 1991), suggesting that erosion, rather than leaching, would cause a threat to water bodies (Zhou and Zhu, 2003). Such a process of leaching is especially effective in soils of Stagnosol type with clear E horizon due to their lower adsorption capacity, with relatively shallowground waters (Tyler, 2004; Väänänen et al., 2008). Fertilization with mineral P in the inorganic pools explains 96 % of the variation in the level of available phosphorus (Beck & Sanchez, 1994).

Great number of researchers studied many aspects of fate and behavior of applied P. However, the chemical processes and following plant availability of soil P remains a big challenge for scientists since it offer wide spectrum of uncertainties, and contradictions. This Chapter is devoted to explanation of mechanisms and distribution of different forms of phosphorus, its transformation and dynamics in the soil based upon the 40-years of experience in phosphate field application.

MATERIALS AND METHODS

Site Description

The investigation was conducted at the Varna experimental station, 44°41'38" and 19°39'10" (near Belgrade, Serbia), where a wide range of different fertilization treatments has been undertaken since 1968. The soil type is Stagnosol (WRB, 2006), a loam textured Pseudogley developed on Pliocene loam and clay materials under aquic conditions at 109 m above sea level. Average annual precipitation of the site is 705 mm, and the average temperature is 12°C. The mineralogical composition of the studied soil was as follows: illite (50–70%), vermiculite (10–30%), and other clay minerals (kaolinite, chlorite, feldspar, quartz and amphibolites) (Aleksandrovic et al., 1965). The cultivated crops were winter wheat (Triticum aestivum L.) and corn (Zea maize L.), with crop residues removed. The soil cultivation was performed by a standard plowing to 25 cm depth.

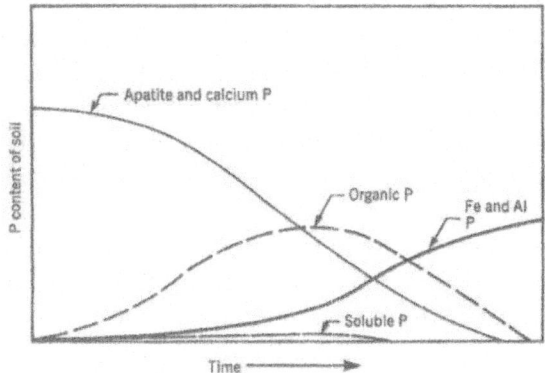

Figure 1: Relative distribution of the major forms of soil P vs time of the soil development (Foth & Ellis, 1997).

Three rates (26, 39, and 52 kg P ha−1) of monoammonium phosphate (MAP) fertilizer (NH4H2PO4) were applied in combination with a consistent rate of N (urea) (60 kg ha−1) and K (KCl) (50 kg ha−1) for 40 yr of the experiment. The fertilized treatments werecompared to the control, with no fertilizer applied. The experiment was arranged as a randomized block design, with each treatment randomized in three blocks for a total of 12 plots. Each plot was 5 by 11 m. Composite samples of five soil subsamples were taken from each plot in the three field replications from two depths: surface (0–30 cm) and subsurface (30–60 cm) layers in spring 2008.

Methods

Soil pH was determined with a glass electrode pH meter in a 1:2.5 water solution. Soil total C and N were measured with an elemental CNS analyzer, Vario model EL III (ELEMENTAR Analysasysteme GmbH, Hanau, Germany; Nelson & Sommers, 1996) Available P and K were determined by the Al-method of Egner–Riehm (Enger & Riehm, 1958), where 0.1 M ammonium lactate (pH = 3.7) was used as an extract. After the extraction, P was determined by spectrophotometry after color development with ammonium molybdate and SnCl2 (Enger & Riehm, 1958). Soil Ca and Mg were extracted by ammonium acetate and determined with a SensAA Dual atomic adsorption spectrophotometer (GBC Scientific Equipment Pty Ltd, Victoria, Australia; Wright & Stuczynski, 1996). Determination of CEC was performed by the steam distillation method after the treatment with 1 M ammonium acetate

(Sumner & Miller, 1996). Exchangeable Al was determined by the titration method by Sokolov: the extraction with 1 M KCl (1:2.5) followed shaking for 1 h and titration with 0.01 M NaOH (Jakovljević et al., 1985).

Trace elements were determined with an ICAP 6300 ICP optical emission spectrometer (Thermo Electron Corporation, Cambridge, UK), after the soils were digested with concentrated HNO3 for extraction of hot acid-extractable forms, and by diethylenetriaminepentaacetic acid (DTPA) for extractable elements (Soltanpour et al., 1996). The F content was determined by ion-selective electrode, after the soil had been fused with NaOH for total F and after extraction with water for available F (Frankenberger et al., 1996). Soil granulometric composition was performed using the pipette method (Day, 1965). All chemical analyses were performed in two analytical replications. The Merck standards were used for the determinations on ICP and SensAA Dual. Before the determination of samples, three blank samples were read, which allowed correcting the results. For the verification of the results, a referent soil sample was determined for all the studied elements (NCS ZC 73005 soil, CNAC for Iron and Steel, Beijing).

Statistical analyses were performed with the SPSS version 16 software. The effects of treatments on all the variables were tested by ANOVA. Statistical differences between the treatments were determined using the t test (95%) Pearson for Fisher's LSD. The significance of their correlations was analyzed via the Pearson correlation matrix (SPSS, 2007).

SEQUENTIAL ANALYSES

The ways phosphates bound to soil particle are the parts of a puzzle whose solution can give many answers concerning their availability to plants and the possible leaching down the soil profile. The best way to obtain the answers is isolation of separate fractions of phosphorus in soil using series of solvents of different strength, i.e. sequential analysis.

Sequential Extraction Procedure

One of the most common phase divisions for sequential extraction was used in the experiment: Exchangeable or sorptive (adsorptive and ion exchange) phase. This phase is used to estimate the maximum quantity of sorbed ions that geological material can release, without visible decomposition of some mineral phases. Neutral solutions of salts (NH_4OAc, $MgCl_2$, $CaCl_2$, $BaCl_2$, KNO_3, etc.) are usually used for this extraction phase. Their concentrations (and ion forces) must be high enough to initiate the most complete ion exchange and desorption from all substrates.

"Easily Reducible" Phase: Weak reduction means (for example, hydroxylamine) are used for selective reduction (solvent) of manganese oxyhydrates, but they are also used for the most mobile fraction of amorphous iron oxides. All the present microelements co-precipitated in these oxides to be detected in the solution.

"Moderate reducible" Phase: For the amorphous iron oxides and the more crystalline manganese oxides, some stronger reduction means are used – oxalic acid, sodium dithionite and similar methods.

Organic-sulfide phase. Distinguishing of organic and sulfide metal fractions in a geological material is one of the disadvantages of sequential extraction. This problem is still unresolved. Pure nitric acid or its combinations with other acids is very effective, but it leads to a noticeable decomposition of silicate material. The use of hydrogen peroxide is acceptable at higher temperatures and low pH (about 2).

Residual Phase: This is the least interesting phase of the ecochemical aspect as it includes silicate and oxide materials as well as incorporated metal ions, i.e. in natural conditions this fraction cannot be mobilized from geological material. Concentrated mineral acids and their mixtures are usually used for decomposition of this crystal matrix (Petrovic, et al., 2009) (Tab. 1).

Table 1: Reagent in the sequential extraction procedure used to study substrates metals (Fe, Al, Mn and Ca) and P

Procedure	Step 1†	Step 2	Step 3	Step 4	Step 5
Petrovic et al. (2009)	1 M CH_3COONH_4 Exchangeable	0.1 M $NH_2OH \cdot HCl$ Bound to carbonates and easily reducible	0.2 M $(NH_4)_2C_2O_4$ and 0.2 M $H_2C_2O_4$ Moderately reducible	30 % H_2O_2 + 3.2 M CH_3COONH_4 Organic-sulphide	6M HCl Residual

Chang and Jackson Sequential Analysis

Fraction of Soil P Extracted by 1M Solution of Ammonium-Chloride (Water-Soluble P)

This fraction of phosphorus is closely linked with the dynamics of P bounding in soil. Such bounding of phosphorous ions can be characterized as an initial reaction. And it represents a non-specific adsorption and ligand exchange on mineral edges or by amorphous oxides and carbonates. This fraction is bound to Mn isolated in step 2 from Table 1 (Mn II) (r=0.994, **), which indicated its sorption on hydrated oxides of manganese. Due to specific further bounding of phosphorus, this fraction is very low in quantity (less than 1 % of the total mineral phosphorus) in acidic soils such as Stagnosol. However, due to application of mineral fertilizers and accumulation of phosphorus (Jaakola, et al. 1997) the processes of saturation of free spaces for adsorption of P in the soil (Vu et al, 2010) result in its significant increase, by about 6 times compared to the control plots in the studied experiment. Considering the low movement of phosphate ions along adepth the soil profile, which isslower than in the processes of bounding them into less soluble forms, such increase of P ions concentration is expressed distinct in the surface soil layer 0-30 cm. Passage of this form of phosphorus into bounded-to-aluminum phosphorus is a process characteristic for acidic soils. The reverse process is also possible (correlation coefficients 0.974 and 0.780 for 0-30 and 30-60 cm, respectively). A very strong correlation between water-soluble P and available P (0.945 and 0.715 for 0-30 and 30-60 cm, respectively) proved that this form of phosphorus was available for plants.

Fraction of Soil P Extracted by 0.5M NH_4F Solution (Al Bound P)

Such an isolated fraction of phosphorus is a characteristic for monodent and bident bounds (Tisdale et al., 1993). Consequently, these compounds are very labile and are described as pseudo sorption (Van der Zee et al., 1987; Van der Zee et al., 1988). In acid mineral soils, such as Podzols, P is mostly retained by Al and Fe oxides by the ligand exchange mechanism where the OH- or H_2O groups of sesquioxides surfaces were are displaced by dihydrogenphosphate anions (Simard et al., 1995). In certain soils, this bound is not strictly confined to Al but can bound bind to Si, as well (Manojlovic, et al. 2007). However, in the studied Stagnosol, the strong correlation of Al-P with Al extracted in the step 2 (Al II) could be attributed to carbonates and alumosilicates (r=0.998**). Somewhat increased content of this fraction versus to the available phosphorus indicates that not only modettant bounds are involved (Fig. 3).

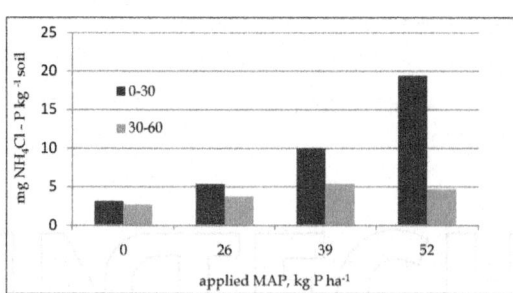

Figure 2: Content of ammonium-chloride extractable P upon 40-years of phosphate application (water-soluble P).

It is obvious that this fraction of soil phosphorus is the most important for plants since there is a high correlation between the Al bound P and the available forms of phosphorus extracted by the Al-method (r=987**). Also, application of mineral phosphorus influences this fraction of soil P the most with the recorded increase of its content from 16.08% in the control to 34.51% in the treatment with 52 kg P ha^{-1}. This fraction of P is responsible for migration of phosphorus along the soil depth, which is confirmed by a significant correlation between the values at two depths (0.876**), as well as for the replenishment of the pool of other fractions of soil P.

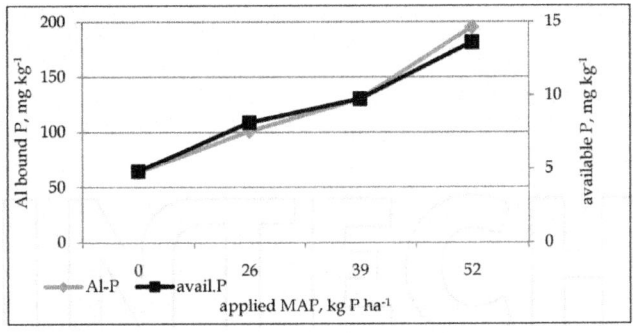

Figure 3: Content of Al bound P and available P upon 40-years of application different amounts of phosphate fertilizer.

Fraction of Soil P Extracted by M NaOH Solution (Fe Bound P)

Fraction of soil P isolated by such a strong reagent may have a high content in soil and mainly is greater than that bound to Al (Manojlovic et al., 2007, Mustapha et al., 2007), ranging between few hundred mg per kilogram. From the chemistry viewpoint, such bounded P is characteristic for the slow-flowing processes involving formation of covalent Fe-P or Al-P bonds on Fe and Al oxide surfaces (Willett et al. 1988) which can be an additional source of available P (Beck & Sanchez, 1994). However, the strength of this bound is quite high. Therefore, its availability is limited. That determines the absence of correlation between the mentioned fraction and the available P. However, in the layers of soils such as Stagnosol, due to constant wetting and alteration of oxidative – reductive conditions, the content of this form of P can be as low as less than 1 mg per kilogram due to passage into other forms (reducible and occluded). Its movement along the soil depth is also limited as indicated by the absence of correlation between the values at different depths. Sequential analysis didn't show marked correlation with the fraction of Fe, but the correlation with DTPA-extractable Fe was recorded (0.665*)

Fraction of Soil P Extracted by M Na Dithionite, Na Citrate Solution (Reducible P)

In contrary to the previous types of bounding of P in soil, this fraction is characterized by the bounds within the particle. Such bounding results in the process of occlusion where the phosphate is adsorbed to the surface of Al hydroxide and is bound by poorly crystalline Fe oxides from that occluded in the crystalline Fe oxides (Delgado & Scalenghe, 2008). In this structure, the phosphate binds the Al- with Fe^{3+} hydroxide so the surface of Al phosphate particle is enveloped by a Fe3+hydroxide skin. Such adsorbed phosphates are only indirectly available to plants. Thus, in the conditions determined by reduction processes the reduction of iron Fe^{2+} and the breakage of the earlier formed bounds take place, which makes this form of P available for plants. Although this fraction is small compared to other fractions of P in soils, under the oxidized conditions Fe-P represents the dominant fraction (Manojlovic et al, 2007, Mustapha et al., 2007). But under the conditions of soilundergoing alterations of wet and dry regimes with high content of available Fe (Cakmak, D. et al., 2010) the reducible-P can be of significant concentration up to 30% from the total mineral P. The high correlation found between the reducible and Al bound P ($r=0.97**$) indicates the indirect availability of this form of P under the alteration reduced conditions in Stagnosol.

Fraction of Soil P Extracted by M NaOH Solution (Occluded P)

Chang & Jackson (1957) noticed that during the sequential extraction some soils, rich in Fe oxides, contain significant amounts of Fe-phosphate occluded within the oxide, which cannot be extracted by sodium dithionite and sodium citrate. This occluded phosphate can be extracted by repeated alkali solution. The P tied in this manner might be increased in quantity by constant addition of mineral phosphate fertilizer where its total content ranges between few milligrams to tens of milligram per kg soil; i.e. in small amounts from 1% to about 10% from the total mineral P (Manojlovic et al, 2007, Mustapha et al., 2007).

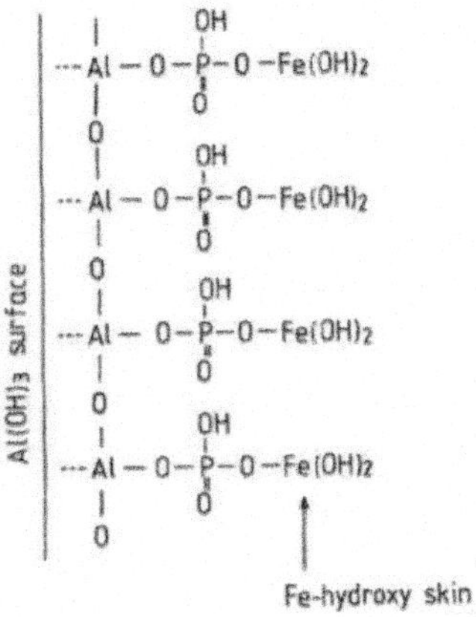

Figure 4: Fe-hydroxy skin covering the phosphate adsorbed to Al oxide/hydroxide (Mengel & Kirkby, 2001).

Under the alteration of reduced and oxidized conditions that predominate in Stagnosol, this form is chemically tied to the reducible form of P, especially, in the upper soil layer (r=0.890**). Also, within such soil particle, Al-phosphate can be present, which can be available under certain conditions within Fe-oxide (Fig.4). Its migration along the soil depth is limited and is of very low mechanic intensity. Absence of correlation with DTPAextractable Fe indicates the un-availability of Fe in such compounds.

Fraction of Soil P Extracted by M H_2SO_4 Solution (Ca Bound P)

In neutral to calcareous soil the concentration of phosphate in soil solution is governed mainly by the formation and dissolution of calcium phosphates. This in turn depends on soil pH and Ca^{2+} concentration in soil solution. The lower are Ca/P ratios in the Ca phosphates -the higher is their solubility in water. However, in acidic soils in spite of significant amount of this fraction (up to 40% from total mineral forms) the Ca-P was widely dispersed in soil minerals and it was weakly changeable. This is supported by the absence of significant correlation between exchangeable Ca and Ca- bound P. Therefore, in such soils fertilization does not result in significant changes in the content of Ca-bound P (Hartikainen, 1989). However, relative increase of this fraction is possible in the subsurface soil layer due to leaching and accumulation of Ca ions, under acidic conditions, in deeper layers where it is transformed into non-labile phosphate fractions. This process of phosphate ageing is especially rapid in acid soils with a high adsorption capacity. The start of this process can also be detected by the negative significant correlation between water-soluble P and DTPAextractable Ca (-0.590*).

BCR ANALYSIS

BCR method according to SMT standard protocol was applied for determination of P in soil (Ruban et al., 1999). BCR is a non-specific extraction procedure for determination of phosphorus in freshwater sediments, developed in the frame of the European Program, Standards, measurements and Testing (SMT) is used for certification campaign for a reference material. The SMT protocol was extended to soil material because bioavaialable forms of phosphorus are important not only for analysis of sediments but also of soils. The detailed description of the SMT protocol is given at Ruban et al., (2001). The Certified Reference material CRM 684 (River Sediment Extractable Phosphorus, from Po River, Italy) was analyzed to verify the results of analyses.

BCR Procedure

Among numerous extraction schemes used, the procedures widely adopted are those developed by Williams (1976), Hieltjes & Lijklem (1980), Rutenberg (1992) and Galterman (1993). Together with the cited procedures, in the literature can be found other sequential schemes (Delaney, et al., 1997; Kleberg et al., 2000). Due to the large number of the existing procedures for extraction of phosphorus and due to the impossibility of comparison of the results from different source samples obtained via different laboratory procedures, the Program of Standardization of extraction scheme was initiated (Ruban et al, 1999).

To overcome the incompatibility of results a Program of the European Commission (SMT: Standards, Measurements and Testing, earlier BCR) initiated the project for selection of sequential extraction procedures for determination of forms of phosphorus in lake sediments. This project targeted the homogenization of extraction schemes to investigate a selected scheme in inter-laboratory investigations that includes expert European laboratories and to certificate traces in the referent material of the sediment. Four methods (Tab. 2) were chosen for testing. These methods were applied in the inter-laboratory investigations and served as the base for development of homogenized procedure for phosphorus extraction from lake sediments. For the determination of phosphorus, all the laboratories used spectrophotometry (Murphy & Riley, 1962). Along with this method, some laboratories used ion-chromatography such as ICP-AES. However, using the laters proved to be unsuitable, since ICP-AES allows determination of total phosphorus, while the method of ion-chromatography determines only orthophosphate.

Table 2: Advantages and disadvantages of the methods for sequential extraction of soil P fractions (adapted from Ruban et al., 1999)

Procedure	Advantages	Disadvantages
Williams	Simple and practical	Partial resorption of phosphorus extracted with NaOH onto CaCO3
Hieltjes - Lijklem	Simple and practical	Dissolution of small amounts of Fe-P and Al-P with NH4Cl; hydrolysis of organic phosphorus is unavoidable; no responce to bioavailability
Golterman	Extraction of specific compounds; allows extraction of organic P; supports information about bioavailability of fraction	Not practical; NTA i EDTA contaminate the determination of phosphorus; complicated preparation of the solution; in some sediments the extraction must be undertaken more times to obtain valid results
Rutenberg	Possibility of differentiation of different types of apatite; no distribution of phosphorus on the residual particle surfaces during extraction	Very long; extraction with butanol is very difficult

Based on the results of inter-laboratory investigations, the modified Williams scheme named SMT (1998) was proposed. The SMT scheme allows definition of the following forms of phosphorus: NaOH-extractable phosphorus (NaOH-P); phosphorus bound to oxides and hydroxides of Fe, Al and Mn; (Fe-Al-Mn-P) HCl-extractable phosphorus (HClP); phosphorus bound to Ca (Ca-P); organic phosphorus (Org-P), inorganic phosphorus (IP); concentrated HCl phosphorus, (conc. HCl-P); total phosphorus(TP) (Ruban et al., 1999).

After the compilation of the extraction scheme, certification of extractable phosphorus from the referent material CRM 684, sampled from Po River (Italy) near the Gorina city, was undertaken within the Project (SEPHOS - sequential extraction of phosphorus from fresh water sediment) (Ruban et al., 2001). After acceptance and certification of SMT procedure, valid comparison of worldwide results became possible, what is of great importance for understanding of biogeochemical cycle of phosphorus in actual systems.

In Tab. 3 there are phases of phosphorus and the reagents used for their extraction in both procedures: modified Chang & Jackson and BCR

Table 3: Phases of soil phosphorus and the reagents used for their sequential extraction

Procedure	Step 1	Step 2	Step 3	Step 4	Step 5	Step 6
Manojlovic (2007) modified from Chang & Jackson (1957)	1M NH4Cl Water soluble P	0.5 M NH4F Al bound P	0.1 M NaOH Fe bound P	0.3 M Na dithionite, Na citrate Reducible P	0.1 M NaOH Occluded P	0.25 M H2SO4 Ca bound P
BCR (Ruban et al., 2001)	1M NaOH NaOH - P	3.5 M HCl HCl - P	1M NaOH Inorganic P	1M HCl + calcination Organic P	3.5 M HCl + calcination Conc HCl - P	

Fraction of Soil P Extracted by 1M NaOH (Al-Fe-Mn Bound P)

This fraction of phosphorus bound to oxides and hydroxides of aluminum, iron and manganese, so called oxide-hydroxide fractions was extracted in step 1 (Tab. 3). The associations of P and Fe are often found in sediments, where phosphorus is tied to complex compounds of iron through changes of ligands (Stumm & Morgan, 1981). In soils such as Stagnosol, this fraction of phosphorus is correlated with Ca from the second phase (Ca II, Table 1) (r=0.951, **) that indicates binding of phosphorus with carbonate fraction. Such bounds are quite labile, what is supported by a strong correlation between Al-Fe-Mn bound P with available P with corresponding coefficients 0.782 and 0.813 (for 0-30 and 30- 60 cm, respectively).

The correlation coefficients between HCl –P and available P were 0.939 and 0.902 (for 0-30 and 30-60 cm, respectively). The good correlation of the Al-FeMn fraction with Al-P fraction from Chang and Jackson method (r=0.775, **) explains the leaching of Al-Fe-Mn bound P fractions from the surface to the subsurface soil layer (r=.901, **) (Fig. 5).

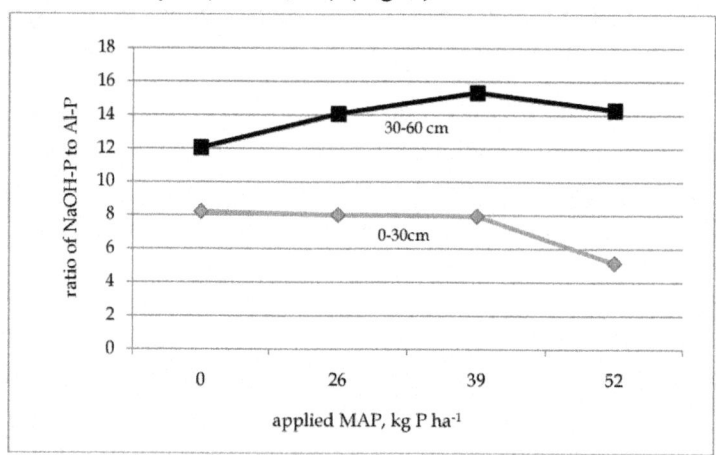

Figure 5: Ratio of NaOH bound P (BCR) to Al bound P (Chang&Jackson) in the 0-30 cm and 30- 60 cm soil intervals.

Fraction of Soil P Extracted by 3.5 M HCl (Ca Bound P)

This fraction generally represents the phosphorus in apatite (Williams et al., 1976; William et al., 1980) and phosphorus bound to Ca (Golterman, 1996, 1982) and was extracted in step 2 (Tab. 3). The adsorption of phosphorus in calcium carbonate is one of the mechanisms of formation of calcium phosphate in sediments. However, apart from the Fe bound phosphorus, formation of

CaPO$_4$ is possible by sedimentation. The behavior and distribution of this fraction is similar to the fraction described above (HCL-P/CaII, r=0965,**). Its distribution along the soil profile is also analogous to the above described fraction of phosphorus (Fig. 6).

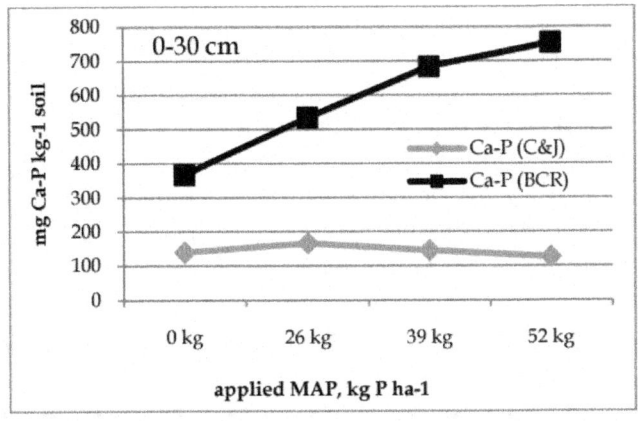

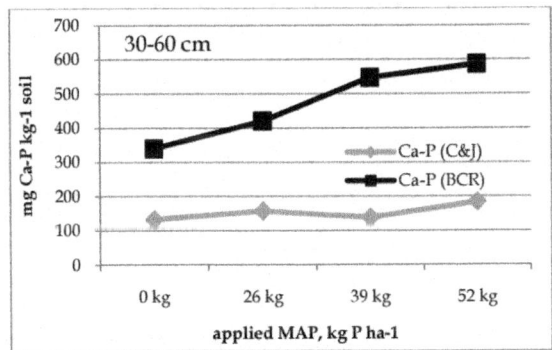

Figure 6: Distribution of Ca bound P in the two applied methods: Chang&Jackson and BCR in the two soil intervals upon 40-years application of phosphate fertilizer.

Fraction of Soil P Extracted by 1M NaOH (Inorganic P; IP)

This fraction is supposed to consist of the later two fractions (Al-Fe-Mn bound P and Ca-P) however, due to the different extracting reagent used for their separation, we have isolated this fraction by extraction in step 3 (Tab. 3). The fraction of inorganic phosphorus in our study was highly correlated with Ca II (Tab. 1), what is the consequence of the decrease of bioavailable phosphorus in the soil. Good correlation of IP with Ca and Al extracted in step 3 of the sequential analysis (Tab. 1) highlights the roles of Ca and Al in the fixation of phosphorus in soil. The decrease of the bioavailability of this fraction is supported by a moderately good positive correlation with the occluded phosphorus (r=.619 *).

Fraction of Soil P Extracted by 1M HCl+Calcinations (Organic P; Org.P)

This fraction of phosphorus is an exact fraction with not precisely defined constitution that partially consists of phitite (De Groot & Golterman, 1993). The extraction of the organic P fraction was performed in step 4 (Table 3). Most of the organic soil phosphorus is present in the form of the inositol phosphate ester while the proportion of phospholipids and nucleic acids in soils is small due to the fact the two groups of phosphate esters are quickly dephosphorylated by microbial phosphatases (Flaig, 1966).

The moderately good correlation of the organic P with Fe extracted in step 4 (Table 1) indicates the importance of this phase in bonding of organic phosphorus. In contrary to other phosphorus fractions determined by BCR method the organic P fraction didn't show increases after 40-years of application of phosphate fertilizer on Stagnosol. The absence of changes in the content of organic phosphorus after the 40-years of application of phosphate fertilizer (Fig.7) obviously was due to its low mobility; the microorganisms easier consumed the applied mineral phosphorus from fertilizer, which resulted in negligible changes of organic phosphorus.

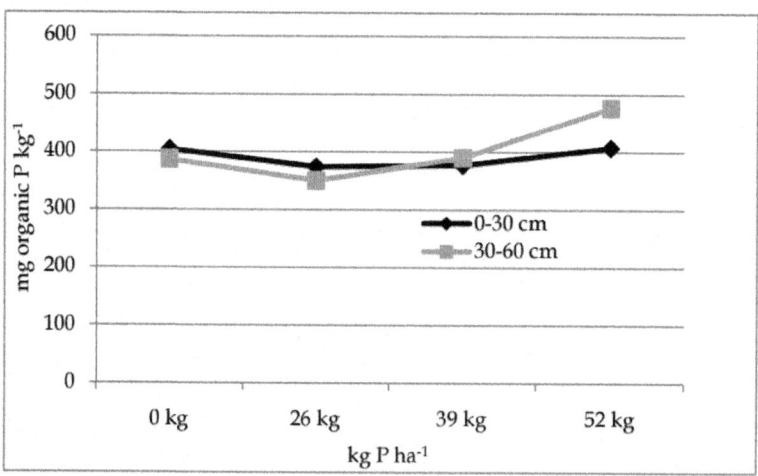

Figure 7: Response of soil organic phosphorus after 40-years application of phosphate fertilizer

MICROWAVE DIGESTION METHOD (ETHOS MILESTONE) USING HNO_3, HCL AND HF (COMPLETELY TOTAL P; TP)

Long-term application of mineral phosphate fertilizer on Stagnosol not only determined the changes in the content of total P extracted by these extraction solutions: known that more than 90% of phosphorus present in soil as insoluble and fixed. Because in soil the process of bounding of phosphate into unavailable forms takes place constantly, the content of total phosphorus can give information about which amount of added phosphate is tied, i.e. unused. Fertilization increases the amount of total forms of P both on surface and subsurface soil layers. But the ratio of the increase in the content of P between the sum of the entire mineral fraction extracted by Chang&Jackson method and the total P by microwave method is 1:6, which indicates the presence of very clear process of fixation and accumulation of P, i.e. formation of the secondary minerals (Fig. 8).

As mentioned earlier, for such type of soil, often due to mineral fertilization there are no increases in the amount of organic P (Sharpley & Smith 1983, Adeptu & Corey, 1976). The process of immobilization of P in deeper layers is closely linked to the mobility of Al-P, where according to the unpublished data, the correlation with the total P was r=0.721** . Therefore, in such soils fertilization by phosphate demands a special caution, i.e. finding the exact ratio between the process of immobilization and the plant demands.

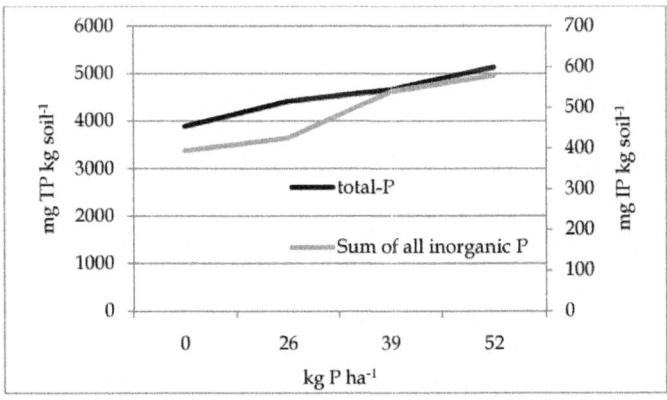

Figure 8: Changes in the concentrations of total P extracted by microwave digestion and the sum of all inorganic P from Chang&Jackson method upon 40-years application of different rates of MAP.

SOIL AGROCHEMICAL PROPERTIES

The change in the basic soil properties and accumulation of micro- and trace elements, some with toxic species, upon the application of phosphate fertilizers are the important factors both for crop yield and for ecological concerns. They can maintain or improve crop yields, but they can also cause changes in the chemical and physical properties of the soil, both directly and indirectly (Hera & Mihaila, 1981; Acton & Gregorich, 1995; Aref & Wander, 1998; Belay et al., 2002). By affecting the basic soil properties (pH, organic C and N, cations, CEC, granulometric composition), phosphate fertilization may influence the solubility of certain elements, such as Al, F, Ca, and Mg (Lindsay, 1979; Kabata-Pendias & Pendias, 2001; Loganathan et al., 2006). On the other hand, mineral phosphate fertilizer could provide an abundance of available phosphorus in soil and increase the efficiency of metal-phosphate mineral formation (Ma et al., 1993; Berti & Cunningham, 1997; Hettiarachhchi et al., 1997; Cooper et al., 1998). Metal-phosphate minerals were shown to control metal solubility in soil suspension when available P was added (Santillian-Medrano & Jurinak, 1975) by inducing the formation of heavy metal phosphate precipitates (Cotter-Howells & Capron, 1996). Additionally, raw materials for P fertilizers contain certain amounts of trace elements and microelements, which may be incorporated into mineral fertilizer (Goodroad & Caldwell, 1979; Adriano, 2001; Kabata-Pendias & Pendias, 2001). The effect of fertilization on soil quality can be best evaluated through the use of long-term experiments (Mitchell et al., 1991; Nel et al., 1996). The general soil-chemical properties are given in Table 4.

Only 15% of the applied phosphorus is consumed by plants quickly after addition (Greenwood, 1981), the rest transforms into insoluble forms and non-labile fractions. Consumption of the applied P by plants becomes more and more difficult each year. However, constant application of fertilizer from year to year can replenish the capacity for the adsorption, and consequently the amount of the available phosphate (Stewart & Sharpley 1987; McCollum, 1991; Maroko, et al., 1999).

As would be expected, the content of P in the soil increased significantly ($p < 0,01$) depending on the amount of applied phosphate. So, in the treatment with 52 kg/Pha-1 an increase by about 3 times was detected in Stagnosol (Figure 9). In the deeper soil layer due to low mobility of phosphates this increase was only by 2 times as compared to the control treatment (Cakmak, D. et al., 2010).

Soil Physical Characteristics

Long-term application of phosphate, besides the direct influence on the content and forms of P in soil, also influences the other soil properties. It can affect the structure of clay minerals where fractions of P can replace Al ions from the tetrahedral structure thus destructing the structure and pulverizing clay minerals (Rajan, 1975).

Table 4: Soil chemical characteristics of Stagnosol in 40-years phosphate fertilization experiment

Treatment	Organic C	Total N	Avail. P†	Exch. Ca²⁺†	Exch. Mg²⁺	Mobile. Al	pH	CEC†	Sum of base
g P ha⁻¹	-- % --			--- mg 100 g⁻¹ ---				-- cmol kg⁻¹ --	
				0-30 cm					
0	0.99 ± 0.03	0.11 ± 0.001	4.8a‡ ± 0.6	237a ± 9	40.7 ± 4.2	3.7a ± 0.9	4.95a ± 0.0	16.4a ± 0.6	8.67a ± 0.2
26	1.06 ± 0.02	0.12 ± 0.003	8.2b ± 0.4	267b ± 2	43.8 ± 2.3	6.9b ± 1.5	4.80b ± 0.1	18ab ± 0.4	9.07b ± 0.2
39	1.03 ± 0.05	0.12 ± 0.004	9.6c ± 0.3	251ba ± 7	45.9 ± 1.7	7.1b ± 1.0	4.84b ± 0.0	18.4b ± 0.2	9.34b ± 0.3
52	1.06 ± 0.04	0.12 ± 0.003	13.5d ± 2.8	241ba ± 6	43.8 ± 2.2	7.9c ± 0.9	4.76c ± 0.0	18.8b ± 0.4	8.94b ± 0.3
P	NS§	NS	***	*	NS	*	*	**	*
				30-60 cm					
0	0.81 ± 0.02	0.10 ± 0.000	3.5a ± 0.7	277 ± 3	48.1 ± 0.9	4.07 ± 0.6	5.03 ± 0.0	17.1 ± 0.5	9.87a ± 0.2
26	0.77 ± 0.05	0.09 ± 0.00	5.2b ± 1.6	262 ± 9	49.9 ± 4.4	5.51 ± 1.1	4.95 ± 0.0	16.9 ± 0.6	9.34a ± 0.5
39	0.83 ± 0.09	0.10 ± 0.01	6.1c ± 1.2	264 ± 16	48.2 ± 4.9	4.37 ± 0.1	5.02 ± 0.1	17.5 ± 0.4	10.44b ± 0.4
52	0.74 ± 0.03	0.09 ± 0.00	7.0d ± 1.5	272 ± 10	50.8 ± 1.8	4.68 ± 0.4	5.03 ± 0.0	17.9 ± 0.9	10.7b ± 0.6
P	NS	NS	***	NS	NS	NS	NS	NS	*

* Significant at $P < 0.05$.
** Significant at $P < 0.01$.
*** Significant at $P < 0.001$.
† Avail., available; Exch., exchangeable; CEC, cation exchange capacity.
‡ Within each depth increment, means with the same letter are not significantly different.
§ NS, not significant.

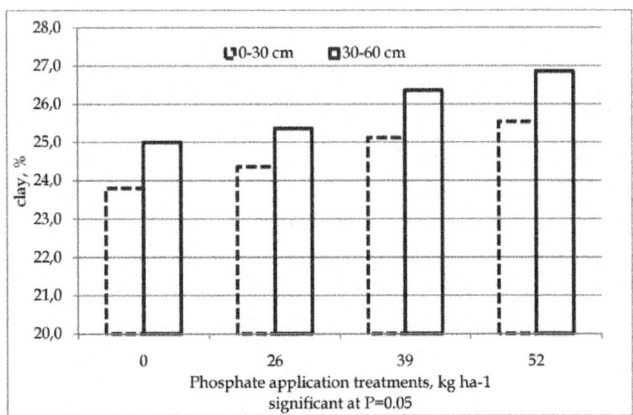

Figure 9: Amount of clay fraction under long-term application of different rates of phosphate fertilizer in two depths on a Stagnosol soil.

Long-term fertilization with MAP distninguished that phenomenon since the changes in soil texture were due to the amount of added P, i.e. in the experiment a significant increase of clay fraction was detected correspondingly to the rates of fertilizer, which is the result of fragmentation of clay particles (Fig. 9). In spite the mineral fertilization causes decreases in soil CEC (Belay et al., 2002) especially if the amount of organic matter did not change, in our case due to the increased content of clay the slightly significant increase of CEC in the treatments with 39 and 52 kg P ha^{-1} ($p < 0,01$) was recorded. The consequence is increases of the value of sum of base in the second depth under the higher rates of fertilizer ($p < 0.05$).

Soil Chemical Characteristics

Acidity (pH): Mineral fertilizers can change the soil pH depending on the dominance of alkali or acidic components. Because in MAP the acidic components predominates, since in soil the process of nitrification leads to formation of nitrate ions, the long-term fertilization unavoidably results in acidification (Magdof et al., 1997; Belay et al., 2002; Saleque et al., 2004). Since the amounts of nitrogen components in the applied fertilizer were small and due to the negligible effect of phosphoric components, the decrease in soil pH after 40-years of application of MAP was slight but significant ($p < 0.05$ (Tab. 4). It should be mentioned that the other phosphoric fertilizers without nitrogen components did not have a negative impact on soil pH even after long application.

Available Phosphorus (Al-Method): Available phosphorus is a fraction of P that is considered available for plants. Chemical extraction is based on solvents, which more or less imitate adsorptive power of plant root. Phosphorus in the first four fractions obtained by the Chang&Jackson method by definition is more or less available to plants but this fact was not confirmed in the 40-year field experiment of application of MAP on Stagnosol. Absence of correlation between the available P and Fe-P is probably due to the low amount of Fephosphate found in this experiment (< 1 mg kg^{-1}). Based on the results of coefficient of correlation, the direct correlation between the available forms of P is recorded for watersoluble P and the P bound to Al at both depths (for 0-30 cm is 945** and 987** respectively; for 30-60 cm 715** 888** respectively). On the other hand, reducible phosphate is indirectly correlated with the available P via water-soluble and Al bound P. The bounds of occluded P in the first depth with water-soluble ($r=.585*$) and in the second depth with water-soluble and Al-P indicates on its possible availability under the extreme conditions of wetting anddrying, characteristic for Stagnosol. Except the Ca-bound phosphorus, all the fractions of soil P showed increasing trends accordingly to the applied phosphate fertilizers where the Al-bound phosphorus showed the clearest increasing tendency (Fig. 10).

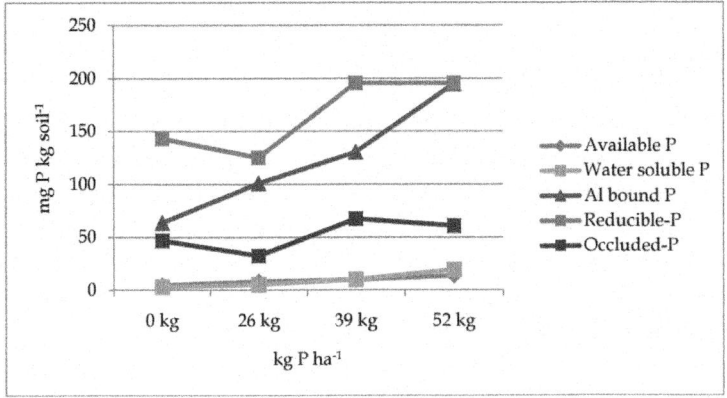

Figure 10: Concentrations soil P extracted by different methods upon 40-years application of phosphate fertilizer.

Mobile Al: One of the indirect consequences of phosphate fertilization that acidifies the soil is the increase in the amount of mobile Al. Soil acidity is caused by the formation of H^+ ions in the soil solution, which is neutralized by Al and Fe oxy/hydroxy complexes (Schwertmann et al., 1987) where the end product of neutralization are species of Al cations such as Al^{3+}, $Al(OH)^{2+}$, $Al(OH)_2^+$. Connection of these two processes is confirmed by a high correlation between soluble Al and decreases in soil pH (r=.897**). It is known that the solubility of Al progressively increases below 5.5 soil pH (H_2O) (Abrahmsen, 1984; McKenzie & Nyborg, 1984; Mrvic et al., 2007). In our experiments, in the 26 kgPha^{-1}, treatment in the subrface soil the amount of mobile Al is doubled versus to control (Tab. 4).

Exchangeable Cataions and CEC: A recent research showed that the decrease of exchangeable Ca and Mg with P fertilization is caused by their replacement with H^+ ions and leaching to layers down the soil profile (Belay et al., 2002), immobilization by phosphates, and by their assimilation by plants. However, destruction of clay minerals as affected by phosphorus and increase in CEC reduces the leaching. In soils with pH above 5.0, increases in acidity results in destruction of minerals, i.e. exchangeable Ca^{2+} and in lesser degree Mg^{2+}, might be derived from structure of soil primary minerals thus increasing the Ca availability in soil solution (McLaughlin & Wimmer, 1999). This process of releasing Ca and Mg is much weaker than that of Al because the release of cations is proportional to charge, which results in higher amounts of cationic species in the soil solution. Under 40-years of phosphate application, the amount of exchangeable Ca is increased (Tab. 4), although the increase was not regular but significant in the surface soil due to the mentioned reverse processes.

MICROELEMENTS

The content of microelements in the soils depends on their amounts in the soil-forming rocks and on the soil-forming processes. The role of microelements in the physiological and the biochemical processes is immeasurably great. Soil is a source of microelements for plants, animals and humans. Deficiency or excess amount of microelements in fodders and food products lead to the disturbance of exchange of substances and the appearance of diseases in plants, animals and people. Basic for the vital activity of the plants and other living organisms are manganese, copper, boron, zinc, molybdenum, nickel, cobalt, fluorine, vanadium, iodine. One of the anthropogenic sources of microelements in soil is agriculture, including the application of phosphate fertilizers.

Iron solubility is controlled both by pH and redox potential. Iron becomes more soluble at lower pH values and under reduction condition. Iron solubility is largely controlled by the solubility of the hydrous Fe^{3+} oxides. At higher pH levels the activity of Fe^{3+} in solution decreased for 1000 time for each pH unit rise (Lindsay, 1974). When soil is waterlogged, the reduction of Fe^{3+} into Fe^{2+} increases the solubility of Fe. On the other hand, application of phosphate and its immobilization by formation of compounds with Fe, results in decrease in the amount of DTPA-extractable Fe, what is supported by the absence of correlation between mobile Fe with reducible and occluded P. Application of phosphate during 40- years determined the antagonistic processes fully took place, i.e. a small increase in DTPAextractable Fe in the treatments where the processes of acidification are active was detected ($p < 0.05$) (Tab. 7).

Zinc solubility is decreased with increasing of soil pH (Cakmak, I. et al., 1996) where phosphates can immobilize Zn as zinc phosphate $Zn_3(PO_4)_3 4H_2O$, although this theory is not fully proved (Jurinak & Inouye 1962). However, application of MAP has shown opposite results for the second depth (30-60 cm) (Tab. 6, 7). Consequently, in the subsurface soil the content of the total Zn increases ($p < 0.05$) versus control. This in turn probably resulted in leaching of Zn due to acidification of the surface soil. On the other hand, the decrease of the amount of DTPA-extractable Zn is evident ($p < 0.05$) what can explain the prevalence of the processes of immobilization of Zn by phosphorus.

Copper similarly to Zn responded to the changes in soil pH. However, its solubility is quite lowered due to very strong bounds of Cu with soil organic matter (Mc Bride 1989). Reduced conditions in Stagnosol ties Cu with iron forming cuprous ferrite, possibly controlling Cu solubility, which depends on the solubility of Fe (Shuman, 1985), which was reflected in significant correlation between soluble Fe and Cu.

The amount of fluorine in phosphate fertilizers depends on geographic origin of the raw material and the degree of its treatment (Gooroad & Caldwell, 1979). In fertilizers such as MAP this amount is not negligible (Tab 5). There was no significant increase in the amount of total F (Tab. 6, 7); the values were near or within the world's average (Helmke, 2000; Glandey & Burns, 1985). Solubility of fluoride ions is controlled by the soil pH and the amount of soil Ca and P (Hurd-Karrer 1950). But it should be noted that P affects the availability of F only in case of very high amounts of P in soil. Therefore, in our experiments a statistically significant increase of soluble F was detected in the both soil depths in accordance with the rates of the applied fertilizer, which presumably were induced by the decrease of soil pH ($r=-0.92$ $p < 0.01$) (Pickering, 1985; Barrow & Ellis, 1986; Loganathan et al., 2006). The significant increase of the amount of soluble F in the subsurface soil might be due to its leaching from the upper soil layer.

Trace Elements

The dominant sources of fertilizer contaminants are the raw materials used to manufacture phosphate fertilizer. The most common contaminant metals and metalloids found in phosphate rock are arsenic (As), cadmium (Cd), chromium (Cr), mercury (Hg), lead (Pb), selenium (Se) and fluoride. Therefore, phosphate fertilizer is considered as one of the most important sources of trace element contamination in agricultural soils. The content of some trace elements used in our 40-year experiment is given in Table 5.

Table 5: Content of trace elements in the applied monoammonium phosphate

Element	As	Hg	F	Cr	Cu	Zn	Cd	Ni	Pb	Fe
					$mg\ kg^{-1}$					
Concentration	4-15	0.003-0.005	19,600-26,700	15-315	10-20	10-38	3-16	4-39	10-20	4000-5000

Lead (Pb)

Amount of Pb in phosphate fertilizers ranges between 7-225 mg kg^{-1} (Kabata-Pendias, 2001). The amount of Pb in the applied phosphate ranged between 10 and 20 mg kg^{-1} (Stevanovic et al., 2009). It is known that lead-containing compounds are weakly soluble (Davies, 1995), but application of phosphates affects their solubility, resulting in their greater immobilization (McGowen et al., 2001; Dermatas et al., 2008).

Hettiarachchi & Pierzynski (1996) had more success with the solubility equilibrium approach. They studied the influence of P additions on Pb solubility in a Pb-contaminated soil. Prior to P additions, the soil was in equilibrium with cerrusite ($PbCO_3$), but after addition of a soluble P the equilibrium was shifted to the Pb controlled by hydroxypryromorphite [$Pb_5(PO_4)_3OH$], a less soluble Pb mineral phase (Sims & Pierzynski, 2005).

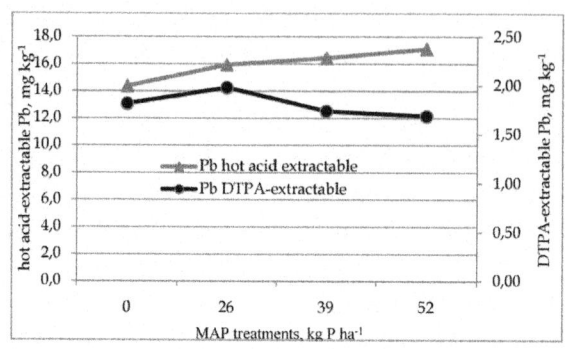

Figure 11: Distribution of hot acid-extractable and diethylenetriaminepentaacetic acid (DTPA)- extractable Pb under different rates of phosphate fertilization on a Stagnosol soil.

In our experiment, the amount of hot acid-extractable Pb in the 0- 30 cm soil interval increased significantly ($p < 0.05$) in accordance with the amount of added P fertilizers (Tab. 6). Significance was observed between the control and all the treatments. The content of hot acid-extractable Pb in the 0-30 cm interval of the control was less than that in the 30-60 cm interval, whereas in the fertilized treatments its content was higher in the upper than in the lower soil depth. Relatively small but significant increase of total P after the 40-years of phosphate application was due to the low content of P in the applied fertilizer, and to the low applied rates of fertilizer. Expectedly, based on the above-mentioned results, in contrast to the hot acid-extractable form, the amount of DTPA-extractable Pb decreased significantly ($p < 0.05$) in the 0-30 cm soil with the increase in fertilization rates (Fig. 11).

Cadmium (Cd)

Cadmium is in fertilized soils as divalent cation Cd_2^+ and the associated organic and inorganic complexes. As the cation, Cd solubility is highest in acidic soils, in soils with low CEC and soils with high anionic ligands (Hahne & Kroontje, 1973; Garcia-Miragaya & Page, 1976; Garcia-Miragaya & Page, 1978; Chubin & Street, 1981). Cadmium present in granular fertilizers rapidly dissolves and the solid-phase Cd in the fertilizer moves into the soil as free Cd_2

+. In contrast to the other trace metals such as Cu and Zn, organic ligands do not bind strongly free Cd_2^+ ions or the Cd complexes. In inorganic ligands the dominant Cd species present in soil solution tend to be more mobile than other trace metals, including Pb and Cu (Alloway, 1995).

Cadmium is a trace metal that is often bound to P fertilizers, especially to rock phosphates and it proved to be less mobile than the P carrier itself in the soil profile (Mulla et al., 1980; Adriano 2001; Kabata-Pendias & Pendias, 2001). However, the fertilizer we used (MAP) contains low concentrations of Cd (3–16 mg kg^{-1}) (Stevanovic et al., 2009). The low measured concentrations of this element and its solubility resulted in almost the same concentrations of Cd in both soil intervals. The minor changes in Cd observed in our studies were clearly due to the low levels of Cd in the source material (Tab. 6). About 71% of the accumulated Cd resided in the surface soil (0 to 15 cm) compared with only 45% for P (Adriano, 2001).

Table 6: Concentrations of hot acid-extractable trace elements in Stagnosol of long-term phosphate fertilization experiment

Treatment					Hot acid-extractable				
	Total F	Cu	Zn	Ni	Pb	Cd	Co	Cr	As	Hg
kg P ha^{-1}					— mg kg^{-1} —				
					0-30 cm				
0	334 ± 16.5†	12.7 ± 0.2	32.0 ± 0.4	26.3 ± 0.5	14.4a† ± 0.7	0.20a ± 0.00	11.5 ± 0.1	18.7a ± 0.6	4.5a ± 0.00	0.049a ± 0.001
26	337 ± 2.6	12.6 ± 0.1	32.3 ± 1.1	27.2 ± 0.3	15.9b ± 0.2	0.20a ± 0.00	12.5 ± 0.7	17.4ab ± 0.5	4.9ab ± 0.17	0.060b ± 0.004
39	339 ± 15.7	13.7 ± 0.8	32.0 ± 0.8	26.8 ± 1.2	16.4b ± 1.0	0.27b ± 0.02	11.3 ± 0.2	16.9ab ± 1.0	5.0b ± 0.20	0.061b ± 0.005
52	358 ± 8.5	12.6 ± 0.3	31.9 ± .0.3	26.2 ± 0.7	17.1c ± 0.1	0.27b ± 0.02	12.0 ± 0.4	15.9b ± 0.5	4.9a ± 0.13	0.092c ± 0.002
P	NS‡	NS	NS	NS	*	**	NS	*	*	***
					30-60 cm				
0	323a ± 10.1	12.5 ± 0.1	30.9 ± 0.1	27.0 ± 0.5	16.7a ± 0.3	0.20 ± 0.03	12.8 ± 0.8	18.6ab ± 0.8	4.7ab ± 0.2	0.04a ± 0.001
26	331a ± 6.4	13.3 ± 0.2	33.0 ± 0.8	28.0 ± 0.7	14.6b ± 0.6	0.20 ± 0.03	13.1 ± 0.4	19.9a ± 1.1	4.6a ± 0.1	0.06b ± 0.002
39	342a ± 3.1	12.8 ± 0.5	31.1 ± 0.6	27.1 ± 0.6	14.0b ± 0.8	0.20 ± 0.00	12.2 ± 0.8	16.8b ± 0.5	4.5a ± 0.1	0.06b ± 0.003
52	351b ± 11.2	13.3 ± 0.5	31.7 ± 0.1	27.3 ± 0.3	15.0ab ± 0.3	0.27 ± 0.02	12.0 ± 0.4	19.0a ± 0.6	5.0b ± 0.1	0.09c ± 0.002
P	*	NS	*	NS	**	NS	NS	*	*	***

Table 7: Concentrations of water-extractable F and DTPA-extractable trace elements in Stagnosol of long-term phosphate fertilization experiment

Treatment	Water-extr. F	DTPA-extractable					
		Cu	Zn	Ni	Pb	Cd	Fe
kg P ha^{-1}				— mg kg^{-1} —			
				0-30 cm			
0	0.32a† ± 0.01	1.60a ± 0.04	0.77 ± 0.09	1.11 ± 0.08	1.81ab ± 0.09	0.07 ± 0.011	158a ± 3.4
26	0.69b ± 0.22	1.67ab ± 0.04	0.73 ± 0.03	1.31 ± 0.11	1.98b ± 0.04	0.07 ± 0.003	196b ± 8.4
39	0.76b ± 0.12	1.70b ± 0.01	0.74 ± 0.07	1.32 ± 0.12	1.74ab ± 0.08	0.06 ± 0.002	195b ± 9.4
52	0.92c ± 0.09	1.65a ± 0.02	0.88 ± 0.09	1.38 ± 0.07	1.65a ± 0.15	0.07 ± 0.008	194b ± 4.3
P	*	*	NS‡	NS	*	NS	**
				30-60 cm			
0	0.14a ± 0.03	1.57a ± 0.01	0.56 ± 0.02	1.22 ± 0.03	1.57 ± 0.02	0.04 ± 0.008	162 ± 6.2
26	0.31b ± 0.10	1.57a ± 0.02	0.47 ± 0.05	1.24 ± 0.07	1.64 ± 0.01	0.03 ± 0.002	173 ± 9.9
39	0.31b ± 0.05	1.57a ± 0.04	0.47 ± 0.06	1.29 ± 0.04	1.43 ± 0.10	0.04 ± 0.005	167 ± 5.4
52	0.38b ± 0.04	1.49b ± 0.02	0.41 ± 0.02	1.19 ± 0.06	1.45 ± 0.12	0.04 ± 0.004	166 ± 5.8
P	*	*	*	NS	NS	NS	NS

Arsenic (As)

Application of phosphates can enhance As mobility, phytoavailability, and phytotoxicity in lead arsenate contaminated soils amended with MAP (Peryea, 1991). The degree of arsenic toxicity depends on the phosphate concentration in the nutrient solution in certain soil-plant environments (Hurd-Karrer, 1939). The main predicators of As solubility are soil mineralogy, organic matter, soil pH and As oxidation state. Arsenic may form insoluble compounds with Fe and Al oxides, be absorbed by the organic matter or hydroxyl groups on clay minerals (Mitchell & Barr, 1995). Among various potassium and sodium salts, potassium phosphate was the most effective in extracting arsenic, attaining more than 40% extraction (Alam, et al., 2001). Phosphate could displace adsorbed or fixed As from sorbing complexes and thereby initially increase the amount of soluble As in soils. (Kabata-Pendias, & Pendias, 2001).

Because As is often a contaminant in phosphate fertilizers (Adriano, 2001; Kabata-Pendias & Pendias, 2001) produced from rock phosphates (O'Neill, 1995) we expected a higher concentration of As in the upper soil interval. However, because of their position in the periodic table of elements, where phosphorus, an essential element capable of being absorbed in relatively large amounts, and arsenic, a highly toxic element, chemically much like phosphorus, these elements often behave as antagonists. Anion antagonism between P and As (Parfitt, 1978), and their similar solubility under certain soil pH (Alloway, 1995), resulted in the movement of As down the soil profile (Tab. 6).

Chromium (Cr)

The content of Cr in phosphate fertilizers ranges between 66-245 mg kg^{-1} (Kabata-Pendias & Pendias, 2001). However, in the fertilizer used in our experiments the content of Cr was wider and ranged from 15-315 mg kg^{-1}. Chromium in soil could be found mainly in two forms: Cr^{3+} and Cr^{6+}. In the majority of soils, the relatively insoluble and less mobile Cr(III) form predominates and it geneally occurs as insoluble hydroxides and oxides (McGrath, 1995). Phosphorus states in soil and soil pH can determine the similarity of adsorption and solubility behavior between Cr^{3+} and Al^{3+}. Although the amount of Cr entering soil via theuse of phosphate fertilizers is uncertain, it is likely to exist as Cr(III). Chromium (Cr III) in soil is not likely to be toxic (McGrath, 1995). Presence of Cr^{6+} in the soil surface horizons is due to its linkage to the oxidizing conditions and redox potentials. Once Cr is oxidized into the hexavalent form in the soil, it becomes more mobile and leachable (Adriano, 2001). Ions of $H_2PO_4^-$ and $HCrO_4^-$ are antagonists what is reflected in the increased mobility and solubility of Cr ion in soil solution. Data in the Table 4 confirms this hypothesis since in the control at both depths the concentration of Cr was nearly equal; however, at the surface depth the concentration of Cr is significantly decreasing correspondingly with the applied rate of phosphate; whilst at the subsurface depth the concentration of Cr significantly increases after the addition of phosphate (Adriano, 2001).

Nickel, Cobalt, Mercury (Ni , Co, Hg)

The content of Ni and Co did not show any differences in any of the treatments. The concentrations of these trace elements in the studied soil were within the range of average world concentrations. Most commercial fertilizers have Hg content below 50 ng per gram, but considerably higher amounts occur in phosphate fertilizers. Hg may be derived partly from the phosphate rocks and partly from the sulphuric acid used for dissolution of phosphates (Steinnes, 1995). Because of the initially low concentrations of hot-acid extractable Hg in the control treatment, the increased concentrations of Hg detected at both soil depths (Tab. 6) under the fertilized treatments are obviously due to the applied MAP. However, in spite of statistically significant increases in the content of Hg, its absolute amount is small and doesn't pose a risk for the environment even after 40-years application of MAP in Stagnosol.

CONCLUSIONS

Formation of the forms of soil phosphorus and their binding largely depends on soil pH. In neutral and alkali soils the main phosphate compounds are calcium phosphates, and their solubility depends mainly on the ratio Ca/PO_4. In such soils, the activity of Ca ions increased resulting in reduced solubility of phosphate while in neutral and acidic soil the adsorption and desorption of phosphate mainly occurs at Al and Fe oxide surfaces. The bounds Al-O-P forms much more labile forms than formations with double bounds of P. Al bound P is the most labile form that supplies the plants with P-nutrient, and is the most responsible form for the movement of P along the soil profile and replenishment of other soil P-fractions. Both sequential analyses used for P fractionation proved that Al bound P fraction is closely related to the mobile Al of soil solution. Application of mineral phosphates results in the increase of Al bound P fraction thus increasing the amount of plant-available phosphorus in soil. In acidic soils, the application of MAP does't results in considerable changes in the amount of Ca-bound P and of organic P in soil. Also, the 40-year application of MAP destroys the structure of the clay minerals what in its turn increases the soil CEC. The long-term application of MAP didn't result in the accumulation of potentially toxic elements in their considerable concentrations, i.e. their concentrations were negligible in Stagnosol.

An important outcome is that soil P can exist in a series of "pools", which can be defined in terms of the extractability of P in different reagents. Also, the P in these pools can be related to the availability of P to plants, recognizing that there is a continuum of both extractabilityand availability. The most important concept is the reversible transfer of P among the most of soil phosphorus pools, what opens a possibility of the effective use of the applied phosphorus. These complex transformations of soil phosphorus grant wide opportunities for further researches.

REFERENCES

1. Abrahamsen, G. & Miller, H.G. 1984. The effects of acid deposition on forest soil and vegetation. Phil. Trans. R. Soc. B., Vol. 305, pp. (369–382). ISSN 1471-2970
2. Acton, D.F. & Gregorich, L.J. (Eds.). (1995). The Health of Our Soils: Toward sustainable agriculture in Canada. 1906/E. Centre for Land and Biological Resources Research, Research Branch, Agriculture & Agri-Food, ISBN 0-660-15947-3, Canada, Ottawa, ON.
3. Adeptu, J.A. & Corey, A.B., 1976: Organic phosphorus as a predictor of plant available phosphorus in soil of southern Nigeria. Soil Sci. Vol. 122, pp. (159-164). ISSN 0038- 075x
4. Adriano, D.C. (2001). Trace elements in terrestrial environments. Sheridan Books, ISBN 0-387- 98678-2, Ann Arbor, MI.
5. Alam, M. G. M., Tokunaga, S., & Maekawa, T. (2001). Extraction of arsenic in a synthetic arsenic-contaminated soil using phosphate, Chemosphere, Vol. 43, No.8 ,(June 2001), pp. (1035-1041)
6. Aleksandrovic, D., Pantovic, M., & Lotse, E., (1965). The minerological composition of some Pseudogley soils in Serbia, Zemljiste i Biljka Vol.14, pp. (333-351).
7. Alloway, B.J. (1995). Heavy metals in soils. Chapman and Hall, ISBN 0-7514-0198-6, London.
8. Aref, S., & Wander, M.M. (1998). Long-term trends of corn yield and soil organic matter in different crop sequences and soil fertility treatments on the Morrow Plots. Adv.
9. Agron. Vol. 62, pp. (153–161). ISSN 0065-2113
10. Aydin, I., Aydin, F., Saydut, A. Hamamci, C. A., 2009. Sequential extraction to determine the distribution of phosphorus in the seawater and marine surface sediment. J Hazard Mater. Vol.168, No.2-3, (Sep.15), pp. (664-669), ISSN 0304-3894

11. Barrow, N.J., & Ellis, A.S., (1986). Testing a mechanistic model: III. The effects of pH on fluoride retention by soil. Eur. J. Soil Sci. Vol.37, pp. (287–293), ISSN 1351-0754

12. Beck, M. A., & Sanchez, P.A. (1994). Soil phosphorus fraction dynamics during 18 years of cultivation on Typic Paledult. Soil. Sci. Soc. Am. J., Vol. 58, pp. (1424–1431), ISSN 0361-5995

13. Belay, A., Claassens, A.S., & Wehner, F.S. (2002). Effects of direct N and K and residual phosphorus fertilizers on soil chemical properties, microbiological components and maize yield under long-term crop rotation. Biol. Fertil. Soils, Vol. 35, pp. (420–427), ISSN 0178-2762.

14. Berti, W.R., & Cunningham, S.D. (1997). In-place inactivation of Pb in contaminated soils. Environ. Sci. Technol, Vol. 31, pp. (2673–2678), ISSN 0013-936X

15. Cakmak, D., Saljnikov, E., Mrvic, V., Jakovljevic, M., Marjanovic, Z., Sikiric, B., & Maksimovic, S. (2010). Soil properties and trace elements under 40-years of phosphate fertilization. J. Environ. Quality, Vol. 39, pp. (541-547), ISSN 0047-2425

16. Cakmak, I., Sari, N., Marschner, H., Ekiz, H., Kalayci, M., Yilmaz, A., & Braun, H.J. (1996). Phytosiderophore release in bread wheat and durum wheat genotypes differing in zinc efficiency. Plant and Soil, Vol.180, pp. (183-189), ISSN 0032-079X.

17. Chang, S.C., &. Jackson, M.L. (1957). Fractionation of soil phosphorus. Soil Sci. Vol. 84, pp. (133–144), ISSN 0038-075X.

18. Cooper, E.M.., Strawn, D.G., Sims, J.T., Sparks, D.L., &. Oneken, B.M. (1998). Effect of chemical stabilization by phosphate amendment on desorption of P and Pb from contaminated soil, In: Agronomy abstracts. pp. 343, ASA, ISSN-L: 00654671 0065 4671. Madison, WI;

19. Cotter-Howells, J., & Capron, S. (1996). Remediation of contaminated land by formation of heavy metal phosphates. Appl. Geochem, Vol. 11, pp. (335–342), ISSN 0883-2927.

20. Davies, B.E. 1995. Lead, In: Heavy metals in soils, B.J. Alloway (Ed.) pp. (206–220), Chapman and Hall, ISBN 0-7514-0198-6, London

21. Delgado, A., & Scalenghe, R. (2008). Aspects of phosphorus transfer from soils in Europe. J. Plant Nutr. Soil Sci. Vol. 171, pp. (552–575), ISSN 0718-9508.

22. De Groot, K.J. & Golterman, H.L. (1993). On the presence of organic phosphate in some Camargue sediments: evidence for the importance of phytate. Hydrobiologia, Vol. 252, pp. (105-116); ISSN 0018-8158

23. Dermatas, D., Chrysochoou, M., Grubb, D.G., & Xu, X. (2008). Phosphate treatment of firing range soils: Lead fixation or phosphorus release? J. Environ. Qual. Vol. 37, pp. (47– 56), ISSN 0047-2425
24. Enger, H., & Riehm, H. (1958). Die Ammoniumlaktatessigsäure-Methode zur Bestimmung der leichtlöslichen Phosphorsäure in Karbonathaltigen Böden. (In German.) Agrochimica Vol. 3, No. 1, pp. (49–65), ISSN 0002-1857
25. Flaig, W. (1966). The Use of Isotopes in Soil Organic Matter Studies. Pergamon, ISBN 0-582- 44245-1, New York, NY;
26. Frankenberger, W.T., Tabatabai, M.A., Adriano, D.C., & Doner, H.E. (1996). Bromine, chlorine, and fluorine, In: Methods of soil analysis, Part 3. Chemical methods, D.L. Sparks (Ed.), pp. 833–867. SSSA, ISBN 96-70096, Madison, WI;
27. Foth, H.D., & Ellis, B.G. (Jan. 31, 1997). Soil fertility, 2nd ed. CRC Press, Boca Raton, ISBN-10: 1566702437, FL.
28. Frossard, F., Brossard, M.., Hedley, M.J., & Metherell, A. (1955). Reaction controlling the cycling of P in soils.. In: Phosphorus in the global environment, H.Tiessen (ed.), pp. (107-137), JohnWiley&Sons, ISBN 0716750791, NY.
29. Garcia-Miragaya, J. & Page, A.L. (1976). Influence of ionic strength and inorganic complex formation on the sorption of trace amounts of Cd by montmorillonite . Soil Sci. Sac.Amer. J. Vol. 40, pp. (658-663), ISSN 0361-5995
30. Garcia-Miragaya, J. & Page, A.L. (1978). Sorption of trace quantities of cadmium by soils with different chemical and mineralogical composition Water, Air & Soil Pollution Vol. 9, No.3, pp. (289-299), ISSN 0049-6979
31. Gee, G.W., & Bauder, J.W. (1996). Particle fractionation and particle-size analysis. In: Methods of soil analysis, Part I, Series 5, C.A. Black (Ed.), pp. (545–567), ASA and SSSA, Madison, WI.
32. Gladney, E.S., & Burns, C.E. (1985). Compilation of elemental concentration data for samples SO-1 to SO-4. Geostand. Newsl, Vol. 9, pp. (38–43), ISSN 0150-5505
33. Golterman, H.L. (1982). Differential extraction of sediment phosphorus with NTA solutions. Hydrobiologia, Vol. 92, pp. (683-687), ISSN 0018-8158
34. Golterman, H.L. (1996). Fractionation of sediment phosphate with chelating compounds: a simplification, and comparison with other methods. Hydrobiologia, Vol. 335, pp. (87- 95), ISSN 0018-8158

35. Goodroad, L.L., & Caldwell, A.C. (1979). Effects of phosphorus fertilizer and lime on the As, Cr, Pb, and V content of soils and plants. J. Environ. Qual. Vol. 8, pp. (493–496), ISSN 0047-2425
36. Greenwood, D.J. (1981). Fertilizer food production: world scene, Fertilizer Research, Vol. 2, pp. (31-51); ISSN 01671731.
37. Hartikainen, H. (1989). Effect of cumulative fertilizer dressing on the phosphorus status of mineral soils. I. Changes in inorganic phosphorus fractions. Journal of Agricultural Science in Finland, Vol. 61, pp. (55-59), ISSN 1795-1895
38. Hedley, M.J., Stewart, J.W.B., & Chauhan, B.S. (1982). Changes in inorganic and organic soil phosphorus fractions induced by cultivation and by laboratory incubations. Soil.Sci.Am.J. Vol. 46, pp. (970-976), ISSN 0361-5995
39. Helmke, P.A. (2000). The chemical composition of soil. In: Handbook of soil science, M.E. Sumner (Ed.), pp. B12–B17, CRC Press, Boca Raton, ISBN 0-8493-3136-6, FL.
40. Hera, C., & Mihaila, V. (1981). The changing of some agrochemical indices of the soil by application of the fertilizers. Analele ICCPT Vol. 47, pp. (319–327), ISSN 0253-1682
41. Hettiarachchi, G.M., & Pierzynski, G.M. (1989). The influence of phosphorus source and rate on soil solution cadmium and lead activities, In: Agronomy abstracts. pp. (333) ASA, CSSA and SSSA, ISSN 006546/1, Madison, WI.;
42. Hettiarachhchi, G.M., Pierzynski, G.M., Zwonitzer, J., & Lambert, M. (1997). Phosphorus source and rate effects on cadmium, lead, and zinc bioavailability in metalcontaminated soil, Extended Abstr., 4th. Int. Conf. on the Biogeochem. Trace Elements (ICOBTE), ISBN 91-576-6626-1, Berkeley, CA., 23–26 June 1997.
43. Hieltjes, A.H.M., and Lijklem, L. (1980). Fractionation of inorganic phosphates in calcareous sediment. J. Environ. Qual., Vol. 9, No. 3, pp. (405-407), ISSN 0047-2425
44. Hurd-Karrer, A. M. (1939). Antagonism of certain elements essential to plants toward chemically related toxic elements,. Plant Physiology, Vol.14, No. 1, pp. (9-29), ISSN 0- 534-15162-0
45. Hurd-Karrer, A.M. (1950). Comparative fluorine uptake by plants in limed and unlimed soil. Soil.Sci. Vol. 70, pp. (153-159), ISSN 0038-075X.
46. Jaakola, A., Hartikainen, H., & Lemola, R. (1997). Effect of fertilization

on soil phosphorus in long-term field experiment in southern Finland. Agricultural and food science in Finland. Vol. 614, pp. (313-322), ISSN 1239-0992.
47. Jakovljević, M., Pantović, M., & Blagojević, S. (1985). Praktikum iz hemije zemljišta i voda. (In Serbian.) Beograd. Poljoprivredni fakultet, Belgrade, Serbia.
48. Jurinak, J.J., & Inouye, T.S. (1962). Some aspects of zinc and cooper phosphate formation in aqueus systems. Soil.Sci. Amer.Proc. Vol. 26, pp. (144-147), 0038-075X
49. Kabata-Pendias, A., & Pendias, H. (2001). Trace elements in soils and plants. 3rd ed. CRC Press, Boca Raton, ISBN 0-8493-1575-1, FL;.
50. Linsdey, W.L. (1974). Role of chelation in micronutrient availability. In: The plant root and its Environment, ed. E.W. Carson, p.507-524, University Press of Virginia,
51. Lindsay, W.L. (1979). Chemical equilibria in soils. Wiley-Interscience, ISBN 10: 0471027049, New York.
52. Lindsay, W.L., Vlek, P.L.G., & Chien, S.H. (1989). Phosphate minerals, In: Minerals in soil environments, J.B. Dixon and S.B. Weed (ed.), SSS Book Ser. I., pp. (1089-1130), ASA, CSSA, and SSSA, ISBN 0891188444, Madison. WI.
53. Loganathan, P., Gray, C.W., Hedley, M.J., & Roberts, A.H.C. (2006). Total and soluble fluorine concentrations in relation to properties of soils in New Zealand. Eur. J. Soil Sci. Vol. 57, pp. (411–421), ISSN 1351-0754
54. Ma, Q.Y., Traina, S.J., & Logan, T.J. (1993). In situ lead immobilization by apatite. Environ. Sci. Technol. Vol. 27, pp. (1803–1810), ISSN 0013-936X
55. Magdoff, F., Lanyon, L., & Liebhardt, B. (1997). Nutrient cycling, transformation and flows: Implications for a more sustainable agriculture. Adv. Agron. Vol. 60, pp. (1–73), ISSN: 0065 -2113
56. Manojlovic, D., Todorovic, M., Jovicic, J., Krsmanovic, V.D., Pfendt, P.A., & Golubovic, P. (2007). Preservation of water quality in accumulation Lake Rovin: the estimate of the emission of phosphorus from inundation area. Desalination, Vol. 213, pp. (104- 109), ISSN 00119164
57. Maroko, J.B., Buresh, R.J., & Smithson, P.C. (1999). Soil phosphorus fractions in unfertilized fallow–maize system on two tropical soils. Soil Sci. Soc. Am. J. Vol. 63, pp. (320–326), ISSN 0361-5995
58. McBride, M.B. (1989). Reaction controlling heavy metal solubility in soils. Adv. Soil Sci. Vol. 10, pp. (1-56), ISSN 0176-9340

59. McCollum, R.E. (1991). Build-up and decline in soil phosphorus: 30-yr trends on a Typic Umbraquult. Agron. J. Vol. 83, pp. (77–85), IOSSN 0002-1962
60. McGowen, S.L., Basta, N.T., & Brown, G.O. (2001). Use of diammonium phosphate to reduce heavy metal solubility and transport in smelter-contaminated soil. J. Environ. Qual. Vol. 30, pp. (493–500), ISSN 0047-2425
61. McGrath, S.P. (1995). Chromium and nickel, In: Heavy metals in soils, B.J. Allowey (Ed.), p. 152–174, Chapman and Hall, ISBN 0-7514-0198-6, London
62. McKenzie, R.C., & Nyborg, M. (1984). Influence of subsoil acidity on root development and crop growth in soils of Alberta and north-eastern British Columbia. Can. J. Soil Sci. Vol. 64, pp. (681–697), ISSN 0008-4271
63. McLaughlin, S.B., & Wimmer, R. (1999). Calcium physiology and terrestrial ecosystem processes. New Phytol. Vol. 142, pp. (373–417), ISSN 0028-646X
64. Mengel, K., & Kirkby, E.A. (2001). Principles of plant nutrition. 5th Ed. Kluwer Academic Publishers, ISBN 1-4020-0008-1, The Netherlands
65. Mitchell, C.C., Westerman, R.L., Brown, J.R., & Peck, T.R.. (1991). Overview of long-term agronomic research. Agron. J. Vol. 83, pp. (24–29), ISSN 0002-1962
66. Mitchell, P., & Barr, D. (1995). The nature and significance of public exposure to arsenic: A review of its relevance to South West England; Environ. Geochem. Health, Vol. 17, pp. (57-82), ISSN 0269-4042
67. Mrvic, V., Jakovljevic, M., Stevanovic, D., & Cakmak, D. (2007). The Forms of Aluminium in Serbia. Plant Soil Environ. Vol. 53, pp. (482–489), ISSN 1214-1178
68. Mulla, D.J., Page, A.L., & Ganje, T.J. (1980). Cadmium accumulation and bioavailability in soil from long-term phosphorus fertilization. J. Environ. Qual. Vol. 9, pp. (408–412), ISSN 0047-2425
69. Murphy. J., & Riley, P. (1962). A modified single solution method for determination of phosphate in natural waters. Anal. Chim . Acta, Vol. 27, pp. (31-36), ISSN 0003-2670
70. Mustapha, S., Yerima, S.I., Voncir, N., & Ahmed, B.I. (2007). Content and Distribution of Phosphorus Forms in Some Halpic Plinthaquults in Bauchi Local Goverment Area, Bauchi State, Nigeria. International J. Soil Sci. Vol. 2, No. 3, pp. (197-203), ISSN 1351-0754

71. Nel, P.C., Barnard, R.O., Steynberg, R.E., De Beer, J.M., & Groeneveld, H.T. (1996). Trends in maize grain yields in long-term fertilizer trial. Field Crops Res. Vol. 47, pp. (53–64), ISSN 1821-3944
72. Nelson, D.W., & Sommers, L.E. (1996). Total carbon, organic carbon, and organic matter, In: Methods of soil analysis Part 3, D.L. Sparks (Ed.), pp. (961–1010), SSSA, ISBN 96- 700096, Madison, WI.
73. O'Neill, P. (1995). Arsenic. In: Heavy metals in soils, B.J. Alloway (ed.), Chapman and Hall, p. 105–119, ISBN 0-7514-0198-6, London
74. Parfitt, R.L. (1978). Anion adsorption by soils and soil materials. Adv. Agron. Vol. 30, pp. (1– 50), ISSN 0065-2113
75. Peryea J. F. (1991). Phosphate-Induced Release of Arsenic from Soils Contaminated with Lead Arsenate, Issue 5, Soil Sci. Soc. Am.J., Vol. 55, No. 5, pp. (1301-1306), ISSN 0361-5995
76. Petrovic Dj., Todorovic M., Manojlovic D., Krsmanovic V.D. (2009). Speciations of Trace Metals in the Accumulation Bogovina on the Crni Timok River. Polish J. Environ. Stud. Vol. 18, pp. (873-884), ISSN 1230-1485
77. Pickering, W. (1985). The mobility of soluble fluoride in soils. Environ. Pollut, Vol. 9, pp. (281–308), ISSN 0971-4871
78. Rajan, S.S.S. (1975). Phosphate adsorption and the displacement of structural silicon in allophane clay. J. Soil Sci. Vol. 26, pp. (250–256), ISSN 1351-0754
79. Ruban, V., Lopez-Sanchez, J.F., Pardo, P., Rauret, G., Muntau, H., & Quevauviller, Ph. (1999). Selection and evaluation of sequential extraction procedures for the determination of phosphorus forms in lake sediments. J. Environ. Monit. Vol. 1, pp. (51-56), ISSN 1464-0325
80. Ruban, V., Lopez-Sanchez, J.F., Pardo, P., Rauret, G., Muntau, H., & Quevauviller, Ph. (2001). Development of a harmonized phosphorus extraction procedure and certification of a sediment reference material. Environ. Monit. Vol. 3, pp. (121-125), ISSN 1464-0325
81. Ruttenberg, K. C. (1992). Development of a sequential extraction method for different forms of phosphorus in marine sediments. Limnol. Oceanogr. Vol. 37, pp. (1460-1482), ISSN 1541-5856
82. Saleque, M.A., Naher, U.A., Islam, A., Pathan, A.B.M.B.U., Hossain, A.T.M.S., & Meisner, A.C. (2004). Inorganic and organic phosphorus fertilizer effects on the phosphorus fractionation in wetland rice soils. Soil Sci. Soc. Am. J. Vol. 68, pp. (1635–1644), ISSN 0361-5995
83. Santillian-Medrano, J., & Jurinak, J.J. (1975). The chemistry of lead and

cadmium in soil: Solid phase formation. Soil Sci. Soc. Am. Proc. Vol. 39, pp. (851–856), ISSN 0361-5995 Schwertmann, U., Süsser, P., & Nätcher, L.G. (1987). Proton buffer system in soils. Pfanzenernähr. Bodenk. Vol. 150, pp. (174-178), ISSN 0718-2791

84. Sharpley, A.N. & Smith, S.J. (1983). Distribution of phosphorus forms in virgin and fertilized soil and potential erosion losses. Soil Sci. Soc. Am. J. Vol. 47, pp. (37-47), ISSN 0361-5995.

85. Sharpley, A.N. (2000). Phosphorus availability. In: Hanbook of soil science, M.E.Sumner (ed.), pp. (D-18-D-38), CRC Press, Boca Raton, ISBN 10: 0849331366, FL.

86. Shuman, L.M. (1985). Fraction method for soil microelements. Soil Sci. Vol. 140, pp. (11-22), ISSN 0038-075X

87. Simard, R.R., Cluis, D., Ganbazo, G., & Beauchemin, S. (1995). Phosphorus status of forest and agricultural soils from a watershed of high animal density. J. Environ. Qual. Vol. 24, pp. (1010-1017), ISSN 0047-2425

88. Sims, J.T., & Pierzynski, G.M. (2005). Chemistry of Phosphorus in Soils, In: Chemical Processes in Soil, SSSA Book series 8, pp. (151-193), ISBN 0891188436, Madison, WI.

89. Soltanpour, P.N., Johnson, G.W., Workman, S.M.., Bentonjones, J.J., & Miller, R.O. (1996). Inductively coupled plasma emission spectrometry and inductively coupled plasma-mass spectrometry, In: Methods of soil analysis. Part 3, D.L. Sparks (ed.), pp. (91-139), SSSA, ISBN 96-700096, Madison, WI. SPSS. (2007). SYSTAT version 16 sofware: Statistics. SPSS, Chicago, IL.

90. Steinnes, E. (1995). Mercury. In: Heavy metals in soils, B.J. Alloway (ed.), pp. (245–257), Chapman and Hall, ISBN 0-7514-0198-6, London

91. Stevanovic, D., Kresovic, M., Stojanovic, M., & Grubisic, M. (2009). Stanje proizvodnje i problemi primene mineralnih djubriva u Srbiji. Proc. of the XXIII Conf. of Agronomist, Veterinarians, and Technologists. Vol. 15. Serbia.

92. Stewart, J.W.B., &. Sharpley, V. (1987). Controls on dynamics of soil and fertilizer phosphorus and sulfur.. In: Soil fertility and organic matter as critical components of production systems, R.F. Follett, J.W.B. Stewart, and C.V. Cole (ed.)., pp. (101–121), SSSA Spec. Publ. 19. SSSA and ASA, ISBN 0 -7923-4833-8, Madison, WI.

93. Stumm, W. & Morgan, J. J. (1981). Aquatic chemistry, 2nd ed. Wiley, 780pp., ISBN-10: 0471511854, New York

94. Sumner, M.E., & Miller, W.P. (1996). Cation exchange capacity and exchange coefficients, In: Methods of soil analysis. Part 3, D.L. Sparks (ed.), pp. (1201–1229), SSSA, ISBN: 96- 700096, Madison, WI.
95. Tisdale, S.L., Nelson, W.L., Beaton, J.D., & Havlin, J.L. (1993). Soil Fertility and fertilizers. 5th ed. Macmillian, ISBN 978-987-24977-1 -2 New York.
96. Tyler G. (2004). Vertical distribution of major, minor and rare elements in Haplic Podzol. Geoderma. Vol. 119, pp. (277-290), ISSN 0016-7061
97. Väänänen, R., Hristov, J., Taskanen, N., Hatikainen, H., Nieminen M., Ilvesniemi, H. (2008). Phosphorus sorption properties in podzolic forest soils and soil solution phosphorus concentration in undisturbed and disturbed soil profiles. Boreal Envi. Research, Vol. 13, pp. (553-567), ISSN 1239-6095
98. Van Der Zee S.E.A.T.M., Nederlof, M. M., Van Riemsdijk, W. H., & De Haan. F.A.M. (1987). Spatial Variability of Phosphate Adsorption Parameters, J. Environ. Qual. Vol. 17, pp. (682-688); ISSN 0047-2425
99. Van Der Zee, S.E.T.M., van Rimsdijk, W.H. (1988). Model for long-term phosphate reaction kinetics in soil J. Envir. Qual. Vol. 17, pp. (35-41), ISSN 0047-2425
100. Vu, D.T., Tang, C., Armstrong R.D. (2010). Transformation and availability of phosphorus in three contrasting soil types from native and farming systems: A study using fractionation and isotopic labeling techniques. J. Soil Sediments. Vol. 10, pp. (18-29), ISSN 1614-7480.
101. Willet, I.R., Chartres., C.J., & Nguyen, T.T. (1988). Migration of phosphate into aggregated particles of ferrihydrate. J.Soil Sci. Vol. 39, pp. (275-282), ISSN 0038-075X
102. Williams, J. D. H., Jaquet, J.M., & Thomas, R. L. (1976). Forms of phosphorus in the surficial sediments of Lake Erie. J. Fish. Res. Bd. Can., Vol. 33, pp. (413–429), ISSN 0015-296X
103. WRB. (2006). World reference base for soil resources. Food and Agriculture Organization of the United Nations, ISBN 92-5- 105511-4, Rome
104. Wright, R.J., & Stuczynski, T. (1996). Atomic absorption and flame emission spectrometry. In: Methods of soil analysis. Part 3, D.L. Sparks (ed.), p. (65–90), ISBN 96-700096, SSSA, Madison, WI.
105. Zhou, Q, & Zhu, Y. (2003). Potential pollution and recommended critical levers of phosphorus in paddy soils of the southern Lake Tai area, China. Geoderma Vol. 115, pp. (45-54), ISSN 0016-7061

Chapter 6

POULTRY LITTER FERTILIZATION IMPACTS ON SOIL, PLANT, AND WATER CHARACTERISTICS IN LOBLOLLY PINE (PINUS TAEDA L.) PLANTATIONS AND SILVOPASTURES IN THE MID-SOUTH USA

Michael A. Blazier[1], Hal O. Liechty[2], Lewis A. Gaston[1] and Keith Ellum[2]

[1]Louisiana State University Agricultural Center USA
[2]University of Arkansas Monticello USA

INTRODUCTION

Increasing global human populations and wealth have resulted in increased demands for animal protein and widespread use of confined animal feeding operations to meet added animal protein consumptive demands. Disposal of animal wastes from these operations can be ecologically and environmentally problematic (Kellogg et al., 2000; Roberts et al., 2004; Shober & Sims, 2003). Poultry production is an important source of this protein and is a major agricultural industry in the United States. The United States is the world's largest producer and second largest exporter of poultry meat (UDSA Economic Research Service, 2009). Four-fifths of the United States poultry industry is comprised of broiler meat production. Broiler meat production is largely concentrated in Southeastern states (Alabama, Arkansas, Florida, Georgia, Kentucky, Louisiana, Mississippi, North Carolina, Oklahoma, South Carolina, Tennessee, Texas and Virginia), with 82% of U.S. broiler production occurring in these states (National Agricultural Statistic Service, 2008).

Broiler production results in the generation of massive amounts of litter, a mixture of feces, feed, feathers and bedding materials such as straw, peanut or rice hulls, and wood shavings (Gupta et al., 1997; Weaver, 1998). The U.S.A. poultry industry produces more than 11 million Mg of litter per year (Cabrera & Sims, 2000). Broiler poultry litter contains several plant macro- and micronutrients (Table 1), which makes it desirable as an agricultural fertilizer

(Sistani et al., 2008). Following removal from poultry production facilities, litter is commonly applied to nearby pastures, hay meadows, and agricultural crops such as corn and cotton to increase crop production and quality (Harmel et al., 2004; Sims & Wolf, 1994). Applications of poultry litter ranging from 4.5 to 11.2 Mg ha^{-1} yr^{-1} are common to supplement or replace inorganic annual fertilizer additions to pastures (Adams et al., 1994). Thus, poultry litter application is an efficient and potentially cost-effective method for improving forage production within the vicinity of production facilities, which helps to sustain non-poultry related agriculture economies in poultry producing regions. Substitution of broiler litter for inorganic fertilizers continues to increase in the southeastern U.S.A. as prices of inorganic fertilizers escalate (Funderberg, 2009).

Table 1: Ranges of reported nutrient concentrations in broiler poultry litter on an oven-dry basis. Adapted from Eichhorn, 2001; Ekinci et al., 2000; Kingery et al., 1994; Mitchell & Donald, 1995; Pote et al., 2003; Sauer et al., 2000; Sims, 1986; Williams et al., 1999

Element	Nutrient Concentration	
	(g kg^{-1})	(mg kg^{-1})
C	280-320	
N	31-49	
P	4-13	
K	2-28	
Ca	2-28	
Mg	0.4-6	
Fe		1950-2395
Mn		277-424
Cu		263-332
Zn		252-404
B		45-55

Poorly planned, excessive, or long-term applications of broiler litter to pastures and other agroecosystems can result in excessive nutrient losses, reductions in surface water quality, and potential risks to human health. Poultry litter is typically applied as a nitrogen fertilizer, but N availability from litter is relatively difficult to predict because only one-third of the N in litter is in exchangeable forms such as NH_4-N and NO_3-N. Two-thirds of N in litter is in organic form, which must be mineralized before it is plant available. Mineralization of N in litter varies from 40 to 90% with edaphic and environmental conditions, particularly conditions at the time of litter application (Mitchell & Donald, 1995). Gaseous losses of N from litter via volatilization can vary from 5 to 20% of total N, which reduces the amount of N available for plant use (Mitchell & Donald, 1995). While the amounts and forms of N in litter can vary considerably, those of other nutrients, particularly P, are relatively stable. As a result, if litter is applied at rates that supply sufficient N to meet crop demand soils can become saturated with P as well as K, Ca, Mg, Cd, Cu, Mn, and Zn (Edmeades, 2003; Kingery et al., 1994). Surface water runoff or soil water leaching associated with these nutrient-saturated soils can reduce water quality in watersheds (Friend et al., 2006; Gallimore et al., 1999; Gaston et al., 2003; Kellogg et al., 2000; Sauer et al., 1999; Sims & Wolf, 1994). Excess nutrients are transported to surface waters via runoff either in particulate forms or sorbed to soil particles which are suspended in surface runoff. Soluble P and C, NO_3-N, NH_4-N, and some organic N species have been demonstrated to be transported by runoff as a solution. NH4-N and P are often sorbed to soil particles and conveyed by runoff through erosion, and organic C, P, and N have been shown to be moved by runoff in particulate form (Edwards & Daniel, 1992). Repeated applications of poultry litter can lead to accumulations of N and P in soil as well as elevated levels of one or both of these nutrients in surface runoff and subsurface water (McLeod and Hegg, 1984; Sharpley and Menzel, 1987; Kingery et al., 1994). The potential for P saturation and leaching may be particularly high for highly-fertilized and sandy soils (Breeuwsma and Silva, 1992; Nair and Graetz, 2004). Large or chronic accumulations of N and P can contribute to accelerated eutrophication of water bodies, impairing their use and potentially leading to fish mortality and growth of algae (Schindler, 1978; Lemunyon and Gilbert, 1993). Elevated concentrations of N and P in surface water and eutrophication of water bodies have been found in areas with high levels of confined poultry and other animal production (Daniel et al., 1998; Sharpley, 1999; Fisher et al., 2000). State and federal environmental protection agencies have responded to these environmental concerns by implementing regulations requiring poultry operations to develop nutrient management plans, which will frequently reduce the allowable amounts of litter that can be applied (Friend et al., 2006). It is estimated that 50% of the litter produced from areas with high concentrations of poultry production facilities cannot be applied to grasslands and croplands in these same areas due to environmental or economic constraints, which has led to surpluses of manure N and P production in some parts of the southeastern U.S. (Kellogg et al., 2000).

The environmental impacts of poultry litter fertilization could be reduced by applying surplus litter to terrestrial ecosystems other than pasture and cropland where nutrient levels in soils are low and which have a low risk of nutrient transport. Disposal of poultry litter to these ecosystems could increase the area of litter dispersal, reduce the spatial concentration of poultry litter application, and decrease risks to water quality in watersheds. Forests may be a viable alternative to pastures and croplands for broiler litter application. Similar to agroecoystems, forests are often limited by soil N and P supplies (Elser et al., 2007). Forests also have a high potential for nutrient uptake

(O'Neill & Gordon, 1994) and have been successfully used to mitigate environmental impacts of municipal waste, municipal effluent, and mill waste disposal (Henry et al., 1993; Polglase et al., 1995; Falkiner & Polglase, 1997; Jackson et al., 2000). In addition, infiltration rates are much higher in forested landscapes than many agricultural landscapes, which could increase potential retention of nutrients and reduce losses through surface runoff in comparison to agricultural crops. Loblolly pine (Pinus taeda L.) has been identified as a practical species to receive poultry litter application (Beem et al., 1998; Friend et al., 2006; Samuelson et al., 1999). Much of the poultry-producing regions of the southeastern U.S.A. are within the natural range of loblolly pine (Pinus taeda L.), so transportation of litter to land with loblolly pine is likely minimal (Friend et al., 2006). Loblolly pine is a prevalent and economically important species in the southeastern U.S.A.; it is used within the region to produce 18% of the world's supply of industrial timber (Allen et al., 2005; Prestemon and Abt, 2002). Loblolly pine growth is often limited by soil supplies of N and P (Binkley et al., 1999), and tree- and forest-level growth responses to N and P fertilization have been well demonstrated (Blazier et al. 2006; Colbert et al., 1990; Haywood et al., 1997; Murthy et al., 1997; Vose & Allen, 1988). Single applications of poultry litter, ranging between 2 to 23 Mg ha^{-1}, have been shown to increase loblolly pine growth rates (Dickens et al., 2004; Friend et al., 2006; Lynch and Tjaden, 2004; Roberts et al., 2006; Samuelson et al., 1999) and economic value (Dickens et al., 2004) of N and P-deficient loblolly pine. Loblolly pine forests have a high capacity for fertilizer retention due to high plant biomass and soil organic matter. Will et al. (2006) reported that 90 to 100% of annually applied N and P was sequestered in aboveground biomass, soil organic matter, or the uppermost 10 cm of soil of a loblolly pine plantation. Due to this high fertilizer retention capacity, minimal offsite movement of nutrients associated with poultry litter application is expected. Furthermore, water runoff potential of forests is lower than that of grasslands and croplands due to their relatively higher infiltration (Zimmermann et al., 2006) and evapotranspiration rates (Farley et al., 2005). Friend et al. (2006) found that nutrients from an application of 4.6 Mg ha^{-1} of poultry litter (on a dry matter basis) was substantially contained within a loblolly pine forest and did not impair water quality.

The principal limitation to fertilizing loblolly pine with poultry litter is the relatively poor accessibility and maneuverability within dense plantations by ground-based fertilizer application equipment. Conventional manure-spreading equipment (manure trucks, tractordrawn spreaders) cannot be driven through typical pine plantations due to close tree spacing, dense understory vegetation, and/or stumps high enough to cause equipment damage. These issues likely reduce the number of spreader contractors willing to operate within forests (Dickens et al., 2003). Silvopasture is an alternative land management system that would allow the application of poultry litter to loblolly pine trees by largely circumventing maneuverability limitations of conventional loblolly pine plantations. Silvopasture management systems consist of forage grasses established and cultivated beneath trees in order to simultaneously produce timber and livestock (Clason & Robinson, 2000; Clason & Sharrow, 2000). Silvopasture regimes are currently the most popular form of agroforestry in the southeastern U.S. (Clason & Sharrow, 2000; Zinkhan & Mercer, 1996). Silvopastures are created by either planting trees in pastures (Robinson & Clason, 2002) or by establishing forage crops in forests (Clason & Robinson, 2000). Forage management in silvopastures is conducted similarly to conventional grasslands in the southeastern U.S.A.; herbicides and/or prescribed burning are used to reduce herbaceous and woody competition and fertilization is carried out to optimize forage yields. Due to land ownership and use patterns, there is high potential for conversion between agriculture and forestry in the southeastern U.S.A. (USDA SCS, 1989). Clason (1995) determined that loblolly pine was compatible with several forage crops in silvopasture systems. The relatively wide spacing of trees and forage understory in silvopastures make navigation of manure-spreading equipment possible (Figure 1).

Figure 1: Applying broiler poultry litter to a silvopasture at the Louisiana State University Agricultural Center Hill Farm Research Station in northwest Louisiana, U.S.A. Picture by Terry Clason, USDA Natural Resource Conservation Service.

A pine plantation in which straw is harvested is another management system where applications of poultry litter could increase commodity production. Pine straw mulch has emerged as a substantial commercial product for horticultural crops and landscaping in urban and suburban areas (Duryea and Edwards, 1989). Adding straw harvesting to conventional timber management regimes has been shown to markedly increase profits, with straw revenue potentially exceeding that of traditional forest products (Haywood et al., 1998; Lopez-Zamora et al. 2001; Roise et al,. 1991). These plantations are typically designed to allow access by conventional agronomic equipment to harvest the straw and are thus well suited for application of the poultry litter by small to mid-size manure or litter spreaders. Harvests of straw on large plantations are usually performed using a hay or pine straw rake, tractor, and mechanical baler (Mills and Robertson, 1991). Understory biomass is typically suppressed in straw harvesting management regimes to improve straw quality by eliminating woody and herbaceous debris (Mills and Robertson, 1991). Coarse and fine woody debris is also removed from the forest floor prior to baling to improve the economic value of baled pine straw (Minogue et al., 2007). This suppression of vegetation and woody debris removal between rows of trees fosters navigation of the plantations with tractor-drawn straw raking and baling equipment (Figure 2) as well as poultry litter application equipment.

Figure 2: Mechanically baled straw in a 19-year-old loblolly pine plantation at the Louisiana State University Agricultural Center Calhoun Research Station in northeast Louisiana. Inset: Tractor-drawn straw baler used for mechanically baling straw in the plantation. Pictures by Keith Ellem, University of Arkansas Monticello.

Poultry litter can be highly beneficial when applied to plantations in which straw is harvested because it can replenish nutrients lost in straw harvesting. The nutrient content in pine needles is substantial, and repetitive harvesting of pine straw removes significant amounts of nutrients from the soil. One metric ton of harvested straw contains approximately 21.3 kg nitrogen (N), 1.8 kg phosphorus (P), 4.5 kg potassium (K), 9.0 kg calcium (Ca), and 1.8 kg magnesium (Mg) (Pote & Daniel, 2008). Since fallen leaves are major sources of nutrient inputs to soils, repeated raking can reduce soil nutrient availability, particularly N, unless nutrients are replenished through management activities (Jorgenson & Wells, 1986; Lopez-Zamora et al., 2001). As such, periodic fertilization has been recommended to remedy nutrient removals that can occur with straw harvesting (Haywood et al., 1998; Lopez-Zamora et al., 2001).

Since poultry litter contains organic matter, it could potentially replenish some of the organic matter removed by pine straw raking. Fallen pine straw is a prominent source of organic matter in the soil organic horizon of pine forests, and it is the major reservoir of labile carbon used by soil microbes in the synthesis of new cells, a process that also mineralizes N (Pritchett & Fisher, 1987; Sanchez et al., 2006; Wagner & Wolf, 1999). Soil microbial biomass and activity are highly sensitive to changes in soil organic matter and are thus used as indicators of soil quality and sustainability (Fauci & Dick, 1994; Harris, 2003; Powlson & Brookes, 1987). Removal of the soil organic horizon decreased soil microbial biomass carbon (C_{mic}) due to reduced substrate availability in a study simulating organic matter removals associated with tree harvesting and site preparation in a boreal forest (Tan et al., 2005). Activities other than soil organic matter removal associated with straw harvesting may also impact soil biological properties. The suppression of understory vegetation prior to straw raking can reduce microbial biomass and activity because understory vegetation provides rhizodeposition important to soil microbes (Donegan et al., 2001; Gallardo and Schlesinger, 1994; Högberg et al., 2001). Inorganic fertilizers do not replenish organic matter essential as microbial substrates and may exacerbate soil microbial biomass and activity declines caused by organic matter removal (Blazier et al., 2005). In contrast, fertilization with poultry litter can increase soil microbial biomass and activity (Canali et al., 2004; Plaza et al., 2004). Soil organic matter also in part determines soil water availability and temperature (Attiwill & Adams, 1993), and poultry litter has been shown to increase soil water content and available water holding capacity and reduce soil temperature (Agbede et al., 2010; Warren & Fonteno, 1993).

Due to the potential influences of poultry litter on soil and tree nutrition, soil microbes, and tree growth, a series of experiments were conducted in the mid-South region of the U.S.A. This chapter will provide a review of the key results of these trials from 1996 through 2011. The focus of this chapter will be on the changes in soil nutrition, physical properties and microbes, tree nutrition and growth, and water nutrient contents in loblolly pine plantations and silvopastures in response to fertilization with conventional fertilizer and poultry litter.

STUDY DESCRIPTIONS

Results of five studies conducted in the mid-South U.S.A are described in this chapter. At least one treatment in each study received surface application of broiler litter as fertilizer, and loblolly pine was the tree component of each study.

All studies occurred in the Western Gulf Coastal Region within areas identified by Friend et al. (2006) as having high occurrence of poultry production and southern pine forests. Two of the studies (SILVO, SWITCH) included poultry litter applied to silvopastures. The silvopasture in the SILVO study consisted as bahiagrass (Paspalum notatum Flüggé) established under thinned loblolly pine, and the silvopasture in the SWITCH study was comprised of switchgrass (Panicum virgatum L.) established under thinned loblolly pine. Two of the studies (AR-FORvsPAST, LA-FORvsPAST) included a comparison of broiler litter in loblolly pine and pastures. The STRAW study included poultry litter applied to a loblolly pine plantation in which straw was annually harvested.

The SILVO and SWITCH studies were conducted at the Louisiana State University Agricultural Center Hill Farm Research Station in Homer, Louisiana, U.S.A. The STRAW and LA-FORvsPAST trials were carried out at the Louisiana State University Agricultural Center Calhoun Research Station in Calhoun, Louisiana, U.S.A. The AR-FORvsPAST study was conducted at the University of Arkansas Southwest Research and Extension Center near Hope, Arkansas, U.S.A. Average annual precipitation of the region in which the studies were carried out is 120 cm, and average temperature is 18°C (Bailey, 1995). Primary study site characteristics are described in Table 2.

Table 2: Location, vegetation, tree and density at study initiation, and soil characteristics for studies of poultry litter fertilization of loblolly pine in the mid-South U.S.A.

Study	Geographical Coordinates	Vegetation	Tree Age (years)	Tree Density (trees ha^{-1})	Soil Classification
SILVO	32°44'N, 93°03'W	loblolly pine-bahiagrass silvopasture	12	247	Loamy, siliceous, thermic Arenic Paleudults
SWITCH	32°44'N, 93°03'W	loblolly pine-switchgrass silvopasture	17	124	Loamy, siliceous, thermic Arenic Paleudults
STRAW	32°31'N, 92°21'W	loblolly pine plantation	10	618	Fine-loamy siliceous thermic Typic Fragiudults
LA-FORvsPAST	32°31'N, 92°21'W	loblolly pine, bermudagrass as vegetation type treatments	5	1586	Fine-loamy siliceous thermic Typic Fragiudults
AR-FORvsPAST	33°42'N, 93°32'W	loblolly pine, bahiagrass as vegetation type treatments	26	201	Fine-loamy, siliceous thermic Typic Fraigudults; Clayey, mixed, thermic Aquic Hapludults

Treatments (Table 3) were replicated four times each in the STRAW, LA-FORvsPAST studies, three times in the SILVO and AR-FORvsPAST studies, and six times in the SWITCH study. Treatments were applied as a one-way treatment structure in the STRAW, LAFORvsPAST, SWITCH, and SILVO studies. Treatments were applied as a split-plot treatment structure in the

AR-FORvsPAST study, with vegetation type (pasture, forest) as a whole-plot treatment and fertilization as a sub-plot treatment. The experimental design was a randomized complete block design for all studies. In statistical analyses of all variables assessed in these treatments, differences among treatments were determined by analysis of variance at $\alpha = 0.05$; correlation among variables was assessed at $\alpha = 0.05$ as well.

Table 3: Treatments conducted in studies of poultry litter fertilization of loblolly pine conducted in the mid-South U.S.A. 1Italicized treatments were applied as sub-plot treatments, underlined treatments were applied in all possible combinations to whole plots, all other treatments were applied as whole-plot treatments

Study	Treatment[1]	Treatment Description
SILVO	CONTROL	No treatment
	IF	Inorganic fertilizer mixture (diammonium phosphate, ammonium nitrate, muriate of potash to annually supply 114 kg N ha^{-1}, 39 kg P ha^{-1}, 20 kg K ha^{-1})
	PL5	Poultry litter applied at 5 Mg ha^{-1} that supplied N, P, K, Ca, Mg, Fe, Mn, Cu, Zn, B at 112, 36, 78, 106, 23, 9, 2, 1, 1.5, 0.2 kg ha^{-1}, respectively
	PL10	Poultry litter applied at 10 Mg ha^{-1} that supplied N, P, K, Ca, Mg, Fe, Mn, Cu, Zn, B at 224, 73, 157, 211, 45, 18, 3, 2, 3, 0.3 kg ha^{-1}, respectively
SWITCH	CONTROL	No treatment
	IF80	Ammonium nitrate applied that supplied 80 kg N ha^{-1}
	IF160	Ammonium nitrate applied that supplied 160 kg N ha^{-1}
	PL1.5	Poultry litter applied at 1.5 Mg ha^{-1} to supply N, P, K, Ca, Mg, Fe, Mn, Cu, Zn, and B at 80, 42, 90, 54, 15, 2, 1.5, 0.15, 1, and 0.1 kg ha^{-1}, respectively.
	PL3	Poultry litter applied at 3 Mg ha^{-1} to supply N, P, K, Ca, Mg, Fe, Mn, Cu, Zn, and B at 160, 84, 180, 108, 30, 4, 3, 0.3, 2, and 0.2 kg ha^{-1}, respectively.
STRAW	CONTROL	No treatment
	RAKE	Straw harvesting
	RAKE-IF	Straw harvesting, diammonium phosphate and urea inorganic fertilizers that supplied N and P at 193 and 102 kg ha^{-1}, respectively
	RAKE-PL	Straw harvesting, poultry litter applied at 8 Mg ha^{-1} that supplied N and P at 193 and 102 kg ha^{-1}, respectively. Other nutrients added by poultry litter not tested due to budget constraints.
LA-FORvsPAST	CONTROL	No treatment
	PL5	Poultry litter applied at 5 Mg ha^{-1} that supplied N, P, K, Ca, Mg, Fe, Mn, Cu, Zn, B at 112, 109, 92, 159, 34, 9, 2, 3, 2, 0.2 kg ha^{-1}, respectively
	PL10	Poultry litter applied at 10 Mg ha^{-1} that supplied N, P, K, Ca, Mg, Fe, Mn, Cu, Zn, B at 224, 218, 184, 318, 68, 18, 3, 6, 3, 0.3 kg ha^{-1}, respectively
	PL20	Poultry litter applied at 20 Mg ha^{-1} that supplied N, P, K, Ca, Mg, Fe, Mn, Cu, Zn, B at 448, 436, 368, 636, 136, 36, 6, 12, 6, 0.6 kg ha^{-1}, respectively
	FOREST	Loblolly pine plantation
	PASTURE	Bermudagrass pasture
AR-FORvsPAST	FOREST	Loblolly pine plantation
	PASTURE	Bahiagrass pasture
	CONTROL	No treatment
	PL9	Poultry litter at 9 Mg ha^{-1} that supplied N, P, K at 30, 14, 22 kg ha^{-1}. Other nutrients added by poultry litter not tested due to budget constraints.

Soil and/or plant responses to treatments were observed in the studies (Table 4). In the SILVO, SWITCH, and LA-FORvsPAST studies, grass clippings were randomly collected within quadrats either at the end of growing seasons (in the SWITCH study) or multiple times during the season and averaged (in the SILVO and LA-FORvsPAST studies) to determine forage yields. In the SILVO, STRAW, and LA-FORvsPAST studies loblolly pine basal area was measured by converting diameter at breast height measurements into basal area for all trees in measurement plots; basal area measures were summed for each plot to estimate stand level basal area. In the SILVO study soil was sampled by a tractor-mounted auger to the bottom of the B_t horizon and separated into A, E, and B_t horizons, which had average depths of 0.15, 0.48, and 0.59 m, respectively. In the SWITCH study, soil was sampled with punch augers to a 15-cm depth for labile C determination and 30 cm for nutrient analyses. Soil in the STRAW study was sampled with punch augers to a 15-cm depth. In the LA-FORvsPAST study, soil was sampled to a 15-cm depth with punch augers pre-treatment and sampled to 0-15, 15-30, 30-45, 60-80, and 80-100 cm depths post-treatment. Soil was sampled to a 15 cm depth in the AR-FORvsPAST study using punch augers. In the SILVO and STRAW studies, loblolly pine foliage was sampled from the upper third of crowns. Organic matter in soil samples was quantified by the Walkley-Black method (Walkley, 1947) in the SILVO, LA-FORvsPAST, and SWITCH studies and by the loss on ignition method (Ball, 1964) in the STRAW and AR-FORvsPAST studies. Soil pH was determined by pH meters in a 2:1 mixture of deionized water to soil in the SILVO and SWITCH studies. Phosphorus in the samples was extracted by Bray 2 P (Bray & Kurtz, 1945) in the SILVO and LA-FORvsPAST studies and by Mehlich 3 (Mehlich, 1984) in the AR-FORvsPAST and SWITCH studies. Nutrients other than P were extracted by ammonium acetate (K, Ca, Mg, Na) and DTPA (Cu, Fe, Mn, Zn) in the SILVO and LA-FORvsPAST studies (Gambrell, 1996; Helmke & Sparks, 1996). Mehlich 3 was used to extract K, Ca, Mg, S, Cu, and Zn in the SWITCH study (Mehlich, 1984). All nutrients from soil samples were quantified via ICP spectrometry (Jones & Case, 1990) in all studies in which soil was analyzed for nutrient concentration. Exchangeable N (NH_4-N, NO_3-N) was extracted by KCl extraction (Mulvaney, 1996) in the SILVO and AR-FORvsPAST studies and measured colorimetrically on Bran Luebbe (Bran-Luebbe, Inc, Delavan, WI) and Lachat autoanalyzers (Lachat Instruments, Loveland, CO, U.S.A.) in the SILVO and AR-FORvsPAST studies, respectively. In the STRAW and SWITCH studies soil labile C was measured by sequential fumigation incubation (Zou et al., 2005). In the STRAW and SWITCH studies microbial biomass C was measured by fumigation incubation (Jenkinson and Powlson, 1976a,b) and microbial activity was measured by an assay of dehydrogenase activity (Lenhard, 1956; Alef, 1995). In the STRAW study N mineralization and nitrification was measured using the buried bag method (Eno, 1960). In the AR-FORvsPAST study, potential N mineralization and nitrification (Hart et al., 1994) were assessed in samples aerobically incubated in the laboratory for 28 days; NH_4-N and NO_3-N used to determine mineralization and nitrification in this procedure were measured by the cadmium reduction method (Mulvaney, 1996) using a Lachat autoanalyzer. Total N in soil samples from the ARFORvsPAST study was determined by dry combustion using an Elementar Vario MAX CN analyzer (Elementar Analysesysteme GmbH, Hanau, Germany).

All nutrients except N in foliage were analyzed by nitric acid digestion and ICP spectrometry; N in these samples was measured by Dumas combustion and thermal conductivity detection using a Leco N/protein analyzer (Leco Inc., St. Joseph, MI, U.S.A) (Helmke & Sparks, 1996; Tate, 1994; Zarcinas et al., 1987). In the SILVO study, N concentrations of bahiagrass samples were ascertained by Kjeldahl method; other nutrients in the samples were determined by Dumas combustion and nitric acid digestion and ICP spectrometry (Helmke & Sparks, 1996; Horneck & Miller, 1998; Tate, 1994; Zarcinas et al., 1987).

Table 4: Timeline of treatments and measurements in studies of fertilization of loblolly pine with poultry litter in the mid-South U.S.A. Shad cells designate years in which treatments or measurements occurred. [1]OM = organic matter, STR = strength, BD = bulk density, WHC = water holding capacity, LABC = labile C, NMIN = N mineralization, CMIC = microbial biomass C, ACT = microbial dehydrogenase activity

Study	Treatment or Measurement[1]	1996	1997	1998	1999	2000	2001	2002	2003	2004	2005	2006	2007	2008	2009	2010	2011
SILVO	Fertilization			■		■											
	Forage yield			■	■	■											
	Soil pH, OM, nutrients			■													
	Soil NH₄-N, NO₃-N			■													
	Pine foliage nutrients			■			■					■					
	Pine basal area			■			■										
SWITCH	Fertilization															■	
	Forage yield															■	
	Soil pH, OM, nutrients															■	
	Soil LABC, CMIC, ACT															■	
STRAW	Straw raking						■	■	■	■	■	■	■				
	Fertilization							■									
	Soil STR, P, BD, WHC							■									
	Soil OM, LABC, NMIN, exchangeable N							■									
	Soil CMIC, ACT												■				
	Pine foliage nutrients												■				
	Pine basal area							■					■				
LA-FORvsPAST	Fertilization		■	■	■	■	■			■							
	Forage yield		■	■	■	■	■										
	Soil pH, OM, nutrients		■	■	■	■	■										
	Pine basal area							■									
	Runoff nutrients											■					
AR-FORvsPAST	Fertilization											■					
	Forage yield											■					
	Soil NH₄-N, NO₃-N											■					
	Soil potential NMIN											■					
	Soil OM, N, P											■					

Bulk density, porosity, soil moisture content, and air-filled porosity of soil samples collected in the STRAW study were analyzed using procedures of Blake & Hartge (1986) and Danielson & Sutherland (1986). Available water holding capacity was determined in the STRAW study using soil moisture retention curves (Brye, 2003; Gee et al., 1992). Soil strength in the STRAW study to 15- and 30-cm depths was measured with a Scout SCT compaction meter (Spectrum Technologies, Inc., Plainfield, IL, USA) (Bradford, 1986). Soil water was collected to a 30-cm depth using tension lysimeters in the AR-FORvsPAST study; NO_3-N and PO_4-P in the water samples was analyzed by ion chromatography and NH_4-N was measured with a Lachat autoanalyzer. Water samples were also digested using a Kjeldahl digestion procedure and analyzed for total Kjeldahl nitrogen (TKN) and total Kjeldahl phosphorus (TKP) using the Lachat spectrophotometer. In the LA-FORvsPAST study, water was collected from runoff troughs after every rain event. Water samples were analyzed for total P by acid persulfate digestion and ICP spectrometry and for dissolved P by ICP spectrometry (Clesceri et al., 1998; Pote & Daniel, 2000).

PLANT BIOMASS AND NUTRITION

Increases in forage yields were observed in response to poultry litter in the SILVO, LAFORvsPAST, AR-FORvsPAST, and SWITCH studies. In the SILVO study poultry litter increased bahiagrass yields, but the magnitude of response was rate-dependent. The PL10 treatment had greater bahiagrass yields than all other treatments, and the PL5 and IF treatments had greater bahiagrass yields than the CONTROL treatment (Evans, 2000). The PL10 treatment also led to greater P, Zn, and Cu concentrations in bahiagrass relative to the CONTROL and IF treatments (Evans, 2000). These results indicated that poultry litter increased yields and nutritional quality of bahiagrass. Gaston et al. (2003) similarly found in the LA-FORvsPAST study that bermudagrass yields increased with increasing litter application rate. In the AR-FORvsPAST study, bahiagrass yields of the PL9 treatment were ~1.5 times greater than those of the CONTROL treatment in the first two years of fertilization. Switchgrass yield response to poultry litter in the SWITCH study was not ratedependent as in the SILVO and LA-FORvsPAST studies, because both application rates led to comparable yields.

Loblolly pine growth was also improved by poultry litter in the SILVO and AR-FORvsPAST studies. In the SILVO study, tree- and stand-level basal area growth was increased by poultry litter at the 10 Mg ha^{-1} rate (Blazier et al., 2008a). As with forage yields, litter application rate affected the level of growth response. After four annual litter applications, the 10PL treatment had greater annual basal area growth per tree than that of all other treatments, and the 5PL treatment had greater annual basal area growth than the CONTROL treatment. Stand-level basal area growth of the 10PL treatment was greater than that of the CONTROL and IF treatments. All fertilizer treatments led to greater foliage N concentrations than the CONTROL treatment, and both poultry litter treatments had greater foliage P concentrations than the CONTROL treatment. These results, which are consistent with other studies (Dickens et al., 2004; Friend et al., 2006; Roberts et al., 2006), show that loblolly pine growth can be increased with poultry litter amendments. The levels of growth responses were somewhat surprising, because all foliage nutrient concentrations were above critical levels (Allen, 1987; Blazier et al., 2008a; Jokela, 2004). Due to the relatively low density of trees in silvopastures, trees may have more readily responded to fertilization by virtue of larger crown mass (which provides a larger nutrient sink per tree) and less competition for applied nutrients compared to that in typical pine plantations. The similarities in growth responses and N and P application rates of the 5LIT and INO treatments suggests that although the 5LIT treatment supplied more K and a wider array of nutrients than the INO treatment, N and P were likely the primary limiting nutrients in the stand (Blazier et al., 2008a). In the AR-FORvsPAST study, annual loblolly pine basal area growth in response to the PL9 treatment was 10.9% greater than that of the CONTROL treatment.

In the STRAW and LA-FORvsPAST studies, no significant loblolly pine growth responses to treatments were observed (Gaston et al., 2003). Before the studies were established, the land was intensively managed for forage production. As such, the decades of fertilization application at these locations had resulted in high nutrient availability. Foliage P and S concentrations were increased by the RAKE-PL treatment relative to the other treatments in the STRAW study, but these increases appeared to have been luxury consumption since these nutrient increases were not accompanied by increased loblolly pine growth.

SOIL PHYSICAL PROPERTIES AND ORGANIC MATTER

In the STRAW study, all treatments that included straw harvesting induced evidence of soil compaction by significantly increasing bulk densities (Table 5) to levels 0.6 to 3.3% greater than the 1.75 g cm^{-3} bulk density defined as a growth-limiting threshold for forests grown on loamy soils (Daddow and Warrington, 1983), whereas soil in the CONTROL treatment remained below this threshold. These bulk density increases were also associated with significant declines in porosity in all treatments that included straw harvesting (Table 5). These findings suggest that annual straw harvesting had potential to reduce tree growth through reduced rooting volume and aeration. Nevertheless, no decreases in loblolly pine growth were observed in response to raking, as described above. It is likely that equipment traffic and increased exposure of mineral soil to rainfall associated with straw harvesting led to these increases in bulk density. Similarities in bulk density and porosity among the RAKE treatment and treatments that included raking and fertilization suggest that the additional trafficking from fertilization equipment each season did not appreciably compact the soil and that straw harvesting was the predominant cause of soil compaction (Blazier et al, 2008b).

Table 5: Soil physical properties and organic matter in response to pine straw harvesting and fertilization with inorganic fertilizer and poultry litter in a loblolly pine plantation in north central Louisiana, U.S.A. Means within columns followed by different letters differ at P < 0.05. Adapted from Blazier et al. (2008b)

Treatment	Bulk Density (g cm^{-3})	Porosity (g kg^{-1})	Air-filled Porosity	Moisture (g kg^{-1})	Soil Strength (MPa)	Organic Matter (g kg^{-1})	Available Water Holding Capacity (g kg^{-1})
CONTROL	1.67 b	369 a	99 a	270 a	1.25 b	27.8 a	427 a
RAKE	1.81 a	318 b	51 b	268 ab	2.31 a	25.8 ab	367 b
RAKE-IF	1.76 a	334 b	86 a	248 b	2.45 a	19.0 b	353 b
RAKE-PL	1.78 a	329 b	48 b	281 a	0.99 b	25.8 ab	384 ab

The RAKE-PL treatment appeared to have ameliorated some of the soil physical impacts of the raking since soil strength, organic matter, and moisture in the RAKE-PL treatment were similar to that in the CONTROL treatment (Table 5). However, the RAKE-PL treatment was characterized by lower air-filled porosity than the CONTROL treatment (Table 5), so there may have been a compaction potential associated with the application of the poultry litter (Tekeste et al., 2007). Poultry litter did not alter soil physical properties in a manner similar to inorganic fertilizers. The RAKE-PL treatment was characterized by soil moisture content, strength, organic matter concentrations, and available water holding capacity similar to the CONTROL treatment (Table 5). The RAKE-PL treatment may have replenished some organic matter lost through straw harvesting and accelerated decomposition associated with increased nutrient levels, because broiler poultry litter typically consists of 44% organic matter (Adeli et al., 2006; Dick et al., 1998). These results suggest that use of poultry litter as a fertilizer source in an annual straw harvest regime was superior to inorganic fertilizers in sustaining soil physical quality.

In contrast with the RAKE-PL treatment, the RAKE and RAKE-IF treatments had detrimental effects on some soil physical properties. The RAKE and RAKE-IF treatments both had soil strengths 46% greater than the CONTROL treatment. Soil strengths of the RAKE and RAKE-IF treatments also exceeded the 2 MPa soil strength threshold defined as highly compacted because of demonstrated root growth restrictions (Taylor & Gardner, 1963;

Tiarks & Haywood, 1996). Available water holding capacity was also reduced by the RAKE and RAKE-IF treatments relative to the CONTROL treatment. These findings suggest that the RAKE and RAKE-IF treatments made soil less amenable for root growth in the uppermost 5 cm of soil, which is the predominant zone in which tree roots, particularly fine roots, grow (Gilman, 1987). Relative to the CONTROL treatment, only the RAKE-IF treatment had greater soil strength, reduced moisture content, and reduced soil organic matter concentrations (Table 5). Repeated fertilization with inorganic nitrogen has been shown to reduce soil organic matter concentrations by increasing decomposition rates (Khan et al., 2007). Increased soil strength in response to the RAKE-IF treatment may have been due to the reductions in soil organic matter concentrations caused by this treatment. Soil strength tends to increase with decreasing soil organic matter concentrations because soil organic matter serves as an organic aggregate binding and bonding material (Munkholm et al., 2002). The relatively lower moisture content and available water holding capacity of the RAKE-IF treatment is consistent with its lower soil organic matter content because organic matter fosters soil moisture retention (Plaza et al., 2004; Powers et al., 2005).

There were no differences in soil organic matter among treatments in the SILVO, SWITCH, LA-FORvsPAST, and AR-FORvsPAST studies (Blazier et al., 2008a; Liechty et al., 2009), in which litter was not removed. As such, increases in forage and/or tree yields from fertilization in these studies were not associated with concomitant increases in soil organic matter. In the SILVO and SWITCH studies, the lack of declines in organic matter in response to inorganic fertilizer application as seen in the STRAW study was likely due to the straw raking done in tandem with fertilization in the STRAW study. As organic matter supply was drastically reduced by annual straw harvesting, stimulating decomposition with inorganic fertilizer led to significant declines in soil organic matter. Additionally, the increases in forage understory biomass of the SILVO and SWITCH studies may have been less prone to lead to increases in organic matter, as evidenced in the LA-FORvsPAST study. In that study no differences in organic matter were found among treatments in the pasture despite the increases in bermudagrass yields described above, whereas organic matter in the loblolly pine plantation differed among treatments as PL20 > PL10, PL5 > CONTROL.

SOIL LABILE C, MICROBIAL BIOMASS C, AND MICROBIAL ACTIVITY

Annual application of inorganic fertilizer had a profound effect on microbial biomass and activity in the STRAW study (Table 6). Microbial biomass C of the RAKE-IF treatment was lower than that of the CONTROL and RAKE treatments, and dehydrogenase activity of the RAKE-IF treatment was lower than all other treatments. The reductions in microbial biomass C and activity were apparently not a result of lower substrate supply, because labile C was similar among treatments (Table 6). Consequently, the higher potential turnover rate of the RAKE-IF treatment relative to all others is likely a result of reduced microbial biomass and activity rather than relatively high recalcitrance of organic matter. The reductions in microbial biomass C and activity in the RAKE-IF treatment were likely associated with the lower pH of this treatment relative to all others. It has been welldemonstrated that intensive fertilization with inorganic N reduces soil pH and that declining pH is associated with reductions in soil microbial biomass and activity (Anderson and Domsch, 1993; Baath et al., 1995; Blazier et al., 2005).

These results thus showed that microbial biomass and activity were reduced by declines in pH from inorganic fertilizer, whereas annual raking and fertilization with poultry litter had no such effects. In contrast to inorganic fertilizer, poultry litter tends to increase soil pH because litter contains calcium carbonate originating from poultry rations (Hue, 1992; Kingery et al., 1993). Although litter did not significantly increase pH in the STRAW study, litter sustained pH at levels comparable to the CONTROL treatment, which fostered microbial biomass C and activity levels comparable to the CONTROL treatment as well.

As in the STRAW study, inorganic fertilizer led to declines in microbial biomass C relative to the CONTROL treatment (Table 6) in the SWITCH study. Fertilization has been shown to reduce soil microbial biomass C in forest soils (Rifai et al., 2010; Wallenstein et al., 2006). Rifai et al. (2010) identified several possible mechanisms for soil microbial biomass declines in response to fertilization, including (1) pH reduction caused by nitrate leaching induced by application of high rates of NH_4NO_3, and (2) inhibition of organic compound decomposition from excess N that reduces organic matter available to soil microbes. In the SWITCH study there were no declines in pH among treatments consistent with declines in soil microbial biomass C, although pH of the inorganic fertilizer treatments were lower than those of the poultry litter treatments.

Dehydrogenase activity decreased as fertilizer application rates increased for both fertilizer types. Since N was the sole nutrient added by inorganic fertilizer treatments in this study, these dehydrogenase activity trends suggest that excess N perturbed microbial decomposition of organic matter in this loblolly pine and switchgrass system.

However, potential C turnover rate was shorter for the lower rate of inorganic fertilizer (IF80) relative to the lower rate of poultry litter (PL1.5) despite the equivalent N rate of the two treatments. Since labile C supply, microbial biomass C, and dehydrogenase activity were similar for the IF80 and PL1.5 treatments, the reason for the higher potential C turnover rate of the PL1.5 treatment was unclear and merited further study.

Table 6: Soil labile C, microbial, and pH responses to fertilization in an annually raked loblolly pine plantation (STRAW) and a loblolly pine and switchgrass silvopasture (SILVO) in the mid-South U.S.A. For each study, means within columns followed by different letters differ at P < 0.05. Adapted in part from Blazier et al. (2008b)

STRAW	Treatment				
	CONTROL	RAKE	RAKE-IF	RAKE-PL	
Labile C (mg kg^{-1})	475.1 a	522.3 a	457.0 a	582.5 a	
Potential C turnover rate (days)	46.0 b	53.2 b	92.9 a	62.8 ab	
Microbial biomass C (mg kg^{-1})	169.2 a	157.2 a	75.3 b	143.5 ab	
Dehydrogenase activity (μg g^{-1})	50.6 a	71.0 a	25.8 b	44.5 a	
pH	4.9 a	4.9 a	4.3 b	5.1 a	
SWITCH	CONTROL	IF80	IF160	PL1.5	PL3
Labile C (mg kg^{-1})	835.6 a	585.2 a	718.1 a	878.7 a	836.7 a
Potential C turnover rate (days)	29.7 ab	24.7 b	30.4 ab	43.8 a	37.3 ab
Microbial biomass C (mg kg^{-1})	410.9 a	341.6 ab	348.4 ab	320.4 ab	377.4 ab
Dehydrogenase activity (μg g^{-1})	11.0 ab	24.2 a	9.9 b	24.3 a	5.9 b
pH	5.5 bc	5.4 c	5.4 c	5.6 ab	5.7 a

SOIL NUTRIENTS

Nitrogen

In all studies in which exchangeable soil N was measured, NO_3-N amounts or proportions of in soil increased in response to poultry litter application (Table 7). In the AR-FORvsPAST study, NO_3-N significantly increased in the loblolly pine plantation and in pasture relative to the CONTROL treatment following two years of poultry litter application. The proportion of NO3-N to total exchangeable N was also greater in response to poultry litter than without litter application (Liechty et al., 2009). There was no difference in NO_3-N concentrations among treatments in the SILVO study, but as in the AR-FORvsPAST study the ratio of NO_3- N to total exchangeable N increased in response to poultry litter additions. This increase in the proportion of NO_3-N in the SILVO study occurred in response to both rates of broiler litter tested; no such increase was observed in response to the inorganic fertilizer mixture (Blazier et al., 2008a). Results similar to the SILVO study were also found in the STRAW study; NO_3-N increased in response to the treatment regime that included poultry litter, whereas no such increase was observed in response to non-fertilized treatments and the treatment regime that included a mixture of inorganic fertilizers (Liechty et al., 2009). Increases in soil NO_3-N in response to poultry litter were attributable to greater nitrification rates (Table 7). Soil in plots treated with broiler litter had greater N mineralization rates in the AR-FORvsPAST study, and a greater proportion of mineralized N was nitrified. There was also a significant positive correlation between NO_3-N in soil and nitrification rates (Liechty et al., 2009). Similar results were observed in the STRAW study, in which both rates of poultry litter had greater N mineralization and nitrification than CONTROL and IF treatments (Blazier et al, 2008b). The greater nitrification and NO_3-N of poultry litter treatments relative to CONTROL treatments in both studies was likely predominately a function of the addition of N to soil by litter. Relatively high NO_3-N in soil after fertilization is in part indicative of low plant sequestration of applied N (Adeli et al., 2006), so consecutive applications of litter at the rates in these studies likely exceeded loblolly pine, bermudagrass, and bahiagrass N demand. The higher nitrification rates seen in response to poultry litter in these studies relative to inorganic fertilizer, even when both fertilizer sources were applied to provide the same N rates, was likely due to the differences in the effects of the fertilizer sources on soil pH. In the SILVO and STRAW studies, soil pH declined in response to inorganic fertilization applications relative to all other treatments (Tables 6 and 7). Likewise, soil pH of the poultry litter treatments in the SWITCH study was greater in response to broiler litter than to CONTROL and inorganic fertilizer treatments (data not shown). Ellum (2010) found in the STRAW study that nitrification was significantly and positively correlated with pH. Nitrification rates have been shown to decline with decreasing pH due to reductions in populations and activity of nitrifying bacteria (Aune & Lal, 1997).

Although differences in soil NO_3-N and nitrification between loblolly pine and bahiagrass pasture in the AR-FORvsPAST study in part reflected the differences in pH and C:N ratios of the soils in these two land uses (Richardson 2006), they also reflected the differences in uptake and use of available N forms by the loblolly pine and pastures. Although N mineralization and nitrification was greater in pasture when fertilized with poultry litter, the increase in NO_3-N remaining in soil per unit increase in potential net nitrification was greater in loblolly pine plantation than in pasture by the second application of poultry litter (Liechty et al., 2009). Conifer tree roots have been shown to preferentially absorb NH4-N rather than NO_3-N (Kronzucker et al., 1997), whereas NO_3-N is preferentially taken up by forage (Blevins and Barker, 2007). Thus, loblolly pine plantation had a greater propensity to retain a higher proportion of NO_3-N than pasture. Given this tendency of loblolly pine to retain proportionally greater NO_3-N, it is likely that less poultry litter should be applied to such plantations than to pastures to minimize NO_3-N pollution in surface and subsurface water.

Table 7: Soil exchangeable N, mineralization, nitrification, and pH in response to fertilization in loblolly pine plantations, silvopasture, and bahiagrass pasture in a series of trials conducted in the mid-South U.S.A. Means within rows followed by different letters differ at P < 0.05. Adapted in part from Liechty et al. (2009)

	Treatment			
SILVO – 5 years post-treatment	CONTROL	IF	PL5	PL10
NO_3-N (mg kg^{-1})	7.2 a	0.1 a	4.8 a	16.2 a
Total exchangeable N (mg kg^{-1})	34.8 a	18.6 a	29.6 a	71.0 a
% NO_3-N	31.0 b	11.1 c	51.7 a	55.6 a
pH	4.9 a	4.5 a	5.0 a	5.0 a
STRAW – 5 years post-treatment	CONTROL	RAKE	RAKE-IF	RAKE-PL
NO_3-N (mg kg^{-1})	0.6 c	0.8 c	1.4 b	10.0 a
Total exchangeable N (mg kg^{-1})	6.5 b	5.1 c	6.3 b	14.4 a
% NO_3-N	15.9 c	17.8 c	25.2 b	65.6 a
N mineralization (mg kg^{-1})	23.6 b	18.5 b	13.7 b	51.2 a
N nitrification (mg kg^{-1})	23.2 b	17.4 b	17.3 b	48.2 a
% N nitrified	98.3 a	94.1 a	126.3 a	94.1 a
AR-FORvsPAST – 2 years post-treatment	PASTURE-CONTROL	PASTURE-PL9	FOREST-CONTROL	FOREST-PL9
NO_3-N (mg kg^{-1})	2.1 b	4.1 a	0.1 b	15.3 a
Total exchangeable N (mg kg^{-1})	8.9 b	11.0 b	6.2 b	24.4 a
% NO_3-N	21.4 bc	37.6 ab	1.1 c	57.1 a
N mineralization (mg kg^{-1})	14.3 b	23.6 a	7.3 c	14.3 b
N nitrification (mg kg^{-1})	15.6 b	26.3 a	3.4 d	13.3 c
% N nitrified	109.0 b	111.4 a	46.5 d	93.3 c

Annual raking in the STRAW study reduced total exchangeable N, and fertilization, regardless of source, replaced at least a portion of the lost N and increased total exchangeable N (Table 7). Interestingly, although both fertilizers increased exchangeable N, poultry litter increased exchangeable N to a greater extent than the inorganic fertilizer, although both fertilizers were applied at the same N rate (Ellum, 2010). The higher exchangeable N concentrations in the RAKE-PL reflected the increases in NO_3-N levels in the RAKE-PL treatment. The NO_3-N concentrations were nearly 7 and 17 times greater in this treatment than those in the RAKE-IF and CONTROL treatments, respectively. This result provides evidence of the propensity of loblolly pine plantations to accumulate NO_3-N in response to annual applications of broiler litter, even when exchangeable N is reduced by annual straw raking. To safeguard against such NO_3-N accumulation, it is likely necessary to fertilize a raked loblolly pine plantation with broiler litter less frequently and at lower rates than in the STRAW study.

Phosphorus

Soil test P accumulation, determined as the annual difference in soil test P concentrations from pre-treatment concentrations, increased in the uppermost soil horizon in all studies in which soil test P was measured (Table 8). In the SILVO study, both litter treatments had significantly greater soil test P accumulation in the uppermost soil horizon than the CONTROL and IF treatments. After the first application, soil test P accumulation was similarly increased by both litter rates. After four annual applications, the PL10 treatment had greater soil test P accumulation than all other treatments. The IF treatment did not result in a significant accumulation of soil test P at any point in the study (Liechty et al, 2009). Soil test P accumulation also increased in the SWITCH study in response to a single application of litter at both rates. In the LA-FORvsPAST and AR-FORvsPAST studies, soil test P accumulation increased in response to broiler litter in loblolly pine plantation and in pasture (Liechty et al., 2009). Increases in soil test P in surface soil in response to litter application have been similarly found in agricultural (Mitchell & Tu, 2006; Sharpley et al., 1993; Sistani et al., 2004) and forest (Friend et al., 2006) soils. In addition to these increases in upper soil horizons, soil test P accumulation was increased to the B_t horizon (an average depth of 0.59 m) by the 10PL treatment after four applications in the SILVO study (Blazier et al., 2008a). Additional evidence of increasing soil test P in lower soil profile was found in the LA-FORvsPAST study, in which soil test P concentrations of the 20PL treatment exceeded that of all others in the 30 to 45 cm depth in loblolly pine and bermudagrass soil in the seventh year of the study (data not shown). These increases in soil test P in surface and subsurface soil in response to annual litter applications suggest that vegetation P demands and soil P sorption capacity were exceeded at all sites irrespective of vegetation type and stand conditions.

Land use type and rate affected soil test P trends in response to broiler litter in the LAFORvsPAST study. Soil test P accumulation in the loblolly pine plantation averaged over all treatments exceeded that of the pasture for six years of the study (Figure 3). Initial soil test P concentrations of the pasture were 1.5 times greater than that of the loblolly pine plantation (data not shown), but in the first three years of treatment soil test P accumulation of the pasture was negative whereas soil test P accumulation of the loblolly pine plantation ranged from 51 to 76 mg kg^{-1} year^{-1} over the same period. Until the final fertilization, soil test P increased more markedly in the loblolly pine plantation than in the pasture. These differences in soil test P accumulation trends between land use types may have been indicative of lower P demand by loblolly pine than bermudagrass, which led to a greater P accumulation in the soils of the loblolly pine plantations than pastures. Litter application rate also influenced soil P accumulation in both land use types in the LA-FORvsPAST study (Table 8). Annual applications of litter at 5 Mg ha^{-1} did not significantly increase soil test P relative to the CONTROL treatment during the study. Soil test P accumulation was greater in response to the 20 Mg ha^{-1} litter application rate relative to the CONTROL and PL5 treatments throughout the study and greater relative to the 10 Mg ha^{-1} rate by the fourth annual fertilization.

Table 8: Soil test P accumulation (mg kg⁻¹) in response to fertilization with poultry litter and inorganic fertilizer in the mid-South U.S.A. For each study site and soil depth, means within columns followed by different letters differ at P < 0.05. ᵃAverage depth of soil samples subdivided into the A horizon, ᵇaverage depth of soil samples subdivided into the E horizon, ᶜaverage depth of soil samples subdivided into the B_t horizon, ᵈsoil test P accumulation reported for study is an average of loblolly pine plantation and pasture soils because analyses did not reveal a treatment x land use type interaction. Adapted in part from Blazier et al. (2008a) and Liechty et al. (2009)

Study	Treatment	Depth (cm)	Year after treatment						
			1	2	3	4	5	6	7
SILVO	CONTROL	0–15ᵃ	13.0 b	------	16.9 b	10.5 c	------	------	------
	IF		19.0 b	------	22.1 b	24.1 c	------	------	------
	PL5		36.7 a	------	68.5 a	87.2 b	------	------	------
	PL10		48.8 a	------	84.2 a	146.5 a	------	------	------
	CONTROL	15–48ᵇ	3.8 a	------	-0.4 a	-5.2 b	------	------	------
	IF		0.9 a	------	-2.0 a	-5.8 b	------	------	------
	PL5		3.3 a	------	1.1 a	0.9 b	------	------	------
	PL10		2.8 a	------	4.4 a	80.2 a	------	------	------
	CONTROL	48–59ᶜ	2.2 a	------	-1.4 a	-4.0 b	------	------	------
	IF		-0.4 a	------	-4.5 a	-6.6 b	------	------	------
	PL5		-0.5 a	------	-3.5 a	-5.7 b	------	------	------
	PL10		-1.8 a	------	-3.5 a	44.6 a	------	------	------
SWITCH	CONTROL	0–15	0.2 c	------	------	------	------	------	------
	IF80		0.1 b	------	------	------	------	------	------
	IF160		0.1 b	------	------	------	------	------	------
	PL90		0.5 a	------	------	------	------	------	------
	PL180		0.5 a	------	------	------	------	------	------
LA-FORvsPASTᵈ	CONTROL	0–15	-19.9 b	-8.67 a	-24.1 b	-11.9 c	------	-11.0 c	-19.6 c
	PL5		3.2 b	9.3 a	-14.2 b	26.4 bc	------	162.5 bc	82.9 bc
	PL10		32.2 ab	29.7 a	65.3 a	103.2 b	------	328.2 b	210.0 b
	PL20		71.2 a	43.5 a	80.8 a	243.4 a	------	760.0 a	447.6 a
AR-FORvsPASTᵈ	CONTROL	0–15	------	10.1 b	------	------	------	------	------
	PL9		------	47.2 a	------	------	------	------	------

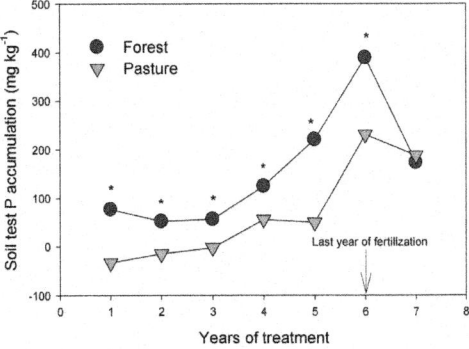

Figure 3: Soil test P accumulation (0 to 15 cm) as affected by annual fertilization with poultry litter in a loblolly pine plantation and a bermudagrass pasture in the mid-South

U.S.A. Asterisks denote years in which soil test P accumulation differed among land use types at $P < 0.05$.

Other Nutrients

Soil K concentrations were increased by broiler litter in the SILVO study (Table 9). A single application of the 10PL treatment increased K concentrations in the A horizon, and subsequent applications led to increases in K concentrations in the E horizon. Increases in K concentrations in lower soil depths have also been observed in response to annual litter fertilization of pastures and agricultural crops on sandy soils (Kingery et al., 1994; Mitchell & Tu, 2006). A similar increase in soil K concentrations in the uppermost 15 cm of soil in response to a single application of broiler litter was found in the SWITCH study (data not shown). In that study soil K increased more in response to the PL3 treatment than all others, and K concentrations of all other fertilizer treatments exceeded that of the CONTROL treatment. Results of both studies indicate that poultry litter can lead to increases in soil K concentrations in these silvopastures, even after a single application.

Although soil K concentrations increased in both the SILVO and SWITCH studies, the amount of poultry litter required to increase the concentrations differed between the two types of silvopastures. An application of only 1.5 Mg ha^{-1} of litter was needed to increase K concentrations in the loblolly pine-switchgrass silvopasture while in the loblolly pinebahiagrass silvopasture K concentrations were observed only after two annual applications of 10 Mg ha^{-1} of poultry litter. Since the soil type was identical for these two studies, these results suggest that loblolly pine-bahiagrass silvopasture had a greater K demand than the loblolly pine-switchgrass silvopasture. The higher demand of the loblolly pine and bahiagrass pasture was likely due in part to loblolly pine density that was nearly double that in the loblolly pine and switchgrass silvopasture.

The switchgrass also likely had a lower K demand than bahaiagrass , because switchgrass is characterized by relatively low nutrient demand despite its relatively high biomass growth potential (Tilman et al., 2006). Nevertheless, annual broiler application at 10 Mg ha^{-1} apparently exceeded vegetation K demand and sorption capacity of the A horizon in the loblolly pine and bahiagrass silvopasture as indicated by increased in K concentrations in the E horizon after four annual applications.

As with K, soil Mg concentrations were increased in the A and E horizons by repeated applications of litter in the SILVO study (Table 9; Blazier et al., 2008a). After two applications soil Mg in the A and E horizons was increased by the 10 Mg ha^{-1} rate relative to the CONTROL and IF treatments, and after four applications the 5 Mg ha-1 rate led to greater soil K concentrations in the A horizon than in the CONTROL and IF treatments. However, the 5 Mg ha^{-1} did not increase soil K concentrations in the E horizon and did not increase soil K concentrations to levels in the A horizon comparable to that of the 10 Mg ha^{-1} rate after the fourth applications. By the fourth application, soil Ca concentrations in the A horizon were also increased by the poultry litter treatments, with that of the PL10 treatment exceeding all other treatments and that of the PL5 treatment greater than the CONTROL and IF treatments.

Table 9: Effects of annually fertilizing a loblolly pine and bahiagrass silvopasture with poultry litter and inorganic fertilizer on soil K and Mg in the A and E soil horizons and on Ca in the A horizon. For each nutrient and horizon, means within a column followed by a different letter differ at $P < 0.05$. Adapted from Blazier et al. (2008a)

Nutrient	Horizon	Treatment	1997	1998	2001	2002
K	A	CONTROL	42.1 a	30.1 b	23.3 a	30.7 a
		IF	33.9 a	42.9 b	23.7 a	26.0 a
		PL5	33.2 a	44.3 b	31.4 a	30.1 a
		PL10	39.8 a	62.6 a	34.9 a	36.1 a
	E	CONTROL	22.8 a	22.2 a	21.3 c	34.1 b
		IF	22.7 a	27.2 a	28.8 bc	37.8 b
		PL5	21.7 a	34.2 a	39.7 ab	43.8 b
		PL10	30.8 a	36.8 a	51.5 a	60.0 a
Mg	A	CONTROL	30.5 b	33.8 b	107.7 bc	96.2 c
		IF	32.0 a	26.9 bc	103.7 c	89.3 c
		PL5	34.8 a	38.4 ab	113.5 ab	100.9 b
		PL10	35.2 a	44.4 a	120.6 a	114.8 a
	E	CONTROL	26.6 a	25.8 b	107.8 b	100.4 b
		IF	29.8 a	25.8 b	112.4 b	104.4 b
		PL5	34.8 a	28.2 ab	126.6 ab	114.3 ab
		PL10	57.6 a	34.9 a	145.1 a	137.6 a
Ca	A	CONTROL	184.6 a	194.6 a	134.4 c	70.2 c
		IF	177.2 a	157.2 a	89.6 c	20.8 c
		PL5	171.2 a	196.0 a	162.9 b	95.5 b
		PL10	186.0 a	226.0 a	229.9 a	174..3 a

WATER NUTRIENTS

Poultry litter applications led to increases in NO_3-N in soil water in the AR-FORvsPAST study. Total N concentrations in soil water were greater for pastures than the loblolly pine plantation and greater for the PL9 treatment than the CONTROL treatment; differences in NO_3-N accounted for the majority of the total N differences between land use types and treatments. In both pasture and loblolly pine plantation, NO_3-N concentrations increased in response to poultry litter application (Figure 4). Soil water NO_3-N concentrations were significantly positively correlated with potential nitrification rates.

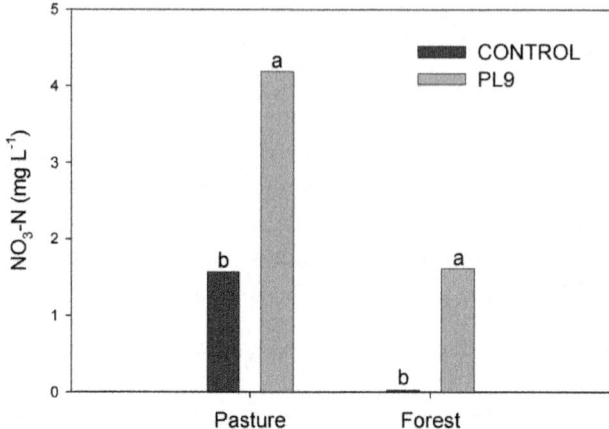

Figure 4: Mean bi-weekly soil water (30 cm) NO3-N concentrations in bermudagrass pasture and loblolly pine plantation treated with poultry litter. For each land use type, means headed by different letters differ at $P < 0.05$. Adapted from Liechty et al. (2009).

Although bi-weekly NO_3-N concentrations in soil water never exceeded the 10 mg L^{-1} drinking water standard of the U.S. Environmental Protection Agency in the loblolly pine plantation, this standard was exceeded in two or more sampling periods in pasture plots fertilized with poultry litter. Soil water N increased 51% more in pastures than in loblolly pine plantation, which suggests the potential for N pollution of water is greater for pastures fertilized with poultry litter than for loblolly pine plantations fertilized with poultry litter. However, because forest soils have an apparently greater propensity than pastures to retain proportionally greater NO_3-N in soil (described above), with long-term litter applications N losses in soil water from forests could be greater than in pastures (Liechty et al., 2009).

Repeated fertilization with poultry litter led to increases in total and dissolved P concentrations in runoff in pasture and loblolly pine plantation in the LA-FORvsPAST study. Total and dissolved P concentrations increased with increasing litter application rate in both land use types, although the P concentrations increased more markedly to 10 and 20 Mg ha^{-1} rates in pasture than in loblolly pine plantation. Total and dissolved P concentrations in runoff were positively correlated with Bray P concentrations in soil. These results indicate potential for losses of P in runoff in response to litter application in pasture and loblolly pine plantation, with modest evidence that P loss potential in loblolly pine plantation was lower. In the AR-FORvsPAST study, there were no significant differences in total P concentrations in soil water among treatments and land use types.

CONCLUSIONS

Poultry litter was a beneficial fertilizer for loblolly pine plantations and silvopastures in this series of studies. Unlike with inorganic fertilizer, soil pH did not decrease with poultry litter application, which sustained microbial biomass and activity at levels comparable to non-fertilized soil. Poultry litter application to soils that had annual pine straw harvesting maintained soil strength, organic matter, and soil moisture similar to those without straw harvesting, whereas applying inorganic fertilizer to soils with straw harvesting negatively impacted these soil attributes. Loblolly pine trees in plantations and silvopastures, as well as the grasses in silvopastures, responded to poultry litter fertilization with increased growth and nutrient concentrations. These increases in plant growth and nutrition provided some buffering against increasing soil nutrient concentrations when these plantations and silvopastures were annually fertilized with poultry litter. Nevertheless, poultry litter was more prone to lead to accumulation of NO_3- N and P in soil than inorganic fertilizer. Loblolly pine plantations were also more prone to increases in soil NO_3-N and P than pastures. Accumulations in soil NO_3-N and P were also associated with increased NO_3-N and P concentrations in soil water and runoff, respectively. As such, poultry litter fertilization of these loblolly pine plantations and silvopastures had the potential to contaminate soil water with N and P. Any poultry litter fertilization regimes for loblolly pine plantations and silvopastures must account for the greater tendencies of N and P accumulation in soil and water of these ecosytems; lower rates and/or frequencies than those used in these trials will likely be necessary for ecologically sustainable fertilization with poultry litter.

REFERENCES

1. Adams, P.L., Daniel, T.C., Edwards, D.R., Nichols, D.J., Pote, D.H. & Scott, H.D. (1994). Poultry litter and manure contributions to nitrate leaching through the vadose zone. Soil Science Society of America Journal, Vol. 58 (No. 4): 1206-1211.
2. Adeli, A., Rowe, D.E. & Read, J.J. (2006). Effects of soil type on bermudagrass response to broiler litter application. Agronomy Journal, Vol. 98 (No. 1): 148-155.
3. Agbede, T.M., Oladitan, T.O., Alagha, S.A., Ojomo, A.O. & Ale, M.O. (2010). Comparative evaluation of poultry manure and NPK fertilizer on soil physical and chemical properties, leaf nutrient concentrations, growth and yield of yam (Dioscorea rotundata Poir) in southwestern Nigeria. World Journal of Agricultural Sciences, Vol. 6 (No. 5): 540-546.

4. Alef, K. (1995). Dehydrogenase activity, In: Methods in applied soil microbiology and biochemistry, Alef, K. & Nannipieri, P. (eds.). pp. 228-231. Academic Press, San Diego, CA.
5. Allen, H.L. (1987). Forest fertilizers: nutrient amendment, stand productivity, and environmental impact. Journal of Forestry, Vol. 85 (No. 2): 37-46.
6. Allen, H.L., Fox, T.R. & Campbell, R.G. (2005). What is ahead for intensive pine plantation silviculture in the South? Southern Journal of Applied Forestry, Vol. 29 (No. 2): 62-69.
7. Alikhani, H.A., Saleh-Rastin, N. & Antoun, H. (2006). Phosphate solubilization activity of rhizobia native to Iranian soils. Plant and Soil, Vol. 287 (No. 1-2): 35-41.
8. Anderson, T.H. & Domsch, K.H. (1993). The metabolic quotient for CO_2 ($qCO2$) as a specific activity parameter to assess the effects of environmental conditions, such as pH, on the microbial biomass of forest soils. Soil Biology and Biochemistry, Vol. 25 (No. 3): 393-395.
9. Attiwill, P.M. & Adams, M.A. (1993). Tansley review no. 50: Nutrient cycling in forests. New Phytologist, Vol. 124: 561-582.
10. Aune, J.B. & Lal, R. (1997). Agricultural productivity in the tropics and critical limits of properties of oxisols, ultisols, alfisols. Tropical Agriculture, Vol.74 (No. 2): 96-103.
11. Baath, E., Frostegard, A., Pennanen, T., & Fritze, H. (1995). Microbial community structure and pH response in relation to soil organic matter quality in wood-ash fertilized, clear-cut or burned coniferous forest soils. Soil Biology and Biochemistry, Vol. 27 (No. 2): 229-240.
12. Bailey, R.G. (1995). Description of the ecoregions of the United States (2nd edition), USDA Forest Service Miscellaneous Publication No. 1391, USDA Forest Service, Washington, D.C.
13. Ball, D.F. (1964). Loss on ignition as an estimate of organic matter and organic carbon in non-calcareous soils. Journal of Soil Science, Vol. 15 (No. 1): 84.92.
14. Beem, M., Turton, D.J., Barden, C.J. & Anderson, S. (1998). Application of poultry litter to pine forests, OSU Extension Factsheet F-5037, Oklahoma Cooperative Extension Service, Oklahoma State University, Stillwater, OK.
15. Binkley, D., Burnham, H. & Allen, H.L. (1999). Water quality impacts of forest fertilization with nitrogen and phosphorus. Forest Ecology and Management, Vol. 121 (No. 3): 191- 213.

16. Blake, G.R. & Hartge, K.H. (1986). Bulk density, In: Methods of Soil Analysis, Part 1: Physical and Mineralogical Methods. (2nd Edition), Klute, A. (ed.). pp. 363-375. Soil Science Society of America, Madison, WI.
17. Blazier, M.A., Hennessey, T.C. & Deng, S.P. (2005). Effects of fertilization and vegetation control on microbial biomass carbon and dehydrogenase activity in a juvenile loblolly pine plantation. Forest Science, Vol. 51 (No. 5): 449-459.
18. Blazier, M.A., Hennessey, T.C., Dougherty, P.M. & Campbell, R. (2006). Nitrogen accumulation and use by a young loblolly pine plantation in southeast Oklahoma: Effects of fertilizer formulation and date of application. Southern Journal of Applied Forestry, Vol. 30 (No. 2): 66-78.
19. Blazier, M.A., Gaston, L.A., Clason, T.R., Farrish, K.W., Oswald, B.P. & Evans, H.A. (2008a). Nutrient dynamics and tree growth of silvopastoral systems: impact of poultry litter. Journal of Environmental Quality, Vol. 37 (No. 4): 1546-1558.
20. Blazier, M.A., Hotard, S.L. & Patterson, W.B. (2008b). Straw harvesting, fertilization, and fertilizer type alter soil microbiological and physical properties in a loblolly pine plantation in the mid-South U.S.A. Biology and Fertility of Soils, Vol. 45 (No. 2): 145- 153.
21. Blevins, D.G. & Barker, D.J. (2007). Nutrients and water in forage crops, In: Forages: The Science of Grassland Agriculture. Volume II (6th edition), Barnes, R.R., Nelson, C.J., Moore, K.J. & Collins, M. (Eds.). pp 67-80. Blackwell Publishing, Ames, IA.
22. Bradford, J.M. (1986). Penetrability, In: Methods of Soil Analysis, Part 1: Physical and Mineralogical Methods. (2nd edition), Klute A. (Ed.). pp. 463-478. Soil Science Society of America, Madison, WI.
23. Bray, R.H. & Kurtz, L.T. (1945). Determination of total, organic, and available forms of phosphorus in soils. Soil Science, Vol. 59 (No. 1): 39-45.
24. Breeuwsma, A. & Silva, S. (1992). Phosphorus fertilization and environmental effects in the Netherlands and the Po region (Italy), Report 57, Winand Staring Centre for Integrated Land, Soil and Water Research, Wageningen, The Netherlands.
25. Brye, K.R. (2003). Long-term effects of cultivation on particle size and water-retention characteristics determined using wetting curves. Soil Science, Vol. 168 (No. 7): 459- 468.
26. Cabrera, M.L. & Sims, J.T. (2000). Beneficial use of poultry by-products: challenges and opportunities, In: Land application of agricultural,

industrial, and municipal by-products (1st edition), Power, J.F. & Dick, W.A. (Eds.). pp. 425-450. Soil Science Society of America, Madison, WI.

27. Canali, S., Tinchera, A., Intrigliolo, F., Pompili, L., Nisini, L., Mocali, S. & Torrisi, B. (2004). Effect of long term addition of composts and poultry manure on soil quality of citrus orchards in Southern Italy. Biology & Fertility of Soils, Vol. 40 (No. 3): 206-210.

28. Clason, T.R. (1995). Economic implications of silvipastures on southern pine plantations. Agroforestry Systems, Vol. 29: 227-238.

29. Clason, T.R. & Robinson, J.L. (2000). From a pine forest to a silvopasture system, Agroforestry Note 18, USDA Forest Service, USDA Natural Resource Conservation Service, Washington, D.C.

30. Clason, T.R. & Sharrow, S.H. (2000). Silvopastoral practices, In: North American Agroforestry: An Integrated Science and Practice, Garrett, H.E., Rietveld, W.J. & Fisher, R.F. (Eds.). pp. 119-147. Agronomy Society of America, Madison, WI.

31. Clesceri, L.S., Greenberg, A.E. & Eaton, A.D. (Eds.). (1998). Standard methods for the examination of water and wastewater (20th edition), American Public Health Association, Washington, DC.

32. Colbert, S.R., Jokela, E.J. & Neary, D.G. (1990). Effects of annual fertilization and sustained weed control on dry matter partitioning, leaf area, and growth efficiency of juvenile loblolly and slash pine. Forest Science, Vol. 36 (No. 4): 995-1014.

33. Daddow R.L. & Warrington, G.E. (1983). Growth-limiting soil bulk densities as influenced by soil texture, Watershed Systems Development Group Report WSDG-TN-00005, USDA Forest Service, Fort Collins, CO.

34. Daniel, T.C., Sharpley, A.N. & Lemunyon, J.L. (1998). Agricultural phosphorus and eutrophication: a symposium overview. Journal of Environmental Quality, Vol. 27 (No. 2): 251-257.

35. Danielson, R.E. & Sutherland, P.L. (1986). Porosity, In: Methods of Soil Analysis, Part 1: Physical and Mineralogical Methods (2nd edition). Klute, A. (Ed.). pp. 443-462. Soil Science Society of America, Madison, WI.

36. Dick, W.A., Eckert, D.J. & Johnson, J.W. (1998). Land application of poultry litter, Ohio State University Cooperative Extension Fact Sheet ANR-4-98, Ohio State University, Columbus, OH.

37. Dickens, E.D., Bush, P.B., & Morris, L.A. (2003). Poultry litter application

recommendations in pine plantations. Warnell School of Forestry and Natural Resources, College of Agricultural and Environmental Sciences, University of Georgia, Athens, GA, Retrieved from http://www.bugwood.org/fertilization/PLARPP.html

38. Dickens, E.D., Richardson, B.W. & McElvany, B.C. (2004). Old-field thinned loblolly pine plantation fertilization with diammonium phosphate plus urea and poultry litter: 4 year growth and product class distribution results, In: Proceedings of the 12th Biennial Southern Silvicultural Research Conference, General Technical Report SRS-48, Outcalt, K. (Ed.), pp. 395-397. USDA Forest Service, Southern Research Station, Asheville, NC.

39. Donegan, K.K., Watrud, L.S., Seidler, R.J., Maggard, S.P., Shiroyama, T., Porteous, L.A. & DiGiovanni, G. (2001). Soil and litter organisms in Pacific Northwest forests under different management practices. Applied Soil Ecology, Vol. 18 (No. 2): 159-175.

40. Duryea, M.L. & Edwards, J.C. (1989). Pine-straw management in Florida's forest, Florida Cooperative Extension Service Institute of Food and Agricultural Science Circular 831, University of Florida, Gainsville, FL.

41. Edmeades, D.C. (2003). The long-term effects of manures and fertilizers on soil productivity and quality: a review. Nutrient Cycling in Agroecosystems, Vol. 66 (No. 2): 165-180.

42. Edwards, D.R. & Daniel, T.C. (1992). Environmental impacts of on-farm poultry waste disposal: a review. Bioresource Technology, Vol. 41 (No. 1): 9-33.

43. Eichhorn, M.M. (2001). Impact of best management practices and organic wastes on water quality and crop production: Poultry litter application demonstration project, Louisiana Department of Environmental Quality Projects CFMS514283 and CFMS554784 Final Report, Hill Farm Research Station, Louisiana Agricultural Experiment Station, Baton Rouge, LA.

44. Ekinci, K., Keener, H.M. & Elwell. D.L. (2000). Composting short paper fiber with broiler litter and additives Part 1: Effects of initial pH and carbon/nitrogen ratio on ammonia emissions. Compost Science & Utilization, Vol. 8 (No. 2): 160-172.

45. Ellum, K. (2010). Pine straw raking and fertilizer source impacts on nitrogen mineralization, pine needle gas exchange, and tree water stress in a loblolly pine plantation, M.S. thesis, School of Forest Resources, University of Arkansas, Monticello, AR.

46. Elser, J. J., Bracken, M.E.S., Cleland, E.E., Gruner, D.S., Harpole, W.S., Hillebrand, H., Ngai, J.T., Seabloom, E.W., Shurin, J.B. & Smith, J.E. (2007). Global analysis of nitrogen and phosphorus limitation of primary producers in freshwater, marine, and terrestrial ecosystems. Ecology Letters, Vol. 10 (No. 12): 1135–1142.

47. Eno, C.F. (1960). Nitrate production in the field by incubating the soil in polyethylene bags. Soil Science Society of America Proceedings, Vol. 24: 277–279.

48. Evans, H.A. (2000). Application of poultry litter and commercial fertilizer in a loblolly pinebahiagrass silvopasture, M.S. thesis, College of Forestry, Stephen F. Austin University, Nacogdoches, TX.

49. Falkiner, R.A. & Polglase, P.J. (1997). Transport of phosphorus through soil in an effluentirrigated tree plantation. Australian Journal of Soil Research, Vol. 35: 385-398.

50. Farley, K.A., Jobbágy, E.G. & Jackson, R.B. (2005). Effects of afforestation on water yield: a global synthesis with implications for policy. Global Change Biology, Vol. 11 (No. 10): 1565-1576.

51. Fauci, F. & Dick, R.P. (1994). Microbial biomass as an indicator of soil quality: effects of long-term management and recent soil amendments, In: Defining soil quality for a sustainable environment (1st edition). Doran, J.W., Coleman, D.C., Bezdicek, D.F. & Stewart, B.A. (Eds). pp. 229-234. Soil Science Society of America, Madison, WI.

52. Fisher, D.S., Steiner, J.L., Endale, D.M., Stuedemann, J.A., Schomberg, H.H., Franzluebbers, A.J. & Wilkinson, S.R. (2000). The relationship of land use practices to surface water quality in the Upper Oconee Watershed in Georgia. Forest Ecology and Management, Vol. 128 (No. 1-2): 39-48.

53. Friend, A.L., Roberts, S.D., Schoenholtz, S.H., Mobley, J.A. & Gerard, P.D. (2006). Poultry litter application to loblolly pine forests: growth and nutrient containment. Journal of Environmental Quality, Vol. 35 (No. 3): 837-848.

54. Funderberg, E. (2009). Poultry litter as fertilizer. In: Ag News and Views, January 2009, Soil & Crops. The Samuel Roberts Noble Foundation. Available from: http://www.noble.org/ag/Soils/PoultryLitter/index.html.

55. Gallardo, A. & Schlesinger, W.H. (1994). Factors limiting microbial biomass in the mineral soil and forest floor of a warm-temperate forest. Soil Biology & Biochemistry. Vol. 26 (No 10): 1409-1415.

56. Gallimore, L.E., Basta, N.T., Storm, D.E., Payton, M.E., Huhnke, R.H. & Smolen, M.D. (1999). Water treatment residual to reduce nutrients in surface runoff from agricultural land. Journal of Environmental Quality,

Vol. 28 (No. 5): 1474-1478.
57. Gambrell, R.P. (1996). Manganese, In: Methods of soil analysis, part 3: Chemical methods (3rd edition). Bartels, J.M. (Ed.) pp. 665-682. Soil Science Society of America, Madison, WI.
58. Gaston, L.A., Clason, T.R. & Cooper, D. (2003). Poultry litter fertilizer on pasture, silvopasture, and forest soils. Louisiana Agriculture, Vol. 46 (No. 3): 22-23.
59. Gaston, L.A., Drapcho, C.M., Tapadar, S. & Kovar, J.L. (2003). Phosphorus runoff relationships for Louisiana Coastal Plain soils amended with poultry litter. Journal of Environmental Quality, Vol. 32 (No. 4): 1422-1429.
60. Gee, G.W., Campbell, M.D., Campbell, G.S. & Campbell, J.H. (1992). Rapid measurement of low soil water potentials using a water activity meter. Soil Science Society of America Journal, Vol. 56 (No. 4): 1068-1070.
61. Gilman, E.F. (1987). Where are tree roots? Florida Cooperative Extension Service, Institute of Food and Agricultural Sciences, University of Florida, Extension Bulletin ENH137, Retrieved from http://edis.ifas.ufl.edu/pdffiles/WO/WO01700.pdf
62. Gupta, G., Borowiec, J. & Okoh, J. (1997). Toxicity identification of poultry litter aqueous leachate. Poultry Science, Vol. 76 (No. 10): 1364-1367.
63. Harmel, R.D., Torbet, H.A., Haggard, B.E., Haney, R. & Dozier, M. (2004). Water quality impacts of converting to a poultry litter fertilization strategy. Journal of Environmental Quality, Vol. 33 (No. 6): 2229-2242.
64. Harris, J.A. (2003). Measurements of the soil microbial community for estimating the success of restoration. European Journal of Soil Science, Vol. 54 (No. 4): 801-808.
65. Hart, S.C., Stark, J.M., Davidson, E.A. & Firestone, M.K. (1994). Nitrogen mineralization, immobilization, and nitrification, In: Methods of Soil Analysis. Part 2. Microbiological and Biochemical Properties. Weaver, R.W., Angele, S., Bottomly, P. (Eds.), pp. 985- 1018. Soil Science Society of America. Madison, WI.
66. Haywood, J.D., Tiarks, A.E. & Sword, M.A. (1997). Fertilization, weed control, and pine litter influence loblolly pine stem productivity and root development. New Forests, Vol. 14 (No. 3): 233-249.
67. Haywood, J.D., Tiarks, A.E., Elliott-Smith, M.L. & Pearson, H.A. (1998). Response of direct seeded Pinus palustris and herbaceous vegetation to

fertilization, burning, and pine straw harvesting. Biomass Bioenergy, Vol. 14 (No. 2): 157-167.
68. Helmke, P.A. & Sparks, D.L. (1996). Lithium, sodium, potassium, rubidium, and cesium, In: Methods of soil analysis, part 3. Chemical methods, Bartels, J.M. (Ed.), pp. 551-574. Soil Science Society of America, Madison, WI.
69. Henry, C.L., Cole, D.W., Hinckley, T.M. & Harrison, R.B. (1993). The use of municipal and pulp and paper sludges to increase production in forestry. Journal of Sustainable Forestry, Vol. 1 (No. 3): 41-55.
70. Högberg, P., Nordgren, A., Buchmann, N., Taylor, A.F.S., Ekblad, A., Högberg, M.N., Nyberg, G., Ottoson-Löfvenius, M. & Read, D.J. (2001). Large-scale forest girdling shows that current photosynthesis drives soil respiration. Nature, Vol. 411: 789-792.
71. Horneck, D.A. & Miller, R.O. (1998). Determination of total nitrogen in plant tissue, In: Handbook of reference methods for plant analysis, Yask, P. (Ed.), pp. 75-83. CRC Press, Boca Raton, FL.
72. Hue, N.V. (1992). Correcting soil acidity of a highly weathered ultisol with chicken manure and sewage sludge. Communications in Soil Science and Plant Analysis, Vol. 23 (No.3- 4): 241-264.
73. Jackson, M.J., Line, M.A., Wilson, S. & Hetherington. S.J. (2000). Application of composted pulp and paper mill sludge to a young pine plantation. Journal of Environmental Quality, Vol. 29 (No. 2): 407-414.
74. Jenkinson, D.S. & Powlson, D.S. (1976a). The effects of biocidal treatments on metabolism in soil-I. Fumigation with chloroform. Soil Biology & Biochemistry, Vol. 8 (No. 3): 167- 177.
75. Jenkinson, D.S. & Powlson, D.S. (1976b). The effects of biocidal treatments on metabolism in soil-V: A method for measuring soil biomass. Soil Biology & Biochemistry, Vol. 8 (No. 3): 209-213.
76. Jokela, E.J. (2004). Nutrient management for southern pines, In: Slash pine: still growing and growing! Proceedings of the slash pine symposium, General Technical Report SRS-76, Dickens, E.D., Barnett, J.P., Hubbard, W.G. & Jokela, E.J. (Eds.), pp. 27-35. U.S. Department of Agriculture, Forest Service, Southern Research Station, Asheville, NC.
77. Jones, J.B. & Case, V.W. (1990). Sampling, handling, and analyzing plant tissue samples. In: Soil testing and plant analysis (3rd edition), Westerman, R.L. (Ed.), pp. 389-447. Soil Science Society of America, Madison, WI.
78. Jorgensen, J.R. & Wells, C.G. (1986). Foresters' primer in nutrient

cycling: a loblolly pine management guide, General Technical Report SE-37, USDA Forest Service, Southeastern Forest Experiment Station, Asheville, NC.
79. Kellogg, R.L., Lander, C.H., Moffitt, D.C. & Gollehon, N. (Eds.). (2000). Manure nutrients relative to the capacity of cropland and pastureland to assimilate nutrients: Spatial and temporal trends for the United States. USDA Natural Resources Conservation Service Publication NPS 00-0579, GSA National Forms and Publication Center, Fort Worth, TX.
80. Khan, S.A., Mulvaney, R.L., Ellsworth, T.R. & Boast, C.W. (2007). The myth of nitrogen fertilization for soil carbon sequestration. Journal of Environmental Quality, Vol. 36 (No. 6): 1821-1832.
81. Kingery, W.L., Wood, C.W., Delaney, D.P., Williams, J.C. & Mullins, G.L. (1993). Implications of long-term application of poultry litter on tall fescue pastures. Journal of Production Agriculture, Vol. 6 (No. 3): 315-395.
82. Kingery, W.L., Wood, C.W., Delaney, D.P., Williams, J.C. & Mullins, G.L. (1994). Impact of long-term land application of broiler litter on environmentally related soil properties. Journal of Environmental Quality, Vol. 23 (No. 1): 139-147.
83. Kronzucker, H.J., Yaeesh Siddiqi, M. & Glass, A.D.M. (1997). Conifer root discrimination against soil nitrate and the ecology of forest succession. Nature, Vol. 385: 59-61.
84. Lemunyon, J. & Gilbert, R. (1993). The concept and need for a phosphorus assessment tool. Journal of Production Agriculture, Vol.6 (No. 4): 483-486.
85. Lenhard, G. (1956). The dehydrogenase activity in soil as a measure of the activity of soil microorganisms. Z. Pflanzenernäh Düng Bodenkd, Vol. 73:1-11.
86. Liechty, H.O., Blazier, M.A., Wight, J.P., Gaston, L.A., Richardson, J.D. & Ficklin, R.L. (2009). Assessment of repeated application of poultry litter on phosphorus and nitrogen dynamics in loblolly pine: implications for water quality. Forest Ecology and Management, Vol. 258 (No. 10): 2294-2303.
87. Lopez-Zamora, I., Duryea, M.L., McCormac, W.C., Comerford, N.B. & Neary, D.G. (2001). Effect of pine needle removal and fertilization on tree growth and soil P availability in a Pinus elliottii Engelm. var. elliottti stand. Forest Ecology and Management, Vol. 148 (No 1-3): 125-134.
88. Lynch, L. & Tjaden, R. (2004). Would forest landowners use poultry manure as fertilizer? Journal of Forestry, Vol. 102 (No. 5): 40-45.

89. McLeod, R.V. & Hegg, R.O. (1984). Pasture runoff quality from application of inorganic and organic nitrogen sources. Journal of Environmental Quality, Vol. 13 (No. 1): 122-126.
90. Mehlich, A. (1984). Mehlich 3 soil test extractant: A modification of the Mehlich 2 extractant. Communications in Soil Science and Plant Analysis, Vol. 15 (No. 12): 1409-1416.
91. Mills, R. & Robertson, D.R. (1991). Production and marketing of Louisiana pine straw, Louisiana Cooperative Extension Service Publication 2340, Louisiana State University Agricultural Center, Baton Rouge, LA.
92. Minogue, P.J., Ober, H.K. & Rosenthal, S. (2007). Overview of Pine Straw Production in North Florida: Potential Revenues, Fertilization Practices, and Vegetation Management Recommendations, School of Forest Resources and Conservation Department, Florida Cooperative Extension Service, Institute of Food and Agricultural Sciences Publication 125. University of Florida, Gainesville, FL.
93. Mitchell, C.C. & Donald, J.O. (1995). The value and use of poultry manures as fertilizers, Alabama Cooperative Extension System Circular ANR-224, Auburn University, Auburn, AL.
94. Mitchell, C.C. & Tu, S. (2006). Nutrient accumulation and movement from poultry litter. Soil Science Society of America Journal, Vol. 70 (No. 6): 2146-2153.
95. Mulvaney, R.L. (1996). Nitrogen-inorganic forms, In: Methods of Soil Analysis. Part 3. Chemical methods, Sparks, D.L. (Ed.), pp. 1123-1184. Soil Science Society of America, Madison, WI.
96. Munkholm, L.J., Schjønning, P., Debosz, K., Jensen, H.E. & Christensen, B.T. (2002). Aggregate strength and mechanical behavior of a sandy loam under long-term fertilization treatments. European Journal of Soil Science, Vol. 53 (No. 1): 129-137.
97. Murthy, R., Zarnoch, S.J. & Dougherty, P.M. (1997). Seasonal trends of light-saturated net photosynthesis and stomatal conductance of loblolly pine trees grown in contrasting environments of nutrition, water, and carbon dioxide. Plant, Cell, and Environment, Vol. 20 (No. 5): 558-568.
98. Nair, V.D. & Graetz, D.A. (2004). Agroforestry as an approach to minimizing nutrient loss from heavily fertilized soils: the Florida experience. Agroforestry Systems, Vol. 60: 269-279.
99. National Agricultural Statistics Service. (2008). Poultry-Production and Value 2007 Summary, United States Department of Agriculture National Agricultural Statistics Service, Washington, D.C.

100. O'Neill, G.J. & Gordon. A.M. (1994). The nitrogen filtering capacity of Carolina poplar in an artificial riparian zone. Journal of Environmental Quality, Vol. 23 (No. 6): 1218-1223.
101. Plaza, C., Hernández, D., García-Gil, J.C. & Polo, A. (2004). Microbial activity in pig slurryamended soils under semiarid conditions. Soil Biology & Biochemistry, Vol. 36 (No. 10): 1577-1585
102. Polglase, P.J., Tompkins, D. & Falkiner, R.A. (1995). Mineralization and leaching of nitrogen in an effluent-irrigated pine plantation. Journal of Environmental Quality, Vol. 24 (No. 5): 911-922.
103. Pote, D.H. & Daniel, T.C. (2000). Analyzing for total phosphorus and total dissolved phosphorus in water samples. In: Methods of Phosphorus Analysis for Soils, Sediments, Residuals, and Waters, Southern Cooperative Series Bulletin No. 396., Pierzynski, G.M. (Ed.), pp. 91-93. Retrieved from: http://www.sera17.ext.vt.edu/Documents/Methods_of_P_Analysis_2000.pdf
104. Pote, D.H. & Daniel, T.C. (2008). Managing pine straw harvests to minimize soil and water losses. Journal of Soil and Water Conservation, Vol. 63 (No.1): 27-28.
105. Pote, D.H., Kingery, W.L., Aiken, G.E., Han, F.X., Moore Jr., P.A. & Buddington, K. (2003). Water-quality effects of incorporating poultry litter in perennial grassland soils. Journal of Environmental Quality, Vol. 32 (No. 6): 2392-2398.
106. Powers, R.F., Scott, D.A., Sanchez, F.G., Voldseth, R.A., Page-Dumroese, D., Elioff, J.D. & Stone, D.M. (2005). The North American long-term soil productivity experiment: findings from the first decade of research. Forest Ecology and Management, Vol. 220 (No. 1-3): 31-50.
107. Powlson, D.S. & Brookes, P.C. (1987). Measurement of soil microbial biomass provides an early indication of changes in total soil organic matter due to straw incorporation. Soil Biology & Biochemistry, Vol. 19 (No. 2): 159-164.
108. Prestemon, J.P. & Abt, R.C. (2002). The southern timber market to 2040. Journal of Forestry, Vol. 100 (No. 7): 16-22.
109. Pritchett, W.L. & Fisher, R.F. (1987). Properties and management of forest soils (2nd edition), John Wiley & Sons, Inc., New York, NY.
110. Richardson, J. (2006). Effects of poultry litter applied to pine plantations and pastures on water quality and soil nitrogen mineralization. M.S. thesis, University of Arkansas at Monticello.
111. Rifai, S.W., Markewitz, D. & Borders, B. (2010). Twenty years of

intensive fertilization and competing vegetation suppression in loblolly pine plantations: impacts on soil C, N, and microbial biomass. Soil Biology & Biochemistry, Vol. 42 (No. 5): 713-723.

112. Roberts, S.D., Friend, A.L. & Gerard, P.D. (2004). The effect of large applications of nutrients from organic waste on biomass allocation and allometric relations in loblolly pine, In: Proceedings of the 12th Biennial Southern Silvicultural Research Conference, General Technical Report SRS-48, Outcalt, K. (Ed.), pp. 398-402. USDA Forest Service, Southern Research Station, Asheville, NC.

113. Roberts, S.D., Friend, A.L. & Schoenholtz, S.H. (2006). Growth of precommercially thinned loblolly pine four years following application of poultry litter, In: Proceedings of the 12th Biennial Southern Silvicultural Research Conference, General Technical Report SRS-92, Outcalt, K. (Ed). pp. 139-142. USDA Forest Service, Southern Research Station, Asheville, NC.

114. Roise, J.P., Chung, J. & Lancia, R. (1991). Red-cockaded woodpecker habitat management and longleaf pine straw production: an economic analysis. Southern Journal of Applied Forestry, Vol. 15 (No. 2): 88-92.

115. Samuelson, L.J., Wilhoit, J. Stokes, T. & Johnson, J. (1999). Influence of poultry litter fertilization on 18-year-old loblolly pine stand. Communications in Soil Science and Plant Analysis, Vol. 30 (No. 3-4): 509-518.

116. Sanchez, F.G., Scott, D.A. & Ludovici, K.H. (2006). Negligible effects of severe organic matter removal and soil compaction on loblolly pine growth over 10 years. Forest Ecology and Management, Vol. 227 (No. 1): 145-154.

117. Sauer, T.J., Daniel, T.C., Nichols, D.J., West, C.P., Moore, P.A. & Wheeler, G.L. (2000). Runoff water quality from poultry litter-treated pasture and forest sites. Journal of Environmental Quality, Vol. 29 (No. 2): 515-521.

118. Schindler, D. (1978). Factors regulating phytoplankton production and standing crop in the word's freshwaters. Limnology and Oceanography, Vol. 23 (No. 3): 478-486.

119. Sharpley, A.N. (1999). Agricultural phosphorus, water quality, and poultry production: are they compatible. Poultry Science, Vol. 78 (No. 5): 660-673.

120. Sharpley, A.N. & Menzel, R.G. (1987). The impact of soil and fertilizer P on the environment. Advances in Agronomy, Vol. 41: 297-324.

121. Sharpley, A.N., Smith, S.J. & Bain, W.R. (1993). Nitrogen and phosphorus fate from longterm poultry litter applications to Oklahoma soils. Soil

Science Society of America Journal, Vol. 57 (No. 4): 1131-1137.

122. Shober, A.L. & Sims, J.T. (2003). Phosphorus restrictions and land application of biosolids: current status and future trends. Journal of Environmental Quality, Vol. 32 (No. 6): 1955-1964.

123. Sims, J.T. (1986). Nitrogen transformations in a poultry manure amended soil: Temperature and moisture effects. Journal of Environmental Quality, Vol. 15 (No. 1): 59-63.

124. Sims, J.T. & Wolf, D.C. (1994). Poultry waste management: Agricultural and environmental issues. Advances in Agronomy, Vol. 52: 1-83.

125. Sistani, K.R., Adeli, A., McGowen, S.L., Tewolde, H. & Brink, G.E. (2008). Laboratory and field evaluation of broiler litter nitrogen mineralization. Bioresource Technology, Vol. 99 (No. 7): 2603-2611.

126. Sistani, K.R., Brink, G.E., Adeli, A., Tewolde, H. & Rowe, D.E. (2004). Year-round soil nutrient dynamics from broiler litter application to three bermudagrass cultivars. Agronomy Journal, Vol. 96 (No. 2): 525-530.

127. Tan, X., Chang, S.X. & Kabzems, R. (2005). Effects of soil compaction and forest floor removal on soil microbial properties and N transformations in a boreal forest longterm soil productivity study. Forest Ecology and Management, Vol. 217 (No. 2-3): 158- 170.

128. Tate, D.F. (1994). Determination of nitrogen in fertilizer by combustion: Collaborative study. Journal of AOAC International, Vol. 77 (No. 4): 829-839.

129. Tekeste, M., Hatzghi, D.H. & Stroonsnijder, L. (2007). Soil strength assessment using threshold probability approach on soils from three agro-ecological zones in Eritrea. Biosystems Engineering, Vol. 98 (No. 4): 470-478.

130. Tilman, D., Hill, J. & Lehman, C. (2006). Carbon-negative biofuels from low-input highdiversity grassland biomass. Science, Vol. 314 (No. 5805): 1598-1600.

131. United States Department of Agriculture (USDA) Economic Research Service. (2009). Poultry and Eggs: Background. Retrieved from: http://www.ers.usda.gov/Briefing/Poultry/Background.htm

132. United States Department of Agriculture Soil Conservation Service (USDA SCS). (1989). Soil Survey of Claiborne Parish, Louisiana. USDA SCS, Washington, D.C.

133. Vose, J.M. & Allen, H.L. (1988). Leaf area, stemwood growth, and nutrition relationships in loblolly pine. Forest Science, Vol. 34: 547-563.

134. Wagner, G.H. & Wolf, D.C. (1999). Carbon transformations and soil

organic matter formation, In: Principles and Applications of Soil Microbiology, Sylvia, D.M., Fuhrmann, J.J., Hartel, P.G. & Zuberer, D.A. (Eds.). pp. 218-258. Prentice Hall Inc., Upper Saddle River, NJ.

135. Walkley, A. (1947). A critical examination of a rapid method for determining organic carbon in soils: Effect of variations in digestion conditions and of inorganic soil constituents. Soil Science, Vol. 63: 251-263.

136. Wallenstein, M.D., McNulty, S., Fernandez, I.J., Boggs, J. & Schlesinger, W.H. (2006). Nitrogen fertilization decreases forest soil fungal and bacterial biomass in three long-term experiments. Forest Ecology and Management, Vol. 222 (No. 1-3): 459-468.

137. Warren, S.L. & Fonteno, W.C. (1993). Changes in physical and chemical properties of a loamy sand soil when amended with poultry litter. Journal of Environmental Horticulture, Vol. 11 (No. 4): 186-190.

138. Weaver, T. (1998). Managing poultry manure nutrients. Agricultural Research, Vol. 46: 12-13. Will, R.E., Markewitz, D., Hendrick, R.L., Meason, D.F., Crocker, T.R. & Borders, B.E. (2006). Nitrogen and phosphorus dynamics for 13-year-old loblolly pine stands receiving complete competition control and annual N fertilizer. Forest Ecology and Management, Vol. 227 (No. 1-2): 155-168.

139. Williams, C.M, Barker, J.C & Sims, J.T. (1999). Management and utilization of poultry wastes. Reviews of Environmental Contamination & Toxicology, Vol. 162: 105-157.

140. Zarcinas, B.A., Cartwright, B. & Spouncer, L.R. (1987). Nitric acid digestion and multinutrient analysis of plant material by inductively coupled plasma spectrometry. Communications in Soil Science and Plant Analysis, Vol. 18 (No. 1): 131-146.

141. Zimmermann, B., Elsenbeer, H. & DeMoraes, J.M. (2006). The influence of land-use changes on soil hydraulic properties: Implications for runoff generation. Forest Ecology and Management, Vol. 222 (No. 1-3): 29-38.

142. Zinkhan, F.C. & Mercer, D.E. (1996). An assessment of agroforestry systems in the southern U.S.A. Agroforestry Systems, Vol. 35: 303-321.

143. Zou, X.M., Ruan, H.H., Fu, Y., Yang, X.D. & Sha, L.Q. (2005). Estimating soil labile organic carbon and potential turnover rates using a sequential fumigation-incubation procedure. Soil Biology & Biochemistry, Vol. 37 (No. 10): 1923-1928.

Chapter 7

PHYSIOLOGICAL AND BIOCHEMICAL MECHANISMS OF PLANT ADAPTATION TO LOW-FERTILITY ACID SOILS OF THE TROPICS: THE CASE OF BRACHIARIAGRASSES

T. Watanabe[1], M. S. H. Khan[2], I. M. Rao[3], J. Wasaki[4], T. Shinano[5], M. Ishitani[3], H. Koyama[6], S. Ishikawa[7], K. Tawaraya[8], M. Nanamori[1], N. Ueki[8] and T. Wagatsuma[8]

[1]Graduate School of Agriculture, Hokkaido University, Kita-ku, Sapporo, Japan

[2]Department of Soil Science, HMD Science and Technology University, Dinajipur, Bangladesh

[3]Centro Internacional de Agricultura Tropical (CIAT), A.A.6713, Cali, Colombia

[4]Graduate School of Biosphere Science, Hiroshima University, Higashi-Hiroshima, Japan

[5]National Agricultural Research Center for Hokkaido Region, Sapporo, Japan

[6]Faculty of Applied Biological Sciences, Gifu University, Gifu, Japan

[7]National Institute for Agro-Environmental Science, Tsukuba, Japan

[8]Faculty of Agriculture, Yamagata University, Tsuruoka, Japan

INTRODUCTION

Brachiaria species are the most widely planted tropical forage grasses in the world (Miles et al., 2004). For example, in Brazil alone, about 80 million hectares are planted to Brachiaria pastures (Macedo, 2005). They increase animal productivity by 5 to 10 times with respect to native savanna vegetation in the tropical areas of Latin America, thus representing a significant contribution to farmer's income (Rao et al., 1993). Although their origin is from the tropical areas of Africa, they are also used for livestock production in South-East Asia and Australia. Among them, Brachiaria decumbens cv. Basilisk, Brachiaria brizantha cv. Marandu, and Brachiaria ruziziensis cv. Kennedy have been more commonly utilized for livestock production in the tropics (Miles et al., 2004). Among the three grasses, B. decumbens is highly adapted to infertile acid soils, i. e., high level of tolerance to high aluminium (Al) saturation, low phosphorus (P) and low calcium (Ca) supply in soil (Louw-Gaume et al., 2010 a,b; Rao et al., 1995, 1996; Wenzl et al., 2001, 2003), but also highly sensitive to a major insect, spittlebugs (Miles et al., 2006) and produces mycotoxin after infection with Pithomyces chartarum (Andrade et al., 1978). B. brizantha cv. Marandu is highly resistant to spittlebugs, adapted to seasonal drought stress, highly responsive to fertilizer application but is not well adapted to low fertility acid soils (Miles et al., 2004, 2006). B. ruziziensis cv. Kennedy is sensitive to spittlebugs, performs better in well-drained fertile soils, has high forage quality but poorly adapted to low fertility acid soils (Ishigaki, 2010; Miles et al., 2004). B. decumbens and B. brizantha are generally tetraploid, apomicts while B. ruziziensis is diploid, sexual (Miles et al., 2004, 2006).

CIAT and its collaborators have an on-going breeding program to combine the desirable attributes from the three grasses (Miles et al, 2004, 2006). B. hybrid cv. Mulato is the product of three generations of crosses between B. ruziziensis, B. decumbens and B. brizantha. This grow well in low P, low fertility acid soils in both wet and dry seasons (Rao et al., 1998), and produces a large numbers of panicles with well synchronized flowering and good caryopsis formation, which leads to good-quality seed. Defining the specific physiological and biochemical mechanisms that are associated with greater adaptation to low fertility acid soils will contribute to developing rapid and reliable methods to select the phenotypes and to develop molecular markers for marker assisted breeding of brachiariagrasses. Developing superior Brachiaria hybrids from the on-going breeding programs that combine the desirable attributes including adaptation to major biotic and abiotic constraints, forage quality, and seed production will facilitate sustainable intensification of crop-livestock systems in the tropics (Miles et al., 2004, 2006; Rao, 2001 a,b). This chapter reviews the progress made in defining the physiological and biochemical mechanisms of adaptation of brachiariagrasses to low fertility acid soils.

There is limited knowledge on the comparative differences in Al resistance among B. decumbens, B. brizantha, B. ruziziensis and B. hybrids (Mulato and Mulato 2) grown in hydroponic system. Identification of plant attributes that contribute to greater ability to acquire nutrients under low pH, low P and high Al conditions is critical to develop brachiariagrases that are productive and persistant under infertile acid soil conditions. There have been some discussions on the validity of short-term screening technique that uses simple solution of Al and Ca to test the effect of Al on relative root elongation of young seedlings: whether the results obtained by this short-term screening technique can apply to the behaviour of older plants is under discussion (Ryan et al., 2011). However, significant positive correlations were observed on Al resistance of 15 cultivars of sorghum (Sorghum bicolor Moench) and 10 cultivars of maize (Zea mays L.)

with data obtained using short-term (1 day) screening and long-term screening technique using hydroponic system (Akhter et al., 2009). Similar results were also observed with 8 rice cultivars (unpublished data). In this chapter, we consider short-term vs long-term responses of several plant species including brachiariagrasses that differed widely in their level of Al resistance.

B. decumbens is known for very high level of Al resistance, however, the mechanisms responsible for this high level of Al resistance was not associated with exudation capacity of organic acid anions from root tips (Wenzl et al., 2001). It is important to define the specific mechanism(s) contributing to the high level of Al resistance in B. decumbens.

Higher level of Al exclusion was found in B. decumbens, however, the specific mechanisms related to Al exclusion are not known. Several mechanisms other than organic acid anion exudation were found and the related Al resistance genes have been reported (Huang et al., 2009; Yamaji et al., 2009; J.L. Yang et al., 2008; Z-B Yang et al., 2010). We found a new mechanism for higher level of Al tolerance of an Al-resistant rice cultivar based on higher abundance of sterols in plasma membrane (PM) lipids of root-tip cells (Khan et al., 2009). Rice is also known for its greater level of Al resistance than other cereal crops and its level of resistance was also not related to organic acid anion exudation (Ishikawa et al., 2000; Ma et al., 2002). Therefore, it is crucial to test the role of sterols in PM lipids of root-tip cells of brachiariagrasses.

High concentrations of phenolic compounds have been reported as one of the promising mechanisms to explain higher level of Al resistance of a forage legume, Lotus pedunculatus Cav. (Stoutjesdijk et al., 2001) and several common woody plants (Ofei-Manu et al., 2001).

This can be explained by higher complexing abilities of phenolic compounds with Al ions (Cornard & Merlin, 2002; Yoneda & Nakatsubo, 1998). Phenolic compounds have also been reported to be solubilized into lipid layer (Boija et al., 2006, 2007). Existence of phenolic compounds in lipid layer was found to make PM less fluid (Arora et al., 2000). Lipid layer with less fluidity will make PM less permeable even in the presence of Al ions (Khan et al., 2009). The contribution of phenolic compounds in root-tip portion to high level of Al resistance in B. decumbens is not known.

A major constraint to agriculture on tropical and subtropical soils is P deficiency (Fairhust et al., 1999). Applying large amounts of fertilizer to correct P deficiency is not feasible for most resource-poor farmers in developing countries. Thus the agricultural productivity becomes limited in the near future. Moreover, P fertilizer is receiving more attention as a nonrenewable resourse (Cordell et al., 2009; Steen, 1998). For sustainable P management in agriculture on tropical and subtropical soils, it is essential to define the mechanisms involved in making plants more efficient in P acquisition and use (Lynch, 2011; Ramaekers et al., 2010; Rao et al., 1999). P is a constituent of phospholipids (PL), nucleic acids, nucleosides, coenzymes, and phosphate esters in plants. P helps regulate plant metabolisms by controlling enzymatic activity through phosphorylation and/or dephosphorylation. To overcome P deficiency, plants develop several strategies, including the well-known one of secreting acid phosphatase (APase), ribonuclease (RNase), and organic acids into the rhizosphere to improve P availability in the soil (Duff et al., 1994; Green, 1994; Jones, 1998; Rao et al., 1999; Tadano et al., 1993; Wasaki et al., 2003a). Besides from acquiring P from the outside of the plant, APase is also recognized for its role in efficient utilization of absorbed P for metabolism (Duff et al., 1989, 1991). One of the mechanisms to increase P recycling ability is dependent on the activity of APase (Duff et al., 1991, 1994) and RNase (Howard et al., 1998) in the cell. The increase of both enzymes is reported and these enzymes are considered to utilize those P compounds that are stored in vacuole. The other mechanism is bypassing several metabolic pathways to reduce the usage of P molecule. Theodorou and Plaxton (1993) showed that P deficiency induces some glycolytic enzymes, such as phosphoenolpyruvate carboxylase (PEPC) and phosphoenolpyruvate phosphatase (PEPP). These catalyze the bypass reaction of pyruvate kinase (PK), which is responsible for regulating carbon flow from glycolysis to the TCA cycle. PEPC replenishes intermediates of the TCA cycle, and may help regulate both carbon and nitrogen metabolism, and P recycling under P deficiency. Kondracka and Rychter (1997) observed that, in P-deficient bean leaves, the rate of malate synthesis increases, and the accumulation of aspartate and alanine (products of PEP metabolism) is also enhanced. In early stages of P deficiency, the increased activity of PEPC and use of PEP in amino acid synthesis are probably the most important reactions for P recycling in bean leaves during photosynthesis. Thus, PEP metabolism by PEPC and PEPP, or PEP transport via PPT (which transports PEP from the cytosol into chloroplasts in leaves) may affect carbon distribution within the plant under P deficiency.

Under P-deficient conditions, the brachariagrasses improve their P acquisition by enhancing root growth, uptake efficiency, and ability to use poorly available plant P (Louw-Gaume et al., 2010b; Rao et al., 1999, 2001a). Although they have much lower internal requirements for P than do other grasses, they also show interspecific differences (Rao et al., 1996).

Recent advances in post-genomic studies have indicated that transcriptomic analysis is a useful tool for understanding gene expression networks. Some transcriptomic studies of low-P adaptation strategies have also been carried out using cDNA arrays (Hammond et al., 2003; Misson et al., 2005; Ramaekers et al., 2010; Uhde-Stone et al., 2003; Wang et al., 2002; Wasaki et al., 2003b, 2006; Wu et al., 2003). Although some aspects of plant strategies for coping with P-deficient conditions are understood, the majority are not; a broad view of gene expression is necessary to fully elucidate all of the mechanisms involved (Ramaekers et al., 2010). In this chapter, we review the progress in understanding the transcriptomic changes by P deficiency in rice plants, which is a model of Gramineae plants and also relatively tolerant of low P and low pH conditions. We use qRT-PCR for quantification of the effects of P deficiency on phosphohydrolases and carbon metabolism in rice leaves. We also review the progress in defining the physiological and biochemical bases of improved P use efficiency in B. hybrid (cv. Mulato).

MATERIALS AND METHODS

We randomly selected seven plant species, i.e., rice (Oryza sativa L. cv. Sasanishiki), maize (Zea mays L. cv. Pioneer 3352), pea (Pisum sativum L. cv. Kinusaya), barley (Hordeum vulgare L. cv. Manriki), tea (Camellia sinensis L. cv. Yabukita), siclepod (Cassia tora L.) and B. brizantha. Seeds of rice, maize, pea were soaked in tap water under aeration for 3 to 24 h depending on plant species. Seeds of siclepod were notched by a razor to facilitate germination and soaked in tap water for 12 h. Seeds of barley and B. brizantha were not soaked. These seeds were germinated on a nylon screen that was put on a polypropylene container filled with tap water under aeration at 27°C in a growth room. Seeds of tea were germinated in quarts sand wetted with deionized water, and then seedlings with roots approximately 1 cm long were transferred to the container described above. All the seedlings with roots approximately 5 cm long were used in the following experiments. In the short-term experiment, ten seedlings of each plant species were pretreated in 600 mL volume of a solution with (Al treatment) or without (control) 50 μ M $AlCl_3$ containing 0.2 mM $CaCl_2$ at pH 4.7 for 1 h. After measurement of the root length with a ruler, all the seedlings were transferred into a Al-free 0.2 mM $CaCl_2$ solution at pH 4.7. Root length was measured again after 24 h. For the tea plant, the root length was measured after 3 d because of the slower root elongation than in the others. Relative growth (%) in the short-term treatment with Al was calculated as the ratio of net root re-elongation of the primary root in the Al treatment to that in the control. In the long-term experiment, twenty seedlings of each plant species were precultured in a 54-L volume of nutrient solution at pH 5.2. The nutrient solution was composed of 1.43 mM NH_4NO_3, 0.7 mM $NaNO_3$, 0.13 mM NaH_2PO_4, 0.78 mM K_2SO_4, 1 mM $CaCl_2$, 0.6 mM $MgSO_4$, 36 μM $FeSO_4$, 9 μ M $MnSO_4$, 0.08 μ M $CuSO_4$, 0.03 μ M $(NH_4)_6Mo7O_{24}$, 18.5 μ M H_3BO_3, 1.5 μ

M $ZnCl_2$. After 1 week, the seedlings were transferred into the nutrient solution with same composition as above (control) or the nutrient solution containing 100 μ M Al and 10 μ M P in soluble form (Al treatment) at pH 5.2 or 4.5, respectively. The solutions were renewed weekly. The concentrations of Al and P, and pH of the solution were monitored every day and adjusted if required. Two months after the Al treatment, the plants were harvested and dried in a draft oven (60°C). Relative growth (%) in the long-term treatment with Al was calculated as the ratio of the dry weight of whole plant in the Al treatment to that in the control. The concentrations of K, P, Mg, and Ca were determined by ICP-AES (inductively coupled plasma-atomic emission spectrometry) after digestion of the plant samples using an acid mixture (HNO_3: $HClO_4$ = 5: 3, v/v). Extraction and determination of the phenolic compounds were carried out as described by Ofei-Manu et al. (2001). Seedlings of Brachiaria hybrid (B. ruziziensis Ger. & Ev. clone 44-06 × B. brizantha (A. Rich.) Stapf CIAT 36061, also known as cv. Mulato), Andropogon gayanus Kunth (CIAT 621), barley (cv. Ryofu) were transferred to 36-L containers containing nutreint solution with or without 0.37 mM Al (as $Al_2[SO_4]_3$). At the end of 10 days of treatment, the roots of seedlings from each treatment were sampled, dried in a forced-air oven at 80°C for 72 h, and then weighed. The dried samples were ground and digested with H_2SO_4-H_2O_2 for Al analysis by ICP-AES. Brachiaria seedlings were prepared as described above, and transferred to 36-L containers carrying the standard nutrient solution, but with 2.8 mM Al at pH 3.7 added, and left to grow for 1 month. The much higher Al concentration was used to ensure clear peaks in the ^{27}Al NMR spectrum. Even so, the Brachiaria seedlings grew well (data not shown). After treatment, roots were removed from the seedlings and washed, first with tap water, then with deionized water. The roots were grouped into three: Fraction (a), roots given the water washings only, and used to determine total amounts of Al and organic acid anions; and Fraction (b), roots were also washed with 0.1 M HCl for 5 min to remove apoplastic, soluble or loosely bound, components. Each fraction of Brachiaria roots was placed in a 10-mm-diameter NMR tube. $AlCl_3$ (0.1 M) solution was used as an external reference to calibrate the chemical shift (0 ppm). ^{27}Al NMR spectra were recorded, using a Bruker MSL400 spectrometer at 104.262 MHz. The spectra were obtained by using a frequency range of 62.5 kHz, a pulse width of 12 μs, a delay time of 0.16 ms, a cycle time of 0.5 s, and 4000 scans. For estimation of low P tolerance, we have selected Brachiaria hybrid (cv. Mulato) and rice (Oryza sativa L. cv. Nipponbare). For the seedling growth nutrient solution was prepared with 2.12 mM N (NH4NO3), 0.77 mM K (K2SO4:KCl = 1:1), 1.25 mM Ca ($CaCl_2 \cdot 2H_2O$), 0.82 mM Mg ($MgSO_4 \cdot 7H_2O$), 35.8 μM Fe ($FeSO_4 \cdot 7H_2O$), 9.1 μM Mn ($MnSO_4 \cdot 4H_2O$), 46.3 μM B (H_3BO_3), 3.1 μM Zn ($ZnSO_4 \cdot 7H_2O$), 0.16 μM Cu ($CuSO_4 \cdot 5H_2O$), 0.05 μM Mo $(NH_4)_6Mo_7O_{24} \cdot$

4H$_2$O), with 6 µM P (NaH$_2$PO$_4$· 2H$_2$O) and pH was maintained at 5.2 for 1 week preculture. Then phosphorus level was changed as 0, 6 and 32 µM for 2 weeks. After sampling, samples were freeze-dried, then P fractionation was carried out based on Schmidt-Thannhauer-Schneider method. Samples were also used for measurements of APase and RNase activities. ^{14}CO$_2$ was generated by adding 30% PCA into NaH$_{14}$CO$_2$ (18.5 kBq), then applied to plants for 5 min in a vinyl package under natural light condition. After sampling, samples were fractionized by using column method to obtain organic acids, amino acids, and sugars fraction, then ^{14}C content in each sample was determined by using scintillation counter. In transcriptomic analyzes and subsequent molecular analysis, we have selected rice (Oryza sativa L. cv. Michikogane) and the growth condition was same as shown before except for the P treatment was done using 0 or 32 µM levels. Total RNA was extracted from frozen samples using a sodium dodecyl sulfate (SDS)- phenol method. The real-time PCR was performed by the LightCycler™ system (Roche) with the LightCycler DNA Master SYBR Green I kit for PCR (Roche), and the TaqStartTM antibody (Clontech) was used for the repression of unspecific amplification.

Mechanisms of Al Resistance in Brachiariagrasses

Differences in Al Resistance and Nutrient Acquisition among Crops and Brachiariagrasses

Differences in Al resistance among crops and B. brizantha are shown in Fig.1 (the left panel). Al resisance was ranked as follows: B. brizantha > rice > tea > maize > pea, siclepod > barley. The order of Al resistance in the short-term experiment was well correlated with that of the long-term experiment in spite of marked difference in the treatment conditions, i.e., duration of Al treatment, Al concentration, composition of co-existing nutrients, pH, etc. ($R2 = 0.785$ [$p < 0.01$], right side figure of Fig.1) (Ishikawa et al., 2000).

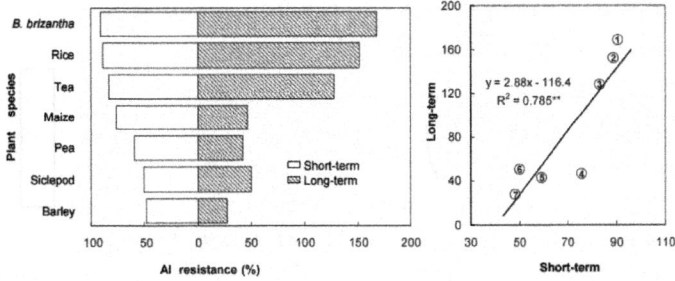

Figure 1: Differences in Al resistance among crops and Brachiaria brizantha. The left panel shows Al resistance values in long-term experiment and

short-term experiment. The right figure shows the relationship between Al resistance in long-term experiment and that in short-term experiment. ① *B. brizantha*, ② rice, ③ tea, ④ maize, ⑤ pea, ⑥ siclepod, ⑦ barley.

Although B. brizantha is relatively less adapted than B.decumbens to infertile acid soils, B. brizantha was found to be superior in its level of Al resistance to the other crops tested.

We also compared nutrient acquisition ability of B. brizantha with that of crops. Nutrient acquisition ability was compared by quantifying relative nutrient status in shoots in Al treatment to that in low pH · low P treatment. Among the 5 species compared, B. brizantha was found to have superior nutrient acquisition abilities for K,P and Mg (Fig. 2). However, B. brizantha showed the lowest acquisition ability for Ca. B. brizantha can acquire highest amounts of K, P and Mg even under high Al, low P and low pH conditions. B. brizantha may have the lowest requirement of Ca for normal growth even in acid soil conditions. This greater ability of B. brizantha to acquire nutrients under simulated acid soil conditions could be responsible for its greater vigour in acid soil conditions during the pasture establishment phase (Rao et al., 1996).

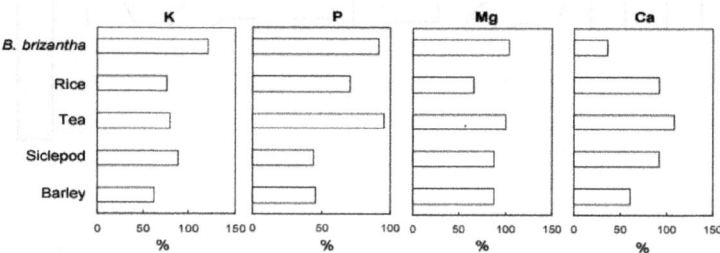

Figure 2: Relative nutrient status of each nutrient concentration in Al treatment to that in control treatment.

Differences in Al Resistance, Al Accumulation and Plasma Membrane Permeability among Brachiariagrasses

Adaptive responses of several brachiariagrasses to infertile acid soils have been identified and described by previous research (Louw-Gaume et al., 2010 a,b; Rao et al., 1995, 1996; Wenzl et al., 2001, 2003). We compared Al resistance, Al accumulation (hematoxylin staining method [Wagatsuma et al., 1995] and PM permeability (FDA-PI fluorescence staining method [Ishikawa et al., 2001]) among 4 brachiariagrasses in short-term experimental conditions that were described in the former section, together with most Al-resistant rice cultivar Rikuu-132 (Khan et al., 2009) as a reference plant species. Al resistance was ranked as follows: B. decumbens, B. hybrid, B. brizantha > B. ruziziensis > Rikuu-132 (Fig. 3). Al resistance of B. hybrid (cv. Mulato) and B. brizantha was found to be comparable to that of B.decumbens which has been ranked as the most Al-resistant brachiariagrass (Wenzl et al., 2001). Although Al resistance of B. ruziziensis was found to be markedly lower than B. decumbens (Wenzl et al., 2001, 2003), its resistance level was higher than that of the most Alresistant rice cultivar. It was suggested that the highest Al resistant phenotype of B. hybrid may be ascribed to the Al resistance genes from B. decumbens or B. brizantha and not from B. ruziziensis. Al accumulation was localized mainly within 1- mm root-tip portion and its concentration corresponds reversely to Al resistance order: the least Al accumulation was recognized for the most Al-resistant B. decumbens. PM lipid layer was less permeable to Al in brachiariagrasses than in the most Al-resistant rice cultivar and its less permeable PM characteristic was localized mainly within 1- mm root-tip portion (Fig.4).

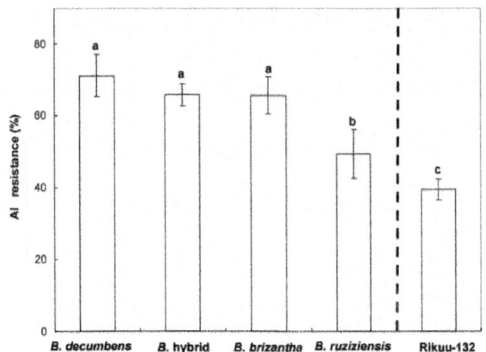

Figure 3: Differences in Al resistance among brachiariagrasses and Al-resistant rice cultivar Rikuu-132.

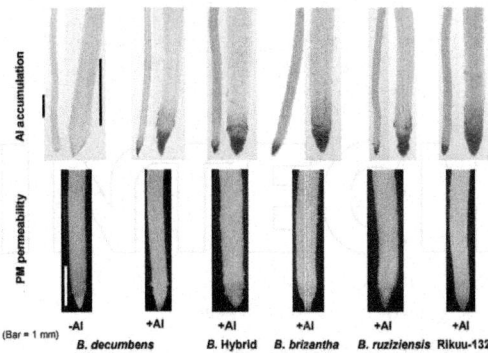

Figure 4: Differences in Al accumulation and PM permeability among brachiariagrasses and Al-resistant rice cultivar (Rikuu-132).

Permeability of PM was negatively associated with Al resistance: the least PM permeabilization was observed with the most Al-resistant B. decumbens. The less permeability of PM and the lower Al accumulation in root-tip portion in Al-resistant brachiariagrass agree well with the former results which have been recognized in Alresistant plant species, cultivars, or lines (Ishikawa et al., 2001; Ishikawa & Wagatsuma, 1998; Wagatsuma et al., 2005).

Lipid Composition and Phenolics Concentration in Root-Tip Portion of Brachiariagrasses in Relation to Al Resistance

The lower ratio of PL to sterols (S) (PL was measured by molybdenum blue spectrophotometric method after extraction with isopropanol-chloroform-H_2O [2:2:1] ; S was measured by ortho-phthalaldehyde colorimetric method after extraction with dichloromethane-methanol [2:1] [Khan et al., 2009]) in root-tip portion was found to be beneficial for the less permeability of PM in the presence of Al, which agrees with the results of rice cultivars (Khan et al., 2009). In the more proximal root region (0-10 mm from root apex), the ratio of PL to S for B. decumbens was higher than that of B. ruziziensis, but on the contrary, it was lower in root-tip portion (0-2 mm from root apex) under Al treatment conditions (Fig.5).

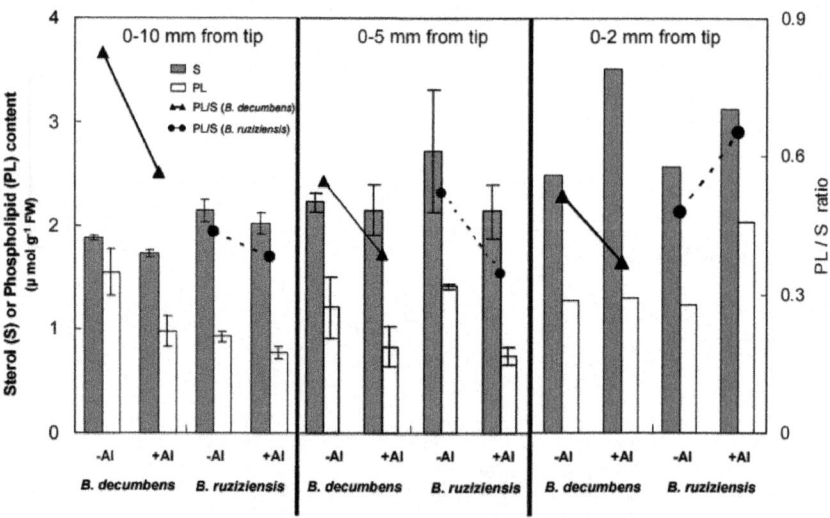

Figure 5: Sterol or phospholipid content in the different segment of root of brachiariagrasses treated with or without Al in solution.

The lower negativity of PM surface that was associated with the lower ratio of PL to S in root-tip portion could contribute to lower permeability of PM to Al. This is highly consistent with Gouy-Chapman-Stern model of Al rhyzotoxicity (Kinraide, 1999). In case of rice cultivars (Khan et al., 2009), wheat lines, triticale lines, maize cultivars (unpublished data), lipid compositional difference in connection with Al resistance were recognized in root-tip portion of 0-10 from root apex. However, in brachiariagrasses, lipid compositional difference was related with Al resistance only in root-tip portion of 0-2 mm from root apex. We suggest that the high level of Al resistance in brachiariagrass is extremely localized at the root tip.

It is known that phenolic compounds can be solubilized into lipid layer, and the lipid layer solubilized with phenolic compounds is transformed into the less fluid layer (Arora et al., 2000; Boija & Johansson, 2006). Higher concentration of phenolic compounds was detected in root-tip portion (0-5 mm and 0-2 mm from root apex) of B. decumbens than in B. ruziziensis (Fig.6).

The concentration of phenolic compounds was lower in the portion of 0-10 mm from root apex than that of the shorter part from root apex (data not shown). Phenolic compounds have been detected basically in the cell wall, vacuole, and to a small extent in the cytoplasm and nucleus (Hutzler et al., 1998). At around neutral pH of cytosol, the binding affinity to Al ions was significantly higher for phenolic compounds than for organic acids (Ofei-Manu et al., 2001). Higher concentration of phenolic compounds is considered to be more effective for greater detoxification of Al ions in cytosol of B. decumbens. Additionally, higher inclusion of phenolic compounds into PM lipid layer may be more favourable for making the PM less permeable in the presence of Al ions, although there are no reports on the inclusion of phenolic compounds in plant lipid layer. Several quantitative and qualitative changes in PM may contribute to superior level of Al resistance in B. decumbens. These include: higher proportion of S relative to PL, higher concentration of phenolic compounds in cytosol, and higher inclusion of phenolic compounds in PM lipid layer in root-tip portion. These changes may contribute to an extremely strong PM lipid layer which plays a key role in exclusion of Al and high level of Al resistance in B. decumbens. Direct demonstration of the existence of phenolic compounds in PM lipid layer will be an important task for the future research.

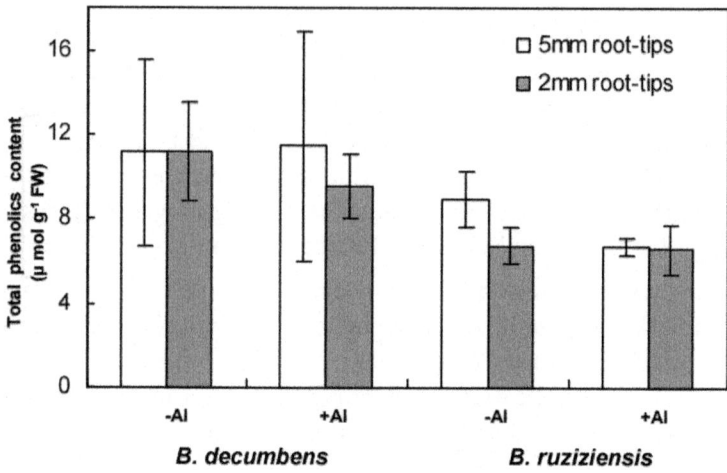

Figure 6: Total phenolic compounds in root-tips of two brachiariagrasses treated with or without Al in solution.

Mechanisms of High Level of Al Resistance in B. Hybrid (cv. Mulato)

B. hybrid showed higher resistance to Al similar to B. decumbens. When B. hybrid seedlings were grown with an extremely high concentration of Al (0.37 mM) for 10 days, no growth inhibition was observed (Fig. 7). Moreover, Al application did not inhibit the uptake of nitrogen (N), P and K in B. hybrid. Andropogon gayanus, a poaceous pasture grass, is also very resistant to Al and Al application significantly increased Al concentration in both leaf and root of this species (Fig. 8). In B. hybrid, by contrast, significant increase in Al accumulation was also observed in root but not in leaf. This indicates that some mechanisms restricting Al translocation from roots to shoots should exist in B. hybrid. The 27Al NMR spectrum obtained from intact roots showed several peaks downfield at 10-20 ppm (Fig.9a), suggesting that most of the soluble Al in roots makes complexes presumably with organic acid anions (Fatemi et al., 1992; Kerven et al., 1995). Since the ^{27}Al NMR spectrum did not change after removing soluble and/or loosely bound apoplastic Al, these Al complexes in roots were likely to be localized in the symplast of cells. In many Al-accumulator species, leaves and roots with high concentration of Al are detoxified by organic ligands, such as Aloxalate in Melastoma malabathricum (Watanabe et al., 1998, 2005). The same mechanisms are considered possible in roots of B. hybrid. It has been reported that Cd translocation from roots to shoots is restricted by Cd isolation in root vacuoles (Miyadate et al., 2010). Al in the B. hybrid may also compartmentalize in root vacuoles and, thus, may not be translocated to shoots.

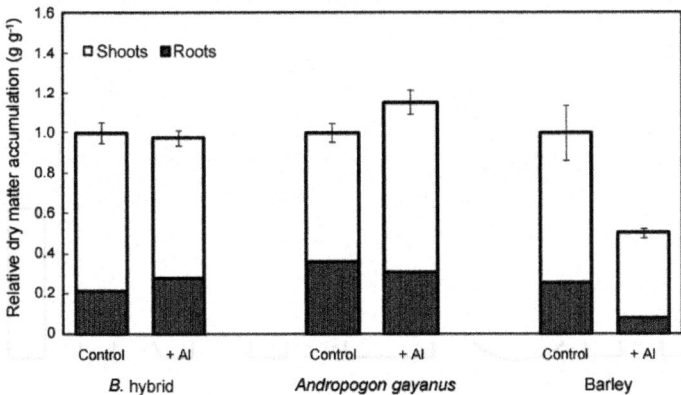

Figure 7: Effects of Al toxicity on growth of Brachiaria hybrid, Andropogon gayanus, and barley. Growth was expressed as the relative dry matter accumulation (i.e. [dry weight after treatment – initial dry weight in each treatment]/[dry weight after treatment-initial dry weight in control treatment]).

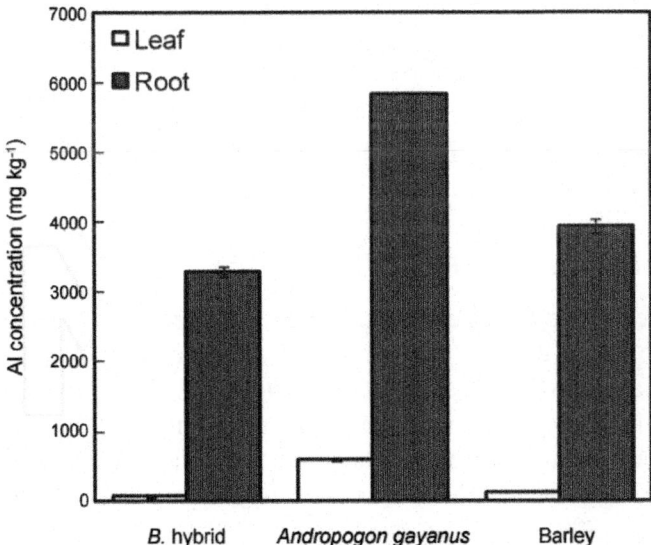

Figure 8: Concetration of Al in Brachiaria hybrid, Andropogon gayanus, and barley after the Al treatment.

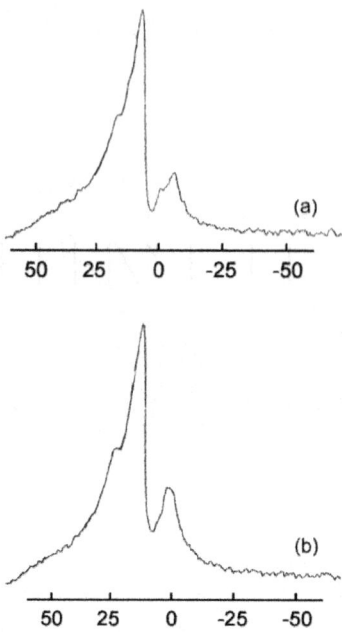

Figure 9: 27Al NMR spectra in intact roots (a) and roots after removing soluble and/or loosely bound apoplastic Al by 0.1 M HCl (b). AlCl3 (0.1 mM) was used as an external reference to calibrate the chemical shift (0 ppm).

Mechanisms of Low P Tolerance in Brachiariagrass Comparing with those in Rice

Low P Tolerance of B. Hybrid

As indicated above, Brachiaria species are well adapted to the low-fertility acid soils of the tropics because they are highly tolerant of high Al and low supplies of P and Ca (LouwGaume et al., 2010b; Rao et al., 1995, 1996, 2001b; Wenzl et al., 2001). They have lower internal requirements for P than other grasses because they are able not only to acquire P with their extensive root systems but also to use the acquired P more efficiently for growth and metabolism (Rao et al., 1996, 1999). However, mechanisms of P-use efficiency are relatively less known in plants, including B. hybrid. Because carbon metabolism is well known to be affected by the P status in plant tissue (Rao, 1996), we studied low-P-tolerance mechanisms, in terms of P recycling and carbon metabolism, in the B. hybrid comparing them with those of rice (Nanamori et al., 2004). B. hybrid and rice plants were cultivated in nutrient solutions with or without 32 μM P.

The data obtained on growth parameters and nutrient status are shown in Fig. 10. When P supply in the nutrient solution was low, root:shoot ratio increased, especially in the B. hybrid. We found that, for the B. hybrid, vigorous root growth is a mechanism for acquiring larger amounts of P from low P conditions. This finding was supported by the high levels of N concentration found in B. hybrid roots, while P concentration in B. hybrid leaves was significantly lower than that of rice leaves. Lower P concentration in B. hybrid leaves may indicate that the B. hybrid uses P more efficiently to sustain active metabolism for dry matter production. The P concentration of B. hybrid was quite low (0.44 and 0.56 mgP/gDW in roots and leaves, respectively) and less than rice, which is also known as a low P tolerant plant. Results on the fractionation of P compounds indicated that acid-soluble Pi accounted for about half of the total P in the B. hybrid (Fig. 11). Results on the Pi:total P ratio in B. hybrid leaves under P deficiency indicate that the B. hybrid can survive with extreme low intracellular Pi concentration. This may be due to rapid turnover of other organic P pools under P-deficient conditions.

Chapin and Bieleski (1982) studied the impact of mild P stress on P fractions in relation to plant growth in barley and low-P-adapted barleygrass. They found that barleygrass had a higher proportion of Pi at each level of P supply. They explained this as a consequence of slower growth in barleygrass and higher P status rather than any inherent difference in mechanism. However, in our study, the higher Pi proportion in the B. hybrid, compared with that of rice, coincided with lower P concentrations, as explained above. We, therefore, speculate that recycling of internal organic P compounds could be an important mechanism of P-use efficiency in the B. hybrid.

Bosse and Köck (1998) have shown activities of APase and RNase were induced during P deficiency, and that this induction is associated with P turnover in plants. In our study, APase and RNase activities were both strongly induced in both rice and B. hybrid by P deficiency (Fig. 12). Induction of APase activity was markedly higher in roots under Pdeficient conditions. Duff et al. (1994) reported the existence of extracellular APase in roots, where it is localized mainly in apical meristems and outer and surface cells. It is involved in hydrolyzing and mobilizing Pi from organic phosphates in the soil for plant nutrition. The induction of APase in roots may also be associated with excretion. Bosse and Köck (1998) suggested that the increase in activity of phosphohydrolases was a specific response to the decline of cellular available Pi in Pi-starved tomato seedlings. Although Pi in roots was lower than in shoots of both test crops, it was impossible to account for the difference of APase induction between roots and shoots only by the difference in intracellular Pi concentration. Thus, we suggest that some other signal transduction pathway must be operating between roots and shoots against P starvation in the cell. APase activity in shoots was greater in the B. hybrid than in rice, suggesting the possibility of rapid P turnover in the B. hybrid. This may enable the B. hybrid to survive under low P conditions. APase may not be a major mechanism for scavenging or acquiring P because differences in APase induction could not sufficiently account for the diverse growth response of genotypes of both common bean and maize plants under P deficiency (Yan et al., 2001; Yun & Kaeppler, 2001).

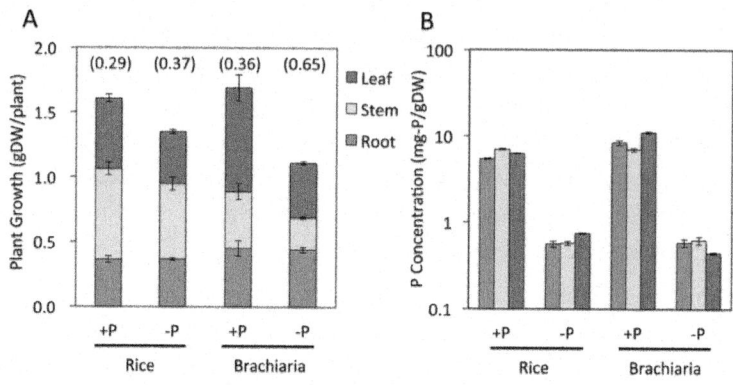

Figure 10: Growth (A) and P concentration (B) of rice and Brachiaria hybrid plants grown under P sufficient and deficient conditions. Error bars indicate S.E. (n = 3).

However, we observed in our study that APase activity was induced by P deficiency and the activity seems to be correlated well with P-use efficiency, as indicated by the lower value of total P concentration, so that the function of APase in adaptation to low-P conditions should not be underestimated. RNase activity was also high in roots under P-deficient conditions (Fig. 12). Nürnberger et al. (1990) and Löffler et al. (1992) showed that both extracellular and intracellular RNase activities were induced in tomato-cell culture under P deficiency. Extracellular RNase could help degrade the RNA from senescing cells that have been either damaged or lysed, and also help degrade any RNA that might be present in the rhizosphere. Thus, the high RNase activity in roots may be associated with secretion similar to APase. RNase activity in shoots was also greater in the B. hybrid than in rice, indicating that RNase also contributes to rapid P turnover. Glund et al. (1990) showed that, in the relationship between Pi concentration and RNase activity, induction of RNase under P starvation occurs when the intracellular content of P is very high.

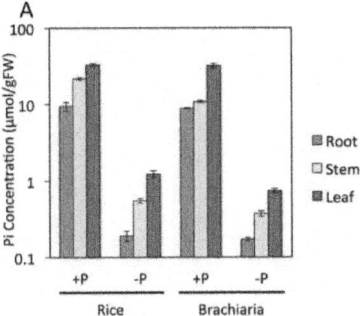

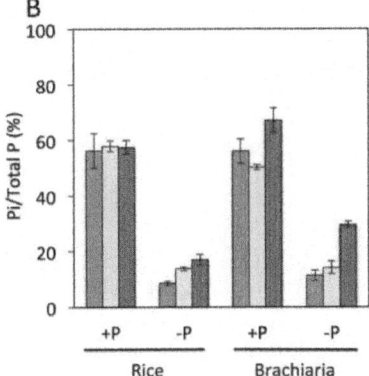

Figure 11: Pi concentration (A) and ratio of Pi to total P (B) of rice and Brachiaria hybrid plants grown under P sufficient and deficient conditions. Error bars indicate S.E. (n= 3).

The above studies indicate that phosphohydrolases, such as APase and RNase, were induced by P deficiency as a P-recycling system. Coinciding with such a mechanism, it is possible that carbon metabolism could also be altered under P deficiency. We therefore studied photosynthate partitioning under P deficiency, tracing photosynthetically fixed ^{14}C in leaves. In rice, photosynthates mainly distributed to sugars, which consist of sucrose, indicating that rice enhanced the sucrose synthesis pathway (Fig. 13). The mRNA accumulation of sucrose phosphate synthase (SPS) also increased as mentioned previously. Hence, sucrose concentration in rice leaves was remarkably high (Fig. 13). The ^{14}C distribution proportion to sugars increased with P deficiency. Enhanced sucrose synthesis in rice leaves through P deficiency may contribute to P recycling because P is liberated during sucrose synthesis (Rao, 1996). However, sucrose catabolism was restricted because the ^{14}C distribution ratio to amino acids and organic acids decreased with P deficiency and with carbohydrate accumulation (Fig. 13).

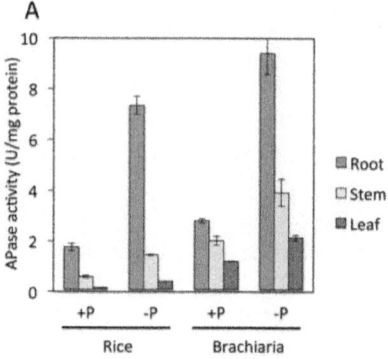
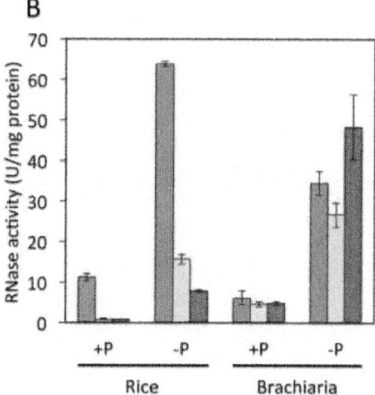

Figure 12: Acid phosphatase (A) and ribonuclease (B) activities in leaves of rice and Brachiaria hybrid grown under P sufficient and deficient conditions. Error bars indicate S.E. (n = 3).

Sucrose synthesis may, therefore, not contribute efficiently to P recycling. However, the 14C distribution proportion to sugars in the B. hybrid was not as marked as in rice (Fig. 13), and the effect of P deficiency was smaller. The ^{14}C distribution ratio to amino acids and organic acids in the B. hybrid was greater than in rice, and slightly affected by P deficiency. The decrease of total organic acids and carbohydrates in B. hybrid leaves under P deficiency suggests that the B. hybrid can sustain active amino acid and organic acid pathways with enhanced sugar catabolism, using P efficiently under P deficiency. PK and its bypassing enzymes catalyze the PEP-consuming reaction in leaves, with PEPP activity increasing by a factor of 5.6 to 6.0 with P deficiency. This induction of PEPP is likely to be associated with P recycling, as Duff et al. (1989) suggested. PK was also induced by P deficiency, but not significantly in the B. hybrid. PEPC activity was slightly induced by P deficiency in rice but not in the B. hybrid. The decrease of PEPC activity in B. hybrid leaves would result from reduced net photosynthesis under P deficiency. Kondracka and Rychter (1997) suggest that facilitating the PEP metabolism may be important in view of the P recycle. PEPC and PEPP are considered to function in P recycling as PK-bypass pathways. If these enzyme activities are induced in P recycling, then the carbon flow to the TCA cycle is expected to increase. The ^{14}C distribution ratio to amino acids and organic acids increased slightly in the B. hybrid with P deficiency (Fig. 13), indicating that these bypassing enzymes may function to facilitate carbon flow to the TCA cycle. However, in rice, the 14C distribution ratio to amino acids and organic acids decreased with P deficiency. Therefore, the PK bypassing mechanism under P deficiency may not contribute to facilitating the carbon flow to the TCA cycle in rice. In addition to the PK-bypassing mechanism, carbon export from chloroplast to cytosol via the triose-phosphate translocator (TPT) may be a process that significantly affects carbon partitioning under P deficiency (Rao, 1996). When plants are starved for P, triose-P exports from chloroplast to cytosol via TPT, and subsequent sucrose synthesis in the cytosol is likely to be restricted (Rao, 1996). The ^{14}C distribution ratio to sugars and to residue, which mainly consists of sucrose and starch respectively, increased with P deficiency in both rice and B. hybrid (Fig. 13), indicating that restriction of triose-P exports from chloroplast to cytosol via TPT may not occur.

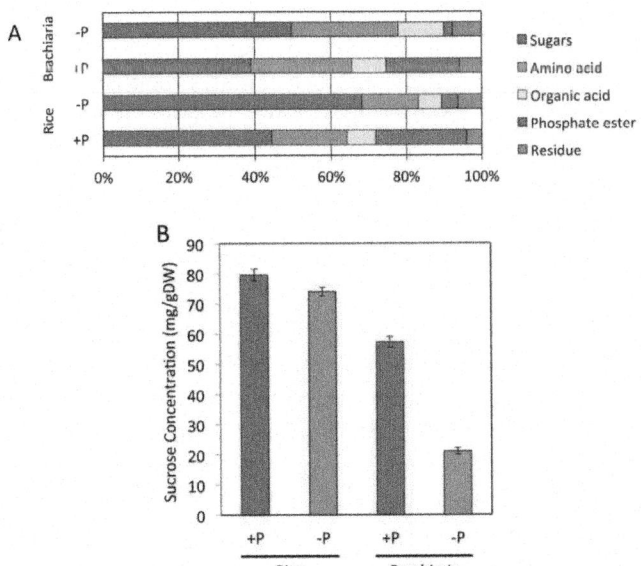

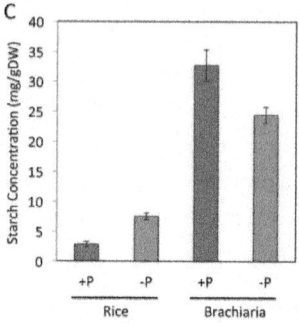

Figure 13: Photosynthetically assimilated 14C distribution (A), sucrose and starch concentration(B and C, respectively) in leaves of rice and Brachiaria hybrid. Error bars indicate S.E. (n = 3).

Transcriptomic Analysis of P Deficient Rice Plants

Rice (Oryza sativa L. ssp. japonica) plants were germinated and cultured in nutrient solutions containing 0 and 32 µM NaH_2PO_4 for –P and +P treatments, respectively. The seedlings were cultivated for 9 days after transplanting. Total RNA of leaves and roots was used for transcriptomic analyzes by using cDNA arrays (Wasaki et al., 2003b, 2006). As the response of rice roots, there were 15 up-regulated genes in the short-term (24 h) and 86 in the longterm (9 d) treatment with –P, whereas there were 23 and 97 down-regulated genes in the two treatments, respectively. The number of genes regulated (especially down-regulated genes) by the P deficiency was lower in leaves than in roots. There was one up-regulated gene in the short-term (24 h) and 48 genes in the long-term (9 days) –P treatments, whereas there were eight and four down-regulated genes in these two treatments, respectively. None of the genes were regulated in a similar manner between the short and long-term –P leaves. This result suggests that the responses in P-deficient rice leaves are different between shortand long-term treatments, whereas those of roots are relatively similar. OsPI1 (Oryza sativa phosphate-limitation inducible gene 1; Wasaki et al. 2003c), showed the most significant increase in its transcription in the long-term –P treatment, in both the roots and leaves. This gene was classified as a member of TPSI1/Mt4 family, which is the P-deficient responsive non-coding RNA. The SqdX-like gene, a homolog of sulfoquinovosyl diacylglycerol (SQDG) synthesis related genes, was up-regulated significantly in the –P roots. P deficiency enhances dynamic lipid reconstruction and causes SQDG or galactolipids accumulation and expression of a related gene in leaves (Essigmann et al., 1998; Nakamura et al., 2009). Because SQDG has the ability to substitute for PL, it was suggested that the increase of SQDG synthesis is available for the efficient use of P in the membrane (Essigmann et al., 1998). Four genes related to P metabolism were induced in leaves by long-term −P treatment. Inorganic pyrophosphatase and a phosphatase probably contributed to the maintenance of Pi concentration in the tissue by the direct production of Pi from organic phosphate compounds. It was concluded that the function of inorganic pyrophosphatase was common in both roots and leaves, because expression was induced in both organs by long-term −P treatment. Both bi-functional nuclease and S-like RNase expression levels were increased by the −P conditions; their contribution is to produce monomeric nucleotides as substrates for phosphatases (Duff et al., 1994; Green, 1994; Palma et al., 2000).

Many genes involved in polysaccharide metabolisms were up- and down-regulated in leaves by long-term −P and P re-supply treatments, respectively. It is probable that the upregulation of ADP-glucose pyrophosphorylase, which is a key enzyme of starch synthesis, and starch synthetic enzymes such as starch branching enzyme and starch synthases, induces the accumulation of starch in leaves under −P conditions. In fact, there are many reports of the accumulation of starch in the chloroplasts of P-deficient rice and other plants (Ciereszko & Barbachowska, 2000; Fredeen et al., 1989; Nomura et al., 1995; Qui and Israel, 1992; Rao et al., 1993; Usuda & Shimogawara, 1991). We concluded that the starch accumulation in leaves grown under P-deficient conditions was caused by the disruption of the export of triose phosphate from the stroma by the Pi translocator (Nátr, 1992). Nátr (1992) also noticed the liberation of Pi by the enhancement of starch synthesis. Because starch synthesis and the induction of Pi utilizing enzymes are synchronized, it is a reasonable speculation that the starch accumulation in the P-deficient leaves is a result of the maintenance of the internal Pi concentration.

Fig. 14 shows a summary of metabolic changes based on the regulation of gene expression in the leaves and roots of rice exposed to −P stress. Some important metabolic changes in roots by −P are suggested, namely: (1) acceleration of carbon supply for organic acid synthesis through glycolysis; (2) alteration of lipid metabolism; (3) rearrangement of compounds for cell wall; and (4) changes of gene expression related to the response for metallic elements such as Al, Fe and Zn. The major responses in leaves were involved in internal P utilization. The response in leaves seems to be less dramatic than that in roots; however, it is probable that an important function is regulated in shoots, such as the regulation of the novel TPSI1/Mt4 gene family (Burleigh & Harrison, 1999), which contains rice OsPI1 (Wasaki et al., 2003c).

Bypass Pathways in Rice for P use Efficiency in Plant

From our previous study using microarray on P deficient rice, we found that several genes relating C and P metabolism in chloroplast changed their expression level. One of them is phosphoenolpyruvate/phosphate translocator (PPT), and it showed enhancement under phosphorus deficient condition. PPT transports PEP into the chloroplast and antiports Pi to cytosol (Hausler et al., 2000), the role of PPT under P deficient condition is considered to supply substrate for the shikimate pathway. There exit another phosphate transporter; triose phosphate translocater (TPT) on chloroplast membrane which loads triose phosphate into cytoplasm and antiports phosphate into chloroplast. These two phosphate translocators are considered to regulate the phosphate level in the chloroplast.

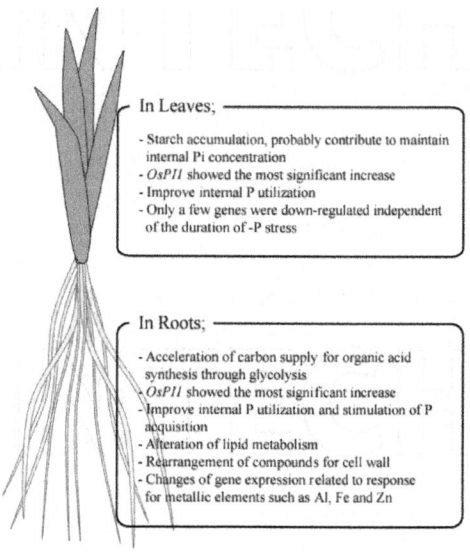

Figure 14: Summary of plant responses to phosphorus deficiency in shoot and root tissue.

As comprehensive analysis of each pathway using intact plant has not reported, we evaluated it by using quantitative real time PCR (qRT-PCR) to determine the expression level of each gene under P deficient condition (Shinano et al., 2005). While the expression level of mRNA is not simply representing the activity of those enzymes corresponding to the gene, obtained information will be very useful to consider plant response to P deficiency.

Gene Coding Key Enzyme of Sucrose Synthesis

The synthesis of sucrose will liberate phosphate from intermediate compounds, thus it is expected that the level of mRNA for SPS increased with -P treatment. SPS exist in cytosol of mesophyll cell and the combined reaction of SPS and sucrose phosphate phosphatase is main route for sucrose synthesis. That is, during sucrose synthesis, one molecule of Pi is liberated from sucrose phosphate. The –P treated plants first uses Pi stored in vacuole but after they used up all Pi in vacuole, cytosolic Pi content became lower. Then the plants with lower Pi concentration may facilitate sucrose synthesis and excrete Pi from sugar phosphate to keep up the Pi concentration in cytosol of mesophyll cell. Sucrose content in phosphate starved plant varies with species. In common bean and sugar beet, leaf sucrose content increased by P deficiency (Ciereszko & Barbachowska, 2000; Rao et al., 1990), although in leaves of Arabidopsis it decreased. Our results indicate that rice increases sucrose synthesis with P deficiency.

Genes Coding Candidates for Glycolytic bypass Enzymes

NADP dependent glyceraldehyde 3-phosphate dehydrogenase (NADP-G3PDH) instead of NAD dependent G3PDH (NAD-G3PDH), and PEPC instead of PK are expected to play alternative pathways to regulate carbon flow under P deficient condition. In rice leaves under P deprivation, we did not see any increase in relative expression of NADP-G3PDH, which is known as P starvation inducible bypassing enzyme for NADG3PDH in Brassica nigra (Duff et al., 1989). On the contrary, NAD-G3PDH relative expression was significantly high in –P plants at 21 days. In the level of gene expression argument, this result may suggest NADP-G3PDH is not working as glycolytic bypass in rice plant. The lack of induction of nonphosphorylating pathway was also seen in other plant species, such as S. minutum (Theodorou et al., 1991) and A. brevipes (Guerrini et al., 2000). Also PEPC and PK have the relationships of glycolytic bypass induced under P deficiency (Li & Ashiharam, 1990). Even though PEPC was thought to be catalyzed with the alternative pathway of PK under P deficiency (Li & Ashiharam, 1990), relative expression of both genes was increased by –P treatment. Increase of PEPC expression by P deficiency is also known in lupin.

Genes Encoding Chloroplast Membrane Transporters

Precise value of Pi concentration in cytosol and chloroplast is not known while it is suggested that the value is between 10 to 15 mM in cytosol (Mimura, 1999) and 20 to 35 mM in chloroplast (Diez & Heber, 1984). This indicates that higher requirement for maintaining Pi level in chloroplast rather than in cytosol, and low P condition is expected to increase the level of TPT and in versa in PPT. While the expression level of TPT was not changed by P deficient condition, the expression level of PPT increased dramatically. When one molecule of P is transported into chloroplast as PEP while exporting one

Pi, the incorporated PEP is decomposed in the chloroplast thereby having no net change in the P level of the chloroplast. We assumed that the role of PPT is increasing the PEP metabolism and makes a cycle from primary photosynthate synthesized in chloroplast and metabolized in cytosol with glycolysis then re-enter chloroplast with PEP then decomposed to release Pi in the chloroplast. From the analysis of rice microarray, PKp (plastid type PK) and shikimate kinase expression were enhanced under P deficiency. These results indicate physiological adaptation to incorporate PEP into chloroplast to support photosynthetic carbon flow and synthesis of secondary metabolic compounds. Recently, another type of phosphate transporter (PHT2; 1, which has high homology with Na^+/Pi symporter of fungi) was reported (Versaw & Harrison, 2002). There is need to evaluate how these transporters are operated to regulate phosphate flux within these subcellular organs.

CONCLUSIONS

Brachiariagrasses are highly adapted to infertile acid soils, however, the physiological and biochemical mechanisms responsible for their superior adaptation have not yet been fully defined. This chapter summarizes the recent progress towards this objective. Comparative differences in Al resistance among 4 brachiariagrasses and 6 reference plant species were analyzed, and the following order of Al resistance was observed: B. decumbens, B. hybrid, B. brizantha > B. ruziziensis > rice (the most Al-resistant cultivar Rikuu-132) > tea (cv. Yabukita) > maize (cv. Pioneer 3352) > pea (cv. Kinusaya) > siclepod > barley (cv. Manriki). The order of Al tolerance in the short-term experiment with exposure to Al (1-h of 50 μ M $AlCl_3$ in 0.2 mM $CaCl_2$ followed by 24-h of Al-free 0.2 mM $CaCl_2$) was well correlated with that in the long-term exposure experiment (2 months of Al treatment with full nutrients) in spite of the differences in the treatment conditions, i.e., duration of Al treatment, Al concentration, composition of co-existing nutrients, and pH. Short-term Al resistance screening technique is accepted to be useful for the evaluation of Al resistance in spite of the simple composition of the treatment solution, considering the positive correlation data obtained formerly among 15 cultivars of sorghum, 10 cultivars of maize, and 8 cultivars of rice. Brachiariagrass showed greater abilities to acquire K, P and Mg, and to tolerate to lower concentration of Ca in shoots in the presence of high concentration of Al in the growing medium including low P at low pH conditions. The level of Al resistance of B. hybrid was ranked to be high as comparable to the most Al-resistant B. decumbens. It was suggested that the highest Al resistance phenotype of B. hybrid may be ascribed to the Al resistance genes from B. decumbens or B. brizantha but not from B. ruziziensis. Extremely high level of Al resistance found in B. decumbens was attributed to localized tip portion of less than 2 mm from root apex due to low amount of Al accumulation, low permeability of PM to Al, lower ratio of PL to S, and higher concentration of phenolic compounds in the tip portion of root as compared with other brachiariagrasses. Thus B. decumbens is considered to possess multiple physiological and biochemical mechanisms to resist high level of Al in soil solution, and its strategy may be extremely localized in the tip portion of the root apex. B. hybrid also exhibited good level of Al resistance. When an extremely high concentration of Al (0.37 mM) was included into the culture solution, significant increase in Al accumulation was observed only in root part. ^{27}Al NMR analysis suggested that the most part of Al in roots was likely to be localized in the cytosol of cells in organically complexed forms and this complexation may inhibit greater upward translocation of Al to shoots, which is beneficial to reduce Al toxicity in shoots.

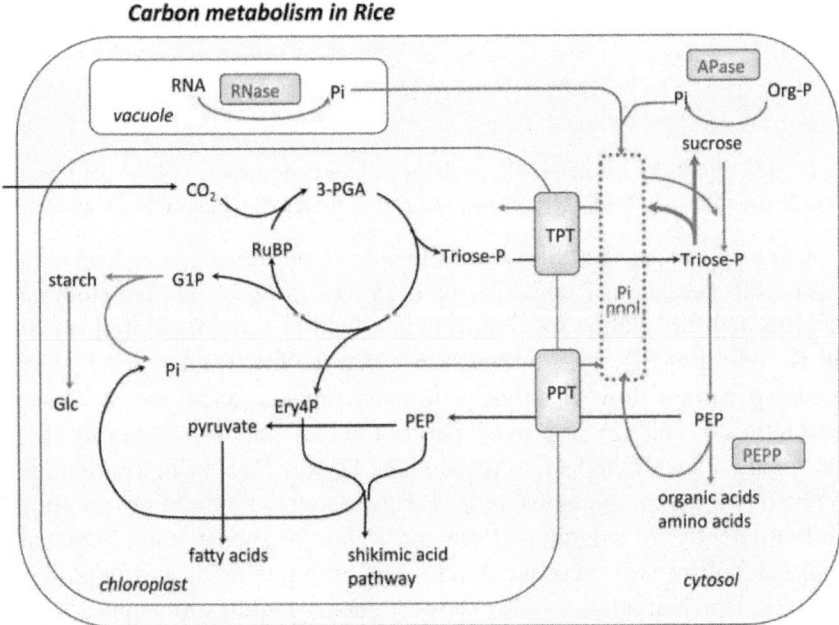

Carbon metabolism in Rice

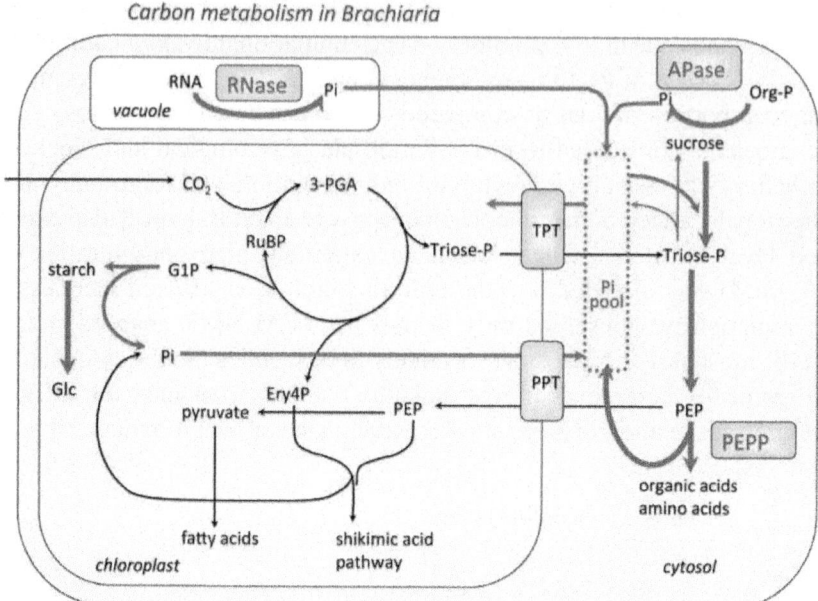

Figure 15: Summary of metabolic changes in leaves and roots with P deficiency: A. Rice; B Brachiaria hybrid. Red arrows indicate P deficiency inducible pathways.

Our study shows that tolerance of low P in both rice and B. hybrid involved marked differences in P recycling and carbon metabolism. We summarized the proposed P recycling mechanisms involved in carbon metabolism of rice and B. hybrid in Fig. 15. For rice, strategies for low-P tolerance include (1) decreased carbon flow to amino acids and organic acids, and decreased N concentration; and (2) improved partitioning of photosynthates to sucrose, combined with restricted sugar catabolism. For the B. hybrid, low-P tolerance involved two major strategies under P deficiency: (1) increasing the ability to use P efficiently by inducing APase and RNase in shoots; and (2) enhancing sugar catabolism and subsequent synthesis of amino acids and organic acids in leaves. Brachiariagrasses also showed greater abilities to acquire K, P and Mg, and to tolerate low concentration of Ca in shoots in the presence of high Al concentration in the growing medium.

In summary, studies on physiological and biochemical mechanism of adaptation of brachiariagrasses grown under simulated conditions of low fertility acid soils indicated their higher level of resistance to Al and tolerance to low supply of P and Ca. This was mainly attributed to their greater ability of Al complexation and Al localization in roots, less upward translocation of Al to shoot tissue, improved P utilization efficiency due to high PPT, and greater acquisition efficiency of K, P and Mg. Identifying the genes responsible for these superior traits of brachiariagrasses is a major objective for future research.

ACKNOWLEDGMENT

This work was supported by a Grant-in-Aid for Scientific Research to Wagatsuma T (18208008 and 23380041) from the Japan Society for the Promotion of Science.

REFERENCES

1. Akhter, A., Khan, M. S. H., Egashira, H., Tawaraya, K., Rao, I. M., Wenzl, P., Ishikawa, S & Wagatsuma, T. (2009). Greater Contribution of Low-nutrient Tolerance to Sorghum and Maize Growth under Combined Stress Conditions with High Aluminum and Low Nutrients in Solution Culture Simulating the Nutrient Status of Tropical Acid Soils. Soil Science and Plant Nutrition, Vol. 55, No. 3, (May 2009), pp. 394-406, ISSN 1747-0765

2. Andrade, S. O., da Sila Lopes, H. O., de Almeida Barros, M., Leite, G. G., Dias, S. M., Saueressig, M., Nobre, D & Temperini, J. A. (1978). Photosensitization in Cattle Grazing on Pasture of Brachiaria decumbens Stapf Infested with Pithomyces chartarum (Berk. & Curt.) M. B. Ellis. Arquivos do Instituto Biologico, Vol.45, No. 2, (April-June 1978), pp. 117-136, ISSN 0020-3653 (In Portuguese with English abstract)

3. Arora, A., Byrem, T. M., Nair, M. G & Strasburg, G. M. (2000). Modulation of Liposomal Membrane Fluidity by Flavonoids and Isoflavonoids. Archives of Biochemistry and Biophysics, Vol.373, No. 1, (Januray 2000), pp. 102-109, ISSN 0003-9861

4. Boija, E & Johansson, G. (2006). Interactions Between Model Membranes and Lignin-related Compounds Studied by Immobilized Liposome Chromatography. Biochimica et Biophysica Acta, Vol. 1758, No. 5, (May 2006), pp. 620-626, ISSN 0005-2736

5. Boija, E., Lundquist, A., Edwards, K. & Johansson, G. (2007). Evaluation of Bilayer Disks as Plant Cell Membrane Models in Partition Studies. Analytical Biochemistry, Vol. 364, No. 2 (May 2007), pp. 145-152, ISSN 0003-2697

6. Bosse, D. & Köck, M. (1998). Influence of Phosphate Starvation on Phosphohydrolases during Development of Tomato Seedlings. Plant, Cell and Environment, Vol.21, No.3, (March 1998), pp. 325-332, ISSN 0140-7791

7. Burleigh, S. H. & Harrison, M. J. (1999). The Down-regulation of Mt4-like Genes by Phosphate Fertilization Occurs Systemically and Involves Phosphate Translocation to the Shoots. Plant Physiology, Vol.119, No.1 (January 1999), pp. 241-248, ISSN 0032- 0889
8. Chapin, F. S. III & Bieleski, R. L. (1982). Mild Phosphorus Stress in Barley and Related Lowphosphorus-adapted
9. Barleygrass: Phosphorus Fractions and Phosphate Absorption in Relation to Growth. Physiologia Plantarum, Vol.54, No.3, (March 1982), pp. 309-317, ISSN 0031-9317
10. Ciereszko, I. & Barbachowska, A. (2000). Sucrose Metabolism in Leaves and Roots of Bean (Phaseolus vulgaris L.) during Phosphate Deficiency. Journal of Plant Physiology, Vol.156, No. 5/6, pp. 640-644, ISSN 0176-1617
11. Cordell, D., Drangert, J. O. & White, S. (2009). The Story of Phosphorus: Global Food Security and Food for Thought. Global Environmental Change, Vol.19, No.2, (May 2009), pp. 292-305, ISSN 0959-3780
12. Cornard, J. P. & Merlin, J. C. (2002). Specroscopic and Structural Study of Complexes of Quercetin with Al(⊂). Journal of Inorganic Biochemistry, Vol. 92, No. 1, (September 2002), pp. 19-27, ISSN 0162-0134
13. Dietz, K. & Heber, U. (1984). Rate-limiting Factors in Leaf Photosynthesis. I. Carbon Fluxes in the Calvin Cycle. Biochimica et Biophysica Acta, Vol.767, No. 3, (December 1984), pp. 432-443, ISSN 0005-2728
14. Duff, S. M., Lefebvre, D. D. & Plaxton, W. C. (1989). Purification and Characterization of a Phosphoenolpyruvate Phosphatase from Brassica nigra Suspension Cells. Plant Physiology, Vol.90, No. 2, (June 1989), pp. 734-741, ISSN 0032-0889
15. Duff, S. M. G., Plaxton, W. C. & Lefebvre, D. D. (1991). Phosphate-starvation Response in Plant Cells: De novo Synthesis and Degradation of Acid Phosphatases. Proceedings of the National Academy of Science of the United States of America, Vol.88, No.21, (November 1991), pp. 9538-9542, ISSN 0027-8424
16. Duff, S. M. G., Sarath, G. & Plaxton, W. C. (1994). The Role of Acid Phosphatase in Plant Phosphorus Metabolism. Physiologia Plantarum, Vol.90, No.4, (April 1994) pp. 791- 800, ISSN 0031-9317
17. Essigmann, B., Güler, S., Narang, R. A., Linke D. & Benning C. (1998). Phosphate Availability Affects the Thylakoid Lipid Composition and the Expression of SQD1, a Gene Required for Sulfolipid Biosynthesis in Arabidopsis thaliana. Proceedings of the National Academy of Sciences of United States of America, Vol.95, No.5, (March 1998), pp. 1950-1955,

ISSN 0027-8424

18. Fairhust, T., Lefrovy, R., Mutert, E. & Batjes, N. (1999). The Importance, Distribution and Causes of Phosphorus Deficiency as a Constraint to Crop Production in the Tropics. Agroforestry Forum, Vol.9, No.1, pp. 2-8, ISSN 0966-8616

19. Fatemi, S. J. A., Williamson, D. J. & Moore, G., R. (1992). A 27Al NMR Investigation of Al3+ Binding to Small Carboxylic Acids and the Proteins Albumin and Transferrin. Journal of Inorganic Biochemistry, Vol. 46, No. 1, (April 1992) pp. 35-40, ISSN 0162- 0134

20. Fredeen, A. L., Rao, I. M. & Terry, N. (1989). Influence of Phosphorus Nutrition on Growth and Carbon Partitioning in Glycine max. Plant Physiology, Vol. 89, No. 1, (January 1989), pp. 225-230, ISSN 0032-0889

21. Glund, K., Nürnberger, T., Abel, S., Jost, W., Preisser, J. & Komor, E. (1990). Intracellular Picompartmentation during Phosphate Starvation–triggered Induction of an Extracellular Ribonuclease in Tomato Cell Culture. In: Progress in Plant Cellular and Molecular Biology. H.J.J. Nijkamp, L.W.H. Van Der Plas, & J. Van Aartrijk, (Eds.), 338–342, Kluwer Academic Publishers, 978-0-7923-0873-7, Boston

22. Green, P. J. (1994). The Ribonuclease of Higher Plants. Annual Review of Plant Physiology, Vol.45, (Jun 1994), pp. 421-425, ISSN 0066-4294

23. Guerrini, F., Cangini, M. & Boni, L. (2000). Metabolic Responces of the Diatom Achanthes brevipes (Bacillariophyceae) to Nutrient Limitation. Journal of Phycology, Vol.36, Nol. 5, (October 2000), pp. 882-890, ISSN 0022-3646

24. Hammond, J. P., Bennett, M.J., Bowen, H.C., Broadley, M.R., Eastwood, D.C., May, S.T., Rahn, C., Swarup, R., Woolaway, K. E. & White , P. J. (2003). Changes in Gene Expression in Arabidopsis Shoots during Phosphate Starvation and the Potential for Developing Smart Plants. Plant Physiology, Vol. 132, No. 2, (June 2003), pp. 578-596, ISSN 0032-0889

25. Hausler, R. E., Baur, B., Scharte, J., Teichmann, T., Eicks, M., Fischer, K. L., Flugge, U. I., Schubert, S., Weber, A. & Fischer, K. (2000). Plastidic Metabooolite Transporters and Their Physiological Functios in the Inducible Crassulacean Acid Metabolism Plant Messembryanthemum crystallinum. The Plant Journal, Vol.24, No.3, (November 2000), pp. 285-296, ISSN 0960-7412.

26. Howard, C. J., LeBrasseur, N. D., Bariola, P. A. & Pamela, J. G. (1998). Control of Ribonuclease in Response to Phosphate Limitation: Induction

of RNS1 in Arabidopsis. In: Phosphorus in Plant Biology: Regulatory Roles in Molecular, Cellular, Organic, and Ecosystem Processes, J.P. Lynch & J. Deikman (Eds.), American Society of Plant Physiology, ISBN 094-3088-38-0, Rockville, MD, U.S.A.

27. Huang, C. F., Yamaji, N., Mitani, N., Yano, M., Nagamura, Y. & Ma, J. F. (2009). A Bacterialtype ABC Transporter is Involved in Aluminum Tolerance in Rice. The Plant Cell, Vol.21, No. 2, (Feburuary 2009), pp. 655-667, ISSN 1040-4651

28. Hutzler, P., Fischbach, R., Heller, W., Jungblut, T. P., Reuber, S., Schmitz, R., Veit, M., Weissenbock, G & Schnitzler, J-P. (1998). Tissue Localization of Phenolic Compounds in Plants by Confocal Laser Scanning Microscopy. Journal of Experimental Botany, Vol. 49, No. 323, pp. 953-965, ISSN 0022-0957

29. Ishigaki, G. (2010). Studies on Establishment of Tissue Culture System and its Utilization for Breeding in Ruzigrass (Brachiaria ruziziensis). (in Japanese with English Tables, Figures, and Summary). Dissertation thesis, University of Miyazaki, Miyazaki, Japan, Retrieved from http://hdl.handle.net/10458/2755

30. Ishikawa, S. & Wagatsuma, T. (1998). Plasma Membrane Permeability of Root-tip Cells Following Temporary Exposure to Al Ions is a Rapid Measure of Al Tolerance among Plant Species. Plant and Cell Physiology, Vol. 39, No. 5, (May 1998), pp. 516- 525, ISSN 0032-0781

31. Ishikawa, S., Wagatsuma, T., Sasaki, R. & Ofei-Manu, P. (2000). Comparison of the Amount of Citric and Malic Acids in Al Media of Seven Plant Species and Two Cultivars Each in Five Plant Species. Soil Science and Plant Nutrition, Vol. 46, No. 3, (September 2000), pp. 751-758, ISSN 1747-0765

32. Ishikawa, S., Wagatsuma, T., Takano, T., Tawaraya, K. & Oomata, K. (2001). The Plasma Membrane Intactness of Root-tip Cells is a Primary Factor for Al Tolerance in Cultivars of Five Plant Species. Soil Science and Plant Nutrition, Vol. 47, No. 3, (September 2001), pp. 489-501, ISSN 1747-0765

33. Jones, D. L. (1998). Organic Acids in the Rhizosphere – A Critical Review. Plant and Soil, Vol. 205, No. 1, (August 1998), pp. 25-44, ISSN 0032-079X

34. Kerven, G. L., Larsen, P. L., Bell, L. C. & Edwards, D. G. (1995). Quantitative 27Al NMR Spectroscopic Studies of Al(⊂) Complexes with Organic Acid Ligand and Their Comparison with GEOCHEM Predicted Values. Plant and Soil, Vol. 171, No. 1, pp. 35-39, ISSN 0032-079X

35. Khan, M. S. H., Tawaraya, K., Sekimoto, H., Koyama, H., Kobayashi, Y., Murayama, T., Chuba, M., Kambayashi, M., Shiono, Y., Uemura, M., Ishikawa, S. & Wagatsuma, T. (2009). Relative Abundance of Δ5-sterols in Plasma Membrane Lipids of Root-tip Cells Correlates with Aluminum Tolerance of Rice. Physiologia Plantarum, Vol. 135, No. 1, (January 2009), pp. 73-83, ISSN 0031-9317

36. Kinraide, T. (1999). Interactions Among Ca^{2+}, Na^+ and K^+ in Salinity Toxicity : Quantitative Resolution of Multiple Toxic and Ameliorative Effects. Journal of Experimental Botany, Vol. 50, No. 338, (September 1999), pp. 1495-1505, ISSN 0022-0957

37. Kondracka, A. & Rychter, A. M. (1997). The Role of Pi Recycling Processes during Photosynthesis in Phosphate-deficient Bean Plants. Journal of Experimental Botany, Vol. 48, No. 7, (July 1997), pp. 1461-1468, ISSN 0022-0957

38. Li, X. N. & Ashiharam H. (1990). Effects of Inorganic Phosphate on Sugar Catabolism by Suspension-cultured Catharanthus roseus. Phytochemistry, Vol. 29, No. 2, (February 1990), pp. 497-500, ISSN 0031-9422

39. Löffler, A., Abel, S., Jost, W., Beintema, JJ. & Glund, K. (1992). Phosphate-regulated Induction of Intracellular Ribonulease in Cultured Tomato (Lycopersicon esculentum) Cells. Plant Physiology, Vol.98, No.4, (April 1992), pp. 1472-1478, ISSN 0032-0889

40. Louw-Gaume, A. E., Rao, I. M., Gaume, A. J. & Frossard, E. (2010a). A Comparative Study on Plant Growth and Root Plasticity Responses of Two Brachiaria Forage Grasses Grown in Nutrient Solution at Low and High Phosphorus Supply. Plant and Soil, Vol. 328, No. 1-2, (March 2010), pp.155-164, ISSN 0032-079X

41. Louw-Gaume, A., Rao, I. M., Frossard, E. & Gaume, A. (2010b). Adaptive Strategies of Tropical Forage Grasses to Low Phosphorus Stress: The Case of Brachiariagrasses. In: Handbook of Plant and Crop Stress. Third Edition, M. Pessarakli, Ed., Taylor & Francis Group, pp. 1111-1144, ISBN 1439813965, USA, Boca Raton

42. Lynch, J.P. (2011). Root Phenes for Enhanced Soil Exploration and Phosphorus Acquisition: Tools for Future Crops. Plant Physiology, Vol. 156, No. 3, (July 2011), pp. 1041-1049, ISSN 0032-0889

43. Ma, J. F., Shen, R., Zhao, Z., Mathias, W., Takeuchi, Y., Ebitani, T & Yano, M. (2002). Response of Rice to Al Stress and Identification of Quantitative Trait Loci for Al Tolerance. Plant and Cell Physiology, Vol. 43, No. 6, (Jun 2002), pp. 652-659, ISSN 0032-0781

44. Macedo, M. A. M. (2005). Pastagens no Ecosistema Cerrados: Evalucao das Pesquisas Para o Desenvolvimento Sustentavel. Reuniao Anual da Siciedade Brasileira de Zootecnia, Vol. 41, pp. 56-84. UFGO, SBZ, Goiania.
45. Miles, J.W.; do Valle, C.B.; Rao, I.M. & Euclides, V.P.B. (2004). Brachiariagrasses. In: Warmseason (C4) grasses (L. Moser, B. Burson & L.E. Sollenberger, Eds.). ASA-CSSA-SSSA, Madison, WI, USA, pp. 745-783, ISSN 0011-183X
46. Miles, J. W., Cardona, C. & Sotelo, G. (2006). Recurrent Selection in a Synthetic Brachiariagrass Population Improves Resistance to Three Spittlebug Species. Crop Science, Vol. 46, No. 3, (June 2006), pp. 1088-1093, ISSN 0011-183X
47. Mimura, T. (1999). Regulation of Phosphate Transport and Homeostasis in Plant Cells. In: International Review of Cytology, K.W. Jeon (Ed.), Vol. 191, pp. 149-200. ISBN 978- 0123-645-95-1
48. Misson, J., Raghothama, K. G., Jain, A., Jouhet, J., Block, M. A., Bligny, R., Ortet, P., Creff, A., Somerville, S., Rolland, N., Doumas, P., Nacry, P., Herrerra-Estrella, L., Nussaume, L. & Thibaud, M. C. (2005). Arabidopsis thaliana Affymetrix Gene Chips Determined Plant Responses to Phosphate Deprivation. Proceedings of the National Academy of Sciences of USA, Vol.102, No.33, (August 2005), pp. 11934-11939, ISSN 0027-8424
49. Miyadate, H., Adachi, S., Hiraizumi, A., Tezuka, K., Nakazawa, N., Kawamoto, T., Katou, K., Kodama, I., Sakurai, K., Takahashi, H., Satoh-Nagasawa, N., Watanabe, A.,Fujimura, T. & Akagi, H. (2010). OsHMA3, P1B-type of ATPase Affects Root-toshoot Cadmium Translocation in Rice by Mediating Efflux into Vacuoles. New Phytologist, Vol. 189, No. 1, (January 2010), pp. 190-199, ISSN 0028-646X
50. Nakamura, Y., Koizumi, R., Shui, G., Shimojima, M., Wenk, M.R., Ito, T. & Ohta, H. (2009). Arabidopsis Lipins Mediate Eukaryotic Pathway of Lipid Metabolism and Cope Critically with Phosphate Starvation. Proceedings of the National Academy of Sciences of USA, Vol. 106, No. 49, (December 2009), pp. 20978-20983, ISSN 0027-8424
51. Nanamori, M., Shinano, T., Wasaki, J., Yamamura, T., Rao, I.M. & Osaki, M. (2004). Low Phosphorus Tolerance Mechanisms: Phosphorus Recycling and Photosynthate Partitioning in the Tropical Forage Grass, Brachiaria Hybrid cultivar Mulato Compared with Rice. Plant and Cell Physiology, Vol. 45, No. 4, (April 2004), pp. 460- 469, ISSN 0032-0781
52. Nátr, L. (1992). Mineral Nutrients – A Ubiquitous Stress Factor for Photosynthesis. Photosynthetica, Vol.27, No. 3, (September 1992), pp.

271-294, ISSN 0300-3604

53. Nomura, M., Imai, K. & Matsuda, T. (1995). Effects of Atmospheric Partial Pressure of Carbon Dioxide and Phosphorus Nutrition on the Ultrastructure of Rice (Oryza sativa L.) Chloroplasts. Japanese Journal of Crop Science, Vol.64, No. 4, pp. 784-793, ISSN 0011-1848

54. Nürnberger, T., Abel, S., Jost, W. & Glund, K. (1990). Induction of An Extracellular Ribonulease in Cultured Tomato Cells upon Phosphate Starvation. Plant Physiology, Vol. 92, No. 4, (April 1990), pp. 970-976, , ISSN 0032-0889

55. Ofei-Manu, P., Wagatsuma, T., Ishikawa, S. & Tawaraya, K. (2001). The Plasma Membrane Strength of the Root-tip Cells and Root Phenolic Compounds are Correlated with Al Tolerance in Several Common Woody Plants. Soil Science and Plant Nutrition, Vol. 47, No. 2, (June 2001), pp. 359-375, ISSN 1747-0765

56. Palma, D. A., Blumwald, E. & Plaxton, W. C. (2000). Upregulation of Vacuolar H+- translocating Pyrophosphatase by Phosphate Starvation of Brassica napus (Rapeseed) Suspension Cell Cultures. FEBS Letter, Vol. 486, No. 2, (December 2000), pp. 155-158, ISSN 0014-5793

57. Qui, J. & Israel, D. W. (1992). Diurnal Starch Accumulation and Utilization in Phosphorusdeficient Soybean Plants. Plant Physiology, Vol.98, No.1, (January 1992), pp. 316- 323., ISSN 0032-0889

58. Ramaekers, L., Remans, R., Rao, I. M., Blair, M. W. & Vanderleyden, J. (2010). Strategies for Improving Phosphorus Acquisition Efficiency of Crop Plants. Field Crops Research, Vol. 117, No. 2-3, (June 2010), pp. 169-175.

59. Rao, I. M., Fredeen, A. L. & Terry, N. (1990). Leaf Phosphate Status, Photosynthesis and Carbon Partitioning in Sugar Beet. III. Diurnal Changes in Carbon Partitioning and Export. Plant Physiology, Vol. 92, No.1, (January 1990), pp. 29-36 , ISSN 0032-0889

60. Rao, I. M. (1996). The Role of Phosphorus in Photosynthesis. In : Handbook of Photosynthesis, M. Pessarakli, (Ed.), pp. 173-194, Marcel Dekker, New York

61. Rao, I. M. (2001a). Adapting Tropical Forages to Low-fertility Soils. In: Proceedings of the XIX International Grassland Congress. (J.A. Gomide.; W.R.S. Mattos. & S.C. da Silva. Eds.), Brazilian Society of Animal Husbandry, Piracicaba, Brazil, pp. 247-254, ISBN 857- 1330-1-7

62. Rao, I. M. (2001b). Role of Physiology in Improving Crop Adaptation to Abiotic Stresses in the Tropics: The Case of Common Bean and Tropical Forages. In: Handbook of Plant and Crop Physiology, M. Pessarakli Ed.,

pp. 583-613, Marcel Dekker, ISBN 0824705467, Inc., New York, USA

63. Rao, I. M., Ayarza, M. A. & Garcia, R. (1995). Adaptive Attributes of Tropical Forage Species to Acid Soils. ⊆. Differences in Plant Growth, Nutrient Acquisition and Nutrient Utilization among C4 Grasses and C3 Legumes. Journal of Plant Nutrition, Vol. 18, No. 10, (October 1995), pp. 2135-2155, ISSN 0176-1617

64. Rao, I. M., Friesen, D. K. & Osaki, M. (1999). Plant Adaptation to Phosphorus-limited Tropical Soils. In: Handbook of Plant and Crop Stress. (M. Pessarakli, Ed.), pp. 61-96, Marcel Dekker, Inc., ISBN 0824719484, New York, USA.

65. Rao, I. M.; Kerridge, P. C. & Macedo, M. C. M. (1996). Nutritional Requirements of Brachiaria and Adaptation to Acid Soils. In: Brachiaria: Biology, Agronomy, and Improvement (J.W. Miles., B. L. Maass. & C. B. Valle, Eds.), Centro Internacional de Agricultura Tropical, Cali, Colombia, pp. 53-71, ISBN 958-9439-57-8

66. Rao, I. M., Miles, J. W & Granobles, J. C. (1998). Differences in Tolerance to Infertile Acid Soil Stress Among Germplasm Accessions and Genetic Recombinants of the Tropical Forage Grass Genus, Brachiaria. Field Crops Reseach, Vol. 59, No. 1, (October 1998), pp. 43-52, ISSN 0378-4290

67. Rao, I. M., Zeigler, R. S., Vera, R. & Sarkarung, S. (1993). Selection and Breeding for Acid-soil Tolerance in Crops: Upland Rice and Tropical Forages as Case Studies. BioScience, Vol. 43, No. 7, (July 1993), pp. 454-465, ISSN 0006-3658

68. Ryan, P. R., Tyerman, S. D., Sasaki, T., Furuichi, T., Yamamoto, Y., Zhang, W. H. & Delhaize, E. (2011). The Identification of Aluminium-resistance Genes Provides Oppotunities for Enhancing Crop Production on Acid Soils. Journal of Experimental Botany, Vol. 62, No. 1, (January 2011), pp. 9-20, ISSN 0022-0957

69. Shinano, T., Nanamori, M., Dohi, M., Wasaki, J. & Osaki, M. (2005). Evaluation of Phosphorus Starvation Inducible Genes Relating to Efficient Phosphorus Utilization in Rice. Plant and Soil, Vol. 269, No. 1-2, (February 2005), pp. 81-87, ISSN 1573-5036

70. Steen, I. (1998). Phosphorus Availability in the 21st Centrury. Phosphorus &Potassium, Vol. 217, (September-October 1998), ISSN 0031-8426

71. Stoutjesdijk, P. A., Sale, P. W. & Larkin, P. J. (2001). Possible Involvement of Condensed Tannins in Aluminium tolerance of Lotus pedunculatus. Australian Journal of Plant Physiology, Vol. 28, No. 11, (November 2001), pp. 1063-1074, ISSN 0310-7841

72. Tadano, T., Ozawa, K., Sakai, H., Osaki, M. & Matsui, H. (1993). Secretion of Acid Phosphatase by the Roots of Crop Plants under Phosphorus-deficient Conditions and Some Properties of the Enzyme Secreted by Lupin Roots. Plant and Soil, Vol. 155/156, No. 1, (October 1993), pp. 95-98, ISSN 1573-5036

73. Theodrou, M.E., Elrifi, I.R., Turpin, D.H. & Plaxton, W.C. (1991). Effect of Phosphorus Limitation on Respiratory Metabolism in the Green Alga Selenastrum minutum. Plant Physiology, Vol. 95, No. 4, (April 1991), pp. 1089-1095. , ISSN 0032-0889

74. Theodorou, M. E. & Plaxton, W. C. (1993). Metabolic Adaptations of Plant Respiration to Nutritional Phosphate Deprivation. Plant Physiology, Vol. 101, No.2, (February 1993), pp. 339-344, ISSN 0032-0889

75. Uhde-Stone, C., Zinn, K. E., Ramirez-Yáñez, M., Li, A., Vance, C. P. & Allan, D. L. (2003b). Nylon Filter Arrays Reveal Differential Gene Expression in Proteoid Roots of White Lupin in Response to Phosphorus Deficiency. Plant Physiology, Vol. 131, No. 3, (March 2003), pp. 1064-1079, ISSN 0032-0889

76. Usuda, H. & Shimogawara, K. (1991). Phosphate Deficiency in Maize. ⊇. Enzyme Activities. Plant and Cell Physiology, Vol. 32, No. 8, (December 1991), pp.1313-1317, ISSN 0032- 0781

77. Versaw, W. K. & Harrison, M. I. (2002) A Chloroplast Phosphate Transporter, PHT2; 1, Influences Allocation of Phosphate within the Plant and Phosphate-starvation Responses. The Plant Cell, Vol. 14, No. 8, (August 2002), pp. 1751-1766, ISSN 1040- 4651

78. Wagatsuma, T., Ishikawa, S., Obata, H., Tawaraya, K. & Katohda, S. (1995). Plasma Membrane of Younger and Outer Cells is the Primary Specific Site for Aluminum Toxicity in Roots. Plant and Soil, Vol. 171, No. 1, (April 1995), pp. 105-112, ISSN 0032-079X

79. Wagatsuma, T., Ishikawa, S., Uemura, M., Mitsuhashi, W., Kawamura, T., Khan, M. S. H. & Tawaraya, K. (2005). Plasma Membrane Lipids are the Powerful Components for Early Stage Aluminum Tolerance in Triticale. Soil Science and Plant Nutrition, Vol. 51, No. 5, pp. 701-704, (September 2005), ISSN 1747-0765

80. Wang, Y. -H., Gravin, D. F. & Kochian, L. V. (2002). Rapid Induction of Regulatory and Transporter Genes in Response to Phosphorus, Potassium, and Iron Deficiencies in Tomato Roots. Evidence for Cross Talk and Root/Rhizosphere-mediated Signals. Plant Physiology, Vol. 130, No. 3, (November 2002), pp. 1361-1370, ISSN 0032-0889

81. Wasaki, J., Shinano, T., Onishi, K., Yonetani, R., Yazaki, J., Fujii, F.,

Shimbo, K., Ishikawa, M., Shimatani, Z., Nagata, Y., Hashimoto, A., Ohta, T., Sato, Y., Miyamoto, C., Honda, S., Kojima, K., Sasaki, T., Kishimoto, N., Kikuchi, S. & Osaki, M. (2006).

82. Transcriptomic Analysis Indicates Putative Metabolic Changes Caused by Manipulation of Phosphorus Availability in Rice Leaves. Journal of Experimental Botany, Vol. 57, No. 9, (September 2006), pp. 2049-2059, ISSN 0022-0957

83. Wasaki, J., Yamamura, T., Shinano, T. & Osaki, M. (2003a). Secreted Acid Phosphatase is Expressed in Cluster Roots of Lupin in Response to Phosphorus Deficiency. Plant and Soil, Vol. 248, No. 1-2, (January 2003), pp. 129-136, ISSN 1573-5036

84. Wasaki, J., Yonetani, R., Kuroda, S., Shinano, T., Yazaki, j., Fujii, F., Shimbo, K., Yamamoto, K., Sakata, K., Sasaki, T., Kishimoto, N., Kikuchi, S., Yamagishi, M. & Osaki, M. (2003b). Transcriptomic Analysis of Metabolic Changes by Phosphorus Stress in Rice Plant Roots. Plant, Cell and Environment, Vol. 26, No. 9, (September 2003), pp. 1515-1523, ISSN 0140-7791

85. Wsaki, J., Yonetani, R., Shinano, T., Kai, M. & Osaki, M. (2003c). Expression of the OSPI1 Gene, Cloned from Rice Roots Using cDNA Microarray, Rapidly Responds to Phosphorus Status. New Phytologist, Vol. 158, No. 2, (May 2003), pp. 239-248, ISSN 0028-646x

86. Watanabe, T., Misawa, S. & Osaki, M. (2005). Aluminum Accumulation in the Roots of Melastoma malabathricum L., an Aluminum-accumulating Plant. Canadian Journal of Botany, Vol. 83, No. 11, (Nobember 2005), pp. 1518-1522, ISSN 0008-4026

87. Watanabe, T., Osaki, M., Yoshihara, T. & Tadano, T. (1998). Distribution and Chemical Speciation of Aluminum in the Al Accumulator Plant, Melastoma marabathricum L. Plant and Soil, Vol. 201, No. 2, pp. 165-173, ISSN 0032-079X

88. Wenzl, P., Mancilla, L. I., Mayer, J. E., Albert, R. & Rao, I. M. (2003). Simulating Infertile Acid Soils with Nutrient Solutions : The Effects on Brachiaria species. Soil Science Society of America Journal, Vol. 67, No. 5, (September-October 2003), pp. 1457-1469, ISSN 1747-0765

89. Wenzl, P., Patino, G. M., Chaves, A. L., Mayer, J. E. & Rao, I. M. (2001). The High Level of Aluminum Resistance in Signalgrass is not Associated with Known Mechanisms of External Detoxification in Root Apices. Plant Physiology, Vol. 125, No. 3, (March 2001), pp. 1473-1484, ISSN 0032-0889

90. Wu, P., Ma, L., Hou, X., Wang, M., Wu, Y., Liu, F. & Deng, X. W.

(2003). Phosphate Starvation Triggers Distinct Alterations of Genome Expression in Arabidopsis Roots and Leaves. Plant Physiology, Vol. 132, No. 3, (July 2003), pp. 1260-1271, ISSN 0032- 0889

91. Yamaji, N., Huang, C. F., Nagao, S., Yano, M., Sato, Y., Nagamura, y. & Ma, J. F. (2009). A Zinc Finger Transcription Factor ART 1 Regulates Multiple Genes Implicated in Aluminum Tolerance in Rice. The Plant Cell, Vol. 21, No. 10, (October 2009), pp. 3339-3349, ISSN 1040-4651

92. Yan, X., Liao, H., Trull, M. C., Beebe, S. E. & Lynch, J. P. (2001). Induction of a Major Leaf Acid Phosphatase Does not Confer Adaptation to Low Phosphorus Availability in Common Bean. Plant Physiology, Vol. 125, No. 4, (April 2001), pp. 1901-1911, ISSN 0032-0889

93. Yang, J. L., Li, Y. Y., Zhang, Y. J., Zhang, S. S., Wu, Y. R., Wu, P., Zheng, S. J. (2008). Cell Wall Polysaccharides Are Specifically Involved in the Exclusion of Aluminum from the Rice Root Apex. Plant Physiology, Vol. 146, No. 2, (February 2008), pp. 602-611, ISSN 0032-0889

94. Yang, Z-B., Eticha, D., Rao, I. M. & Horst, W. J. (2010). Alteration of Cell-wall Porosity is Involved in Osmotic Stress-induced Enhancement of Aluminium Resistance in Common Bean (Phaseorus velgaris L.). Journal of Experimental Botany, Vol. 61, (July 2010), pp. 3245-3258, ISSN 0022-0957

95. Yoneda, S. & Nakatsubo, F. (1998). Effect of the Hydroxylation of Patterns and Degrees of Polymerization of Condensed Tannins on Their Metal-chelating Capacity. Journal of Wood Chemistry and Technology, Vol. 92, No. 2, (April 2008), pp. 193-205, ISSN 0277- 3813

96. Yun, S. J. & Kaeppler, S. M. (2001). Induction of Maize Acid Phosphatase Activities under Phosphorus Starvetion. Plant and Soil, Vol. 237, No. 1, (November 2001), pp. 109- 115, ISSN 1573-5036

Chapter 8

SOIL INDICATORS OF HILLSLOPE HYDROLOGY

Johan van Tol, Pieter Le Roux and Malcolm Hensley

Department Soil, Crop and Climate Sciences, University of the Free State South Africa

INTRODUCTION

The demand for water doubles every 20 years which is more than twice the rate of the world's population growth. New water resources are becoming scarcer and to treat and remediate existing sources more expensive (Clothier et al., 2008). The protection and management of surface and groundwater resources, especially in the highly variable water regime of semi-arid areas, requires accurate analysis of hydrological processes. This involves the identification, definition and quantification of the pathways, connectivities, thresholds and residence times of components of flow making up stream discharge. It is essential that these aspects be efficiently captured in hydrological models for accurate water resource predictions, estimating the hydrologic sensitivity of the land for cultivation, contamination and development, and for quantifying low flow mechanisms (Lorentz et al., 2007; Uhlenbrook et al., 2005; Wenninger et al., 2008). Ideally these hydrological models can best be developed using measurements of the surface and subsurface lateral flow paths, water table fluctuations, connectivity of the various water bodies and the residence flow time of water through the landscape. The landscape unit that is of particular importance is the hillslope (Karvonen et al., 1999; Lin et al., 2006; Ticehurst et al., 2007), hence the accent here on this landscape unit. The measurements named are however expensive and time consuming since these processes are dynamic in nature with strong temporal and spatial variation (McDonnell et al., 2007; Park & Van de Giesen, 2004; Ticehurst et al., 2007). The need for predictions of the named hydrological processes is becoming increasingly important and led to the launch an International Association of Hydrological Sciences (IAHS) initiative called Predictions in Ungauged Basins or PUB

(Sivapalan 2003; Sivapalan et al., 2003) encouraging researchers and modellers to focus their efforts on predicting the hydrological behaviour of catchments based on physical principles without relying on calibrations of hydrological models.

Soils integrate the influences of parent material, topography, vegetation/ land use, and climate and can therefore act as a first order control on the partitioning of hydrological flow paths, residence time distributions and water storage (Park et al., 2001; Soulsby et al. 2008).

The influence of soil on hydrological processes is due to the ability of soil to transmit, store and react with water (Park et al., 2001). Hydrologists agree that the spatial variation of soil properties significantly influences hydrological processes but that hydrologists lack the skill to gather and interpret soil information (Lilly et al., 1998). The relationship between soil and hydrology is interactive. Water is a primary agent in soil genesis, resulting in the formation of soil properties containing unique signatures of the way they formed. Almost every hydrological process of interest to hydrologists is difficult to observe and measure (Sivapalan 2003a). Soil properties are not dynamic in nature in the short term, and their spatial variation is not random (Webster, 2000). The correct interpretation of spatially varying soil properties associated with the interactive relationship between soil and hydrology can serve as an indicator of the dominant hydrological processes (Ticehurst et al., 2007; Van Tol et al, 2010a), and improve the understanding of hydrological behaviour of different hillslopes (Lin et al., 2006). The hillslope scale is considered ideal as it is the smallest unit on which most hydrological processes can be observed. The improved understanding can facilitate the development of conceptual, qualitative 2-dimensional descriptions of hillslope hydrological behaviour. These descriptions reflect the physical processes and should be used in the configuration and structure of hydrological models in order to mimic the hydrological behaviour of hillslopes and thereby aid in PUB.

HYDROPEDOLOGY

The science of hydrology has been poorly serviced by soil science in the past. Soil science started as a natural science in the late 19th century but changed to an applied science in agriculture in the 20th century. This was mainly due to the quest for food security, heightened by the 1st, 2nd world wars and the 'cold' war which followed. During the postwar years attitudes changed internationally as fears about food shortages decreased while apprehension regarding environmental sustainability grew. Agriculture occupies more than 90% of the land surface in most countries and accounts for around 70% of fresh water utilization in many developed countries. Rapidly expanding industrialization, increased urbanization, intensified agricultural production and population growth have all contributed during the last six decades towards drastically increasing the environmental pollution hazard. A large fraction of this pollution ends up in the soil, making it a vital role player in pollution control and rehabilitation, thereby broadening its contribution to holistic environmental studies.

The term hydropedology was introduced in 1966 by Kutilek (Kutilek & Nielson, 2007), and can be defined as the "…synergistic integration of pedology with hydrology to enhance the holistic study of soil-water interactions and landscape-soil-hydrology relationships across space and time, aiming to understand pedologic controls on hydrologic processes and properties, and hydrologic impacts on soil formation, variability, and functions" (Lin et al., 2008). This field aims to bridge gaps between pedology, soil physics, hydrology and geomorphology and also between micro and macroscopic scales of water interactions. Issues covered by hydropedology include: i) hydrology as factor of soil formation, ii) soil as essential component of hydrological cycle and filter of water, iii) soil morphology as signatures of soil hydrology and iv) landscape-soil-water relationships across scales. For excellent reviews and comprehensive discussions on hydropedology see Lin (2003), Kutilek et al. (2007), Van Huyssteen (2008) and Lin (2010).

Hydrological Flowpaths

Three major flow pathways exist in a typical hillslope: overland flow, subsurface lateral flow and bedrock flow (Karvonen et al. 1999; Ticehurst et al., 2007). Subsurface lateral flow can be divided into: subsurface macropore flow, subsurface lateral flow at A-B horizon interface, return flow at the footslope and toeslope and flow at the soil-bedrock interface (Lin et al., 2006). These flowpaths are not mutually exclusive, and water tends to move between them. Some paths are only connected when the hillslope is wet. The relative importance of the various pathways is determined by soil characteristics, the macropore network and the parent material at the base of the soil (Mosley, 1982). Hydrologic conditioning is influenced by soil depth, pore size and organic matter distribution, tortuosity and the surface and subsurface topography (Sidle et al., 2001).

The role of topography varies with the moisture content of the soil. In drier periods the main controlling factor of movement is soil characteristics. In wetter periods, the topography becomes increasingly important (Lin et al., 2006; McGlynn et al., 2002; Park & van de Giesen, 2004).

Overland Flow

Overland flow occurs either as infiltration excess or as saturation excess. In general steeper slopes generate large volumes of overland flow with significant erosive energy. Thinner A horizons usually indicate that the overland flow is dominant, in thicker soils more infiltration due to the greater volume of water needed to saturate the soil is expected. The assumption can be made that thicker soils support more vegetation and this causes a decrease in the overland flow proportion (Ticehurst el al., 2007).

Breaks in slope (normally between midslopes and valley bottoms) reduce the velocity of water and enhance infiltration. The soils in this region are generally thicker, due to deposition of alluvial material and organic matter, which further enhance infiltration. In valley bottoms the runoff rate tends to slow down because of the smaller gradient. These soils are however the wettest in typical hillslopes and the saturated conditions reduce the infiltration rate promoting overland flow.

Where overland flow occurs the water may downslope encounter an area where the soil water deficit has not yet been satisfied, the water then infiltrates. This is called the run-on pathway and is often ignored in rainfall and runoff studies. The water available for infiltration then includes the precipitation as well as water supplied from the upperslope (Nahar et al., 2004).

The amount of overland flow is greatly affected by the texture of the soil, specifically the percentage clay and sand. Sandy soil is generally more permeable and has a greater hydraulic conductivity than clay rich soil, and therefore infiltration excess induced overland flow seldom occurs in sandy soils. In a study by Karnoven et al. (1999) the conductivity of sandy loam soils was 15 times higher than clayey soils.

Subsurface Lateral Flow

Macropores conduct a considerable amount of water during large storms in forested catchments. Water moves through tree root channels, pores created by organisms (earthworms), as well as cracks. Cracks are usually present in soils with a high 2:1 clay content (vertic soils), especially in drier periods (Lin et al., 2006). There are three factors determining the contribution of subsurface macropore flow of water namely; size of the macropores, the accessibility and continuity of the pores. The continuity of these pores seems to increase with an increase in soil moisture (Nieber et al., 2000). Soil pipes are usually flow pathways parallel with the slope and are formed by soil fauna (moles & mice) as well as dead root channels. They contribute a significant amount of subsurface water to streamflow and are usually quick to respond to rainfall.

Pipe flow has a smaller influence in hillslopes with high drainable porosity because water table response is lower due to the high storage potential of the profile (Uchida et al., 2006).

Lateral flow occurs at A/B horizon interfaces due to differences in the structures, densities and hydraulic conductivities of the horizons. Vertical flow will be hindered and water will tend to move laterally if the A horizon is more permeable than the B horizon. In a study by Lin et al. (2006), the lowest water content was recorded below the interface between the A and B horizons due to the great amount of lateral flow. The lower gradient of lower slope terrain units would limit this lateral flow and promote water logging, as well as overland flow due to excess saturation (return flow). Similar mechanisms might result in a flowpath at the bottom of the profile at the interface between the soil and the underlying parent material. Continuous flow after a storm even with little water in the top of the profile suggested that the water moved vertically in the upperslopes and then laterally at or near the soil-bedrock interface in a study by Lin et al, (2006). The permeability, the depth as well as the differentiation between horizons would affect the amount of water moving through this flowpath. Since the clay content of the B horizon in lower slopes usually shows an increase due to luviation, this pathway would generally originate in the upper slopes.

Bedrock Flow

Ticehurst et al. (2007) found in their study that the soils from the summit area, which were sandy and shallow, provided and important water intake area for water supply to the bedrock flowpath. The general movement of water in this region is vertical and soils are usually well drained. Due to the age of the soils and the small amount of deposition, little differentiation between horizons is generally present and water drains vertically through the B horizon into the C horizon. The water that doesn't move on top of the bedrock moves through cracks in the bedrock or on solid bedrock within the saprolite. The bedrock flowpath is extremely important for recharge of lower slopes, groundwater levels and generating baseflow in some catchments (Fanning & Fanning 1989; Ticehurst et al., 2007).

Signatures and control mechanisms

Soil properties, soil horizons, soil profiles and soil patterns are not randomly distributed (Webster, 2000). These soil features are influenced by five soil forming factors i.e. climate, topography, geology, organisms and time. The combination of these factors results in unique soil properties with distinctive vertical and horizontal distributions. These properties and their distribution not only influence hydrological processes but can also serve as verification on the way they were formed. A few examples are presented.

Redox Morphology

Soil morphology developed by oxidizing, reducing and redox conditions serves as signatures of flowpaths and storage mechanisms in soils, hillslopes and catchments. Reducing conditions increase down the profile, down slope and with increasing rainfall if all other factors playing a role in the redox process, remains the same. Redox features in soils involve localities where there is depletion in Fe3+ and Mn2+ concentrations and localities where there is accumulation of Fe^{3+} and Mn^{2+} (Soil Survey Staff, 1992). Depletion in Fe^{3+} and Mn^{2+} is associated with low chroma values (grey colours), and accumulation of Fe^{3+} and Mn^{2+} is associated with high chroma colours (yellow, red and black) in the form of mottles and concretions (Le Roux, 1996).

Micro organisms utilize O_2, NO_3^-, Mn^{2+}, Fe3+ and SO_4^{2-} as oxidation agents (electron acceptors) and easily oxidisable organic matter as reduction agent. These reactions occur sequentially from most likely to least likely to be reduced. In the oxidized state Fe^{3+} and Mn^{2+} are insoluble, and in the reduced state, very soluble. Reduced localities have high Fe^{2+} and Mn^+ concentrations in solution. They diffuse to oxidised localities where the concentration in solution is low to be oxidized again (Van Breedeman & Brinkman, 1976). Grey colours of the silicate clay, quartz and feldspar soil minerals are grey and therefore grey colours appear where the Fe coatings are removed (Vepraskas & Bouma, 1976). Yellow, red and black colours occur juxtapositioned where Fe^{3+} and Mn^{2+} accumulate in sequence with an increase in Fe^{3+} and Mn^{2+} concentration (Le Roux, 1996).

Redox features are easily observed in plinthic soils. Plinthic horizons have an accumulation of iron in the form of oxides and hydroxides and are localized in the form of high chroma mottles and concretions. The simple processes leading to the formation of such a horizon are eluviation (removal of constituents), illuviation (accumulation of eluviated material), oxidation and reduction (Fanning & Fanning, 1989). Fe^{3+} is reduced and together with sesquioxides eluviated from the upper lying horizons and Fe^{2+} oxidized and accumulates in the lower horizon. A fluctuating water level is necessary for this to take place.

Plinthite normally occurs in highly weathered soils of the regions with rainfall exceeding 500 mm and where a fluctuating water table is active. High temperatures and a high evaporative demand favour plinthite formation since they influence the fluctuation of water levels. The formation of plinthite on different topographical positions corresponds to the climate. In the drier climates plinthite forms in the lower lying areas. Redoximorphic features occur in soils of semi-arid climates and wetter. The key factor is a ratio of rainfall/evapotranspiration resulting in water flowing to the deep subsoil and impermeable deep subsoil preventing water loss to the fractured rock and result in subsoil saturation. These conditions typically occur in semi-arid climates and wetter.

The relationship between wetness and position in the landscape is reflected through the variation in soil colour. In a typical catena the red Fe rich soils are typically found on the higher lying drier positions of the hillslope, whereas grey gleyed soils can generally be found in the wetter lower-slope terrain positions. Long term data proves that yellow soils are normally better drained than grey soils, but wetter than red soils (Van Huyssteen & Ellis, 1997; Van Huyssteen et al., 2005).

Presence of Calcium Carbonate (CaCO₃)

The dependence of calcareous precipitates on the presence and behaviour of water makes it a good indicator of hillslope hydrology. Calcretes are materials formed by cementation or selective replacement of the soil particles by carbonate. Calcareous layers in soils are controlled by the soil water regime and are typically found in arid to sub-humid regions. Lime precipitates in the soil due to limited leaching which can be brought about by two processes: leaching that can be limited due to low rainfall/high evapotranspiration, or restricting subsoil layers and associated saturated conditions (Driessen & Deckers., 2001; Netterberg, 1978).

In sub-humid to arid regions, calcification is one of the main processes in soils with carbonate rich parent materials. Weathering of the parent material results in the formation of soils with calcium as the major cation on the cation exchange complex. $CaCO_3$, the dominant carbonate in these soils, is pedogenically formed as follows:

$$Ca^{2+} + CO_2 + H_2O \rightarrow CaCO_3 + H_2 \quad (1)$$

Weathered Ca^{2+} dissolves in water leaches towards lower soil horizons and flows downslope, and filling voids and pores. Plant roots extract water and precipitation in the form of $CaCO_3$ occurs due to the presence of CO_2. The CO_2 are present in the soil as a consequence of diffusion from the atmosphere, but CO_2 generated by oxidation of plant roots enhance this process, especially when the natural vegetation consists of grasses and shrubs. This process is the first stage of the formation of a calcic horizon (Fanning et al., 1989; Shankar & Achyuthan, 2007).

According to Netterberg (1978), the presence of calcretes can serve as indicators of previous (fluctuating) ground water levels as well as of preferred flow paths (faults). When the parent material contains small amounts of $CaCO_3$ and the amount of $CaCO_3$ in the profile exceeds the amount that could be released by weathering, the presence of $CaCO_3$ in the soil can be ascribed to the second process namely, deposition of $CaCO_3$ rich dust from coastal shelves (Bockheim & Douglass, 2006).

Soil Depth and Porosity

Soil depth is the result of the balance between the rate of weathering and the rate of erosion. Under similar climate and hydro-topographical conditions the rate of weathering of the parent material is controlled by the nature of the rock. Porosity (f) is a measure of the total void space in a porous material and is measured, either as a percentage (between 0 and 100%), or as a fraction (between 0 and 1) of the bulk volume. It is defined by the formula:

$$f = 1 - \rho_d \div \rho_s \qquad (2)$$

Where ρ_d is the bulk density (Mg. m^{-3}) and ρ_s is the particle density (Mg. m^{-3}, generally taken as 2.65 in soils low in organic matter). Soil depth together with the porosity determines the storage capacity of the soil.

In a South African case study, the storage capacity of two semi-arid catchments was determined (Van Tol et al, 2010a). The average soil depth of catchment B3 was 450 mm and that of B4 & B5 was 190 mm (due to similarities in B4 and B5 they were considered as one catchment). The average porosity of B3 was 301.5 mm compared to 130.6 mm of B4 & B5. Although the area of B3 is smaller (40.7 km^2) than that of B4 & B5 (49.4 km^2), it can store almost twice the volume of water (12.5 × 10^6 m^3 compared to 6.7 × 10^6 m^3). This facilitates more water infiltration, greater water holding capacity, a greater volume of water contained at saturation and at drained upper limit. This results in more interflow at the A/B-horizon interface and at the soil/bedrock interface, more water contributing to groundwater bodies and consequently a longer duration of streamflow. More water is available for transpiration resulting in a denser vegetative cover.

The thickness of the soil further influences the residence times of water in catchments. This influence can be more important than that of the slope length, or upslope contribution area (Asano et al., 2002). Shallow soils, with a small storage capacity tend to saturate quickly favouring the generation of overland flow due to saturation excess, resulting in short residence times and high peak flows.

HYDROLOGY OF SOIL TYPES

"Theory development will advance if we can develop simple models (which may be caricatures of the basin system but, nevertheless, contain within them the basic properties of the actual basins), provided, importantly, that they can be verified with large-scale patterns extracted from the observed data" (Sivapalan, 2003). In order to develop simple conceptual hydropedological models (and to improve our understanding of the role of hydropedology in both the natural environment and agriculture), it is necessary to understand key hydrological processes, the impact of soil on these processes and the influence of these processes on soil formation.

This relationship between soil and water is however difficult to comprehend at hillslope or catchment scale. For example; water may drain from the soil into the rock and then return to the soil. It may also exit the soil again as return flow. Where a water table occurs in the soil it is often uncertain whether the soil is feeding the rock aquifer or vice versa. This interaction between soil and hydrology can be simplified by firstly studying this interaction at a pedon scale. In this section soils are divided into different soil types based on their hydrological behaviour, similar to the Hydrology of Soil Types (HOST) classification system. In HOST the soils of the UK were divided into 29 classes based on their hydrological response (Boorman et al., 1995). In this section we only focus on three main response mechanisms of soils and use six years of soil moisture content measurements to support the classification. Because hydropedology is a rather young and complex subject, with relatively few quantitative measurements worldwide to verify hypotheses, we considered it wise to include only local case studies about which we have sufficient knowledge and as much quantitive information as possible.

Hydrology of Soil Types

It is hypothesized that soils can be grouped in three main hydropedological types based on their hydrological response: recharge soils, interflow soils and responsive soils. Data from the Weatherley research catchment (31°06'6"S/28°20'13"E) in South Africa (Van Huyssteen et al., 2005) was used to distinguish between these soil types using the degree of water saturation (s), measured over six years. The degree of water saturation is the volume of water relative to the (f) (Hillel, 1980). Porosity can be calculated using equation 2 and the degree of water saturation as:

$$s = V_W \div V_f \qquad (3)$$

Where s is the degree of saturation (as fraction), V_w is the water content (mm^3 mm^{-3}) and V_f is the total pore volume (mm3 mm-3). Complete saturation (s = 1) is seldom reached since air is usually trapped in pores by water (Hillel, 1980). The drained upper limit (DUL) i.e. the water content below which drainage due to gravity virtually ceases is expected to be around 0.65 in most soils. The term "annual duration of degree of water saturation above 0.7 of porosity" ($AD_{s>0.7}$) is the first approximated threshold value for the onset of reduction (Van Huyssteen et al., 2005). The degree of saturation before the start of reduction will however differ between areas, soil forms and horizons since numerous factors influence redox conditions in soils. It is because redox reactions of significant extent in soils leave well defined morphological footprints e.g. mottling and/or grey colours, that $AD_{s>0.7}$ is considered to be a useful parameter in hydropedological studies. $AD_{s>0.7}$ was measured in days per year. The mean annual duration in days of events with s>0.7 ($D_{s>0.7}$) was calculated as follows (Van Huyssteen et al., 2005):

$$D_{s>0.7} = AD_{s>0.7} \div F_{s>0.7} \quad (4)$$

Where $F_{s>0.7}$ is the mean annual frequency of events where s>0.7 (events year^{-1}). Soils were classified according to Soil Classification - A taxonomic system for South Africa (Soil Classification Working Group, 1991) although equivalent classification accordance with WRB (IUSS Working Group WRB, 2006).

Recharge Soils

Several soils in the Weatherley catchment qualify as recharge soils. The average annual duration of saturation above 0.7 of porosity ($AD_{s>0.7}$), expressed as a % of 365 days, is not significant in these soils (Fig. 1) as conditions near saturation only occur when drainable water accumulates. The short degree of saturation in the subsoil shows that water draining through the soil exits the solum to enter the fractured rock underneath.

In recharge soils, the hypothesis is that dominant flow direction is vertical. These soils typically occur on the crest or midslope positions on hillslopes with gentle slopes. Precipitation infiltrates the soil and water flows vertically through the pedon under gravitational forces. The underlying permeable bedrock facilitates infiltration of water.

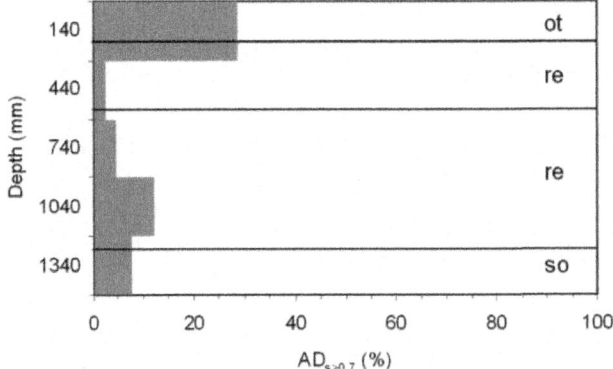

Figure 1: Mean ADs>0.7 (%) values in a typical recharge soil: P221, Hutton 2100 (WRB - Orthidistric Cambisol), Weatherley (after Van Huyssteen et al., 2005).

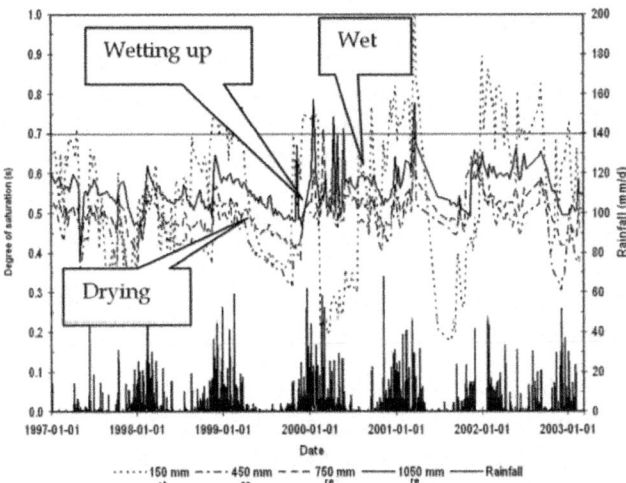

Figure 2: Degree of saturation vs. rainfall over 6 years of a recharge soil: P221, Hutton 2100 (WRB - Orthidistric Cambisol) in the Weatherley catchment (after Van Huyssteen et al., 2005).

From a hydrological perspective the formation and distribution of recharge soils is therefore dependant to a large extent on the permeability of the underlying material. Depending on the nature of the underlying material the infiltrated water can either recharge regional water tables directly, or in the case of aquicludes or aquitards, move laterally after leaving the soil. This lateral moving water can then recharge the stream through transient or perennial groundwater. Its contribution to transient groundwater may be uncertain. Since these flowpaths through the bedrock are usually the longest, recharge soils are important for generating base flow. Recharge soils show no evidence of saturation in any part of the profile. The annual rainfall and potential evapotranspiration should however be considered when classifying a soil as a recharge soil. In arid areas, precipitation is insufficient for redoximorphic features to form and the soils would be classified as recharge soils based on morphological properties even though they are not freely drained.

The contribution of recharge soils to catchment hydrology by implication stops when the soil water balance is negative (i.e. ET>P). This limits its activity to the wet part of the rain season (Fig. 2). Three phases are clearly visible in the graph namely a wetting up cycle with the start of the rain season, a wet phase during the rainy season and a drying phase in the waning portion of the rain season. The drying phase is only stopped by the start of the wetting up phase of the following rain season or when the water content is lower than the lower limit of plant available water.

The wetting up cycle depends on the precipitation, atmospheric demand (ET) and the size of the reservoir. As the grass vegetation of the Weatherley catchment mainly extracts its water from the upper 900 mm (Zere, 2005) of soil, a relative large volume of soil has to be brought to drained upper limit (DUL) before draining starts. In the majority of years (four out of six) this, cycle is two weeks in duration. In the wet cycle the water content of the recharge soils depends mainly on the distribution of rainfall events. Profile water exceeding DUL drains beyond reach of the grass roots.

Interflow Soils

The $(AD_{s>0.7})$ values in the subsoils of interflow soils is distinctive (Fig. 3). Conditions of water contents near saturation (drainable water) occur in all horizons but typically increase with depth. Interflow soils are associated with subsurface lateral flowpaths. For interflow to occur a layer with lower hydraulic conductivity must be present (B horizon or bedrock with restricted permeability) as well as a slope favouring lateral movement down the slope. Interflow soils are therefore typically found in midslope positions with fairly steep gradients. Water starts moving laterally when infiltrated water encounters a layer with lower hydraulic conductivity (A/B horizon interface; soil – bedrock interface or a saturated layer) or when water, fed from upslope recharge soils, encounters such a layer and may return to the soil.

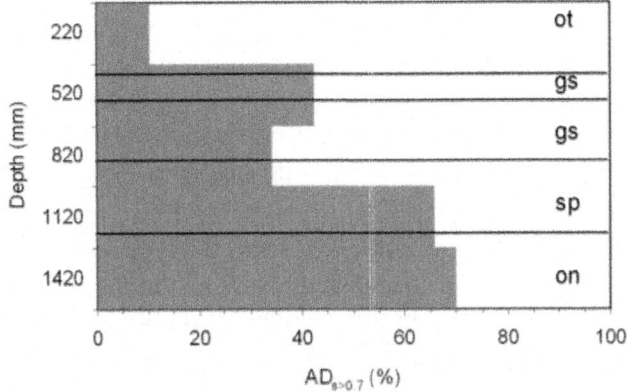

Figure 3: Mean ADs>0.7 (%) values in a typical interflow soil: P225, Longlands 1000 (WRB - Ferric-Endoeutric Albeluvisol) in the Weatherley catchment (Van Huyssteen et al., 2005).

Interflow soils have in contrast with the three phases of recharge soils, a distinctive drainage phase, above DUL (Fig. 4). The duration of $AD_{s>0.7}$ in the soft plinthic (sp) horizon of ± 67% i.e. 244 days or 8 months (Fig. 3) is an indication that this soil body generally releases water up to the end of August. This implies a 5 month draining phase i.e. stretching from the end of the rain season (early April) to the end of the dry season in August.

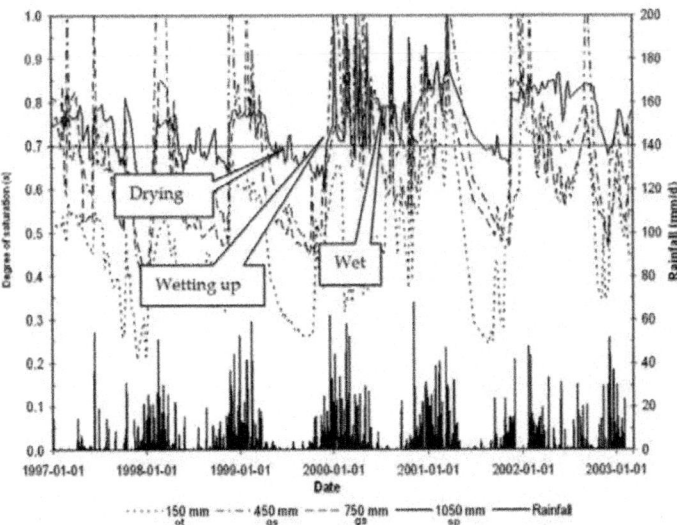

Figure 4: Degree of saturation vs. rainfall over 6 years of an interflow soil: P225, Longlands 1000 (WRB - Ferric-Endoeutric Albeluvisol) in the Weatherley catchment (after Van Huyssteen et al., 2005).

During the wet phase losses of water by drainage and ET are sometimes slower than additions of water by precipitation, and interflow results in a rise of the transient groundwater into the plinthic, E and A horizons. Such a fluctuating water table is typical of subsoils with plinthic and E horizons (Soil Classification Working Group, 1991). These fluctuations are event driven and can be related to rainfall events. The catchment must first fill up before transient groundwater can occur.

Sub soil flowpaths are associated with a residence time shorter than the bedrock flowpaths and longer than overland flow. Interflow soils would therefore contribute mainly to the shoulder of the hydrograph, and to some extent to baseflow. Interflow soils normally have morphological indications of periodic saturation in the profile. If dominant flow exists on the A/B horizon interface, eluvial horizons form. These horizons show marked removal of colloidal material and organic matter. When interflow occurs at the soil/bedrock interface, the transitional horizon usually show indications of periodic saturation.

Responsive Soils

Responsive soils can either be very shallow soils with low infiltration capacity, saturated soils which prohibit water infiltration or soils prone to form crusts resulting in low infiltration rates and generating Hortonian overland flow. In the Weatherley catchment responsive soils generate overland flow due to saturation excess. The overland flow component contributes to peak flow as the first part of the peak of the hydrograph. The influence of the water content of the topsoil on the generation of overland flow is illustrated in Fig. 5. The results show that overland flow only becomes significant when the topsoils are close to saturation. Overland flow from responsive soils is therefore expected in the wettest positions in landscapes i.e. valley bottoms and wetlands.

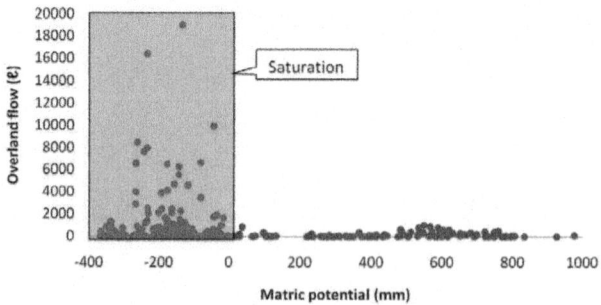

Figure 5: Volume of overland flow measured at five runoff plots vs. topsoil matric potential in the Weatherley catchment.

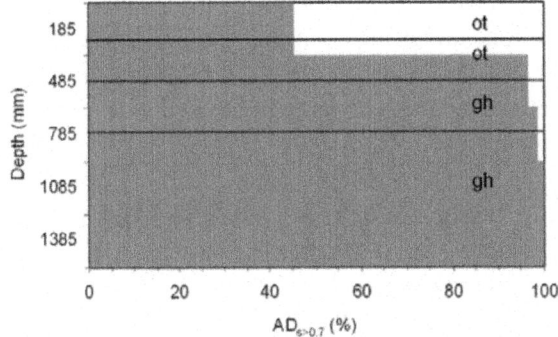

Figure 6: Mean ADs>0.7 (%) values in a typical responsive soil: P235, Katspruit 1000, (WRB - Hyperdistric Gleysol) in the Weatherley (Van Huyssteen et al., 2005).

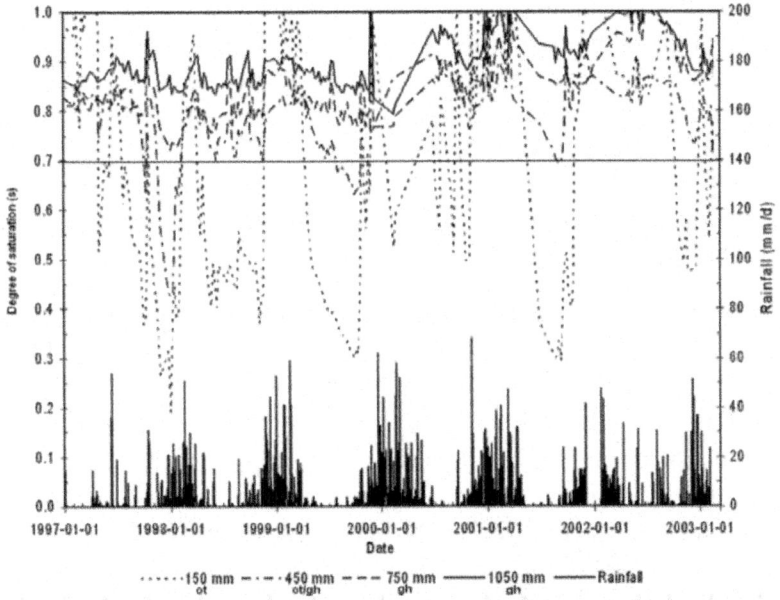

Figure 7: Degree of saturation vs. rainfall over 6 years of a responsive soil: P235, Katspruit 1000, (WRB - Hyperdistric Gleysol) in the Weatherley catchment (after Van Huyssteen et al., 2005).

In the Weatherley catchment these soils are at or near saturation for long periods (Fig. 6), resulting in conditions called saturation excess overland flow in the rain season. Due to long periods of saturation the subsoil (gh horizon) lacks an obvious wetting and draining phase since it is saturated or close to saturation throughout the year. Only the topsoil horizon loses water to ET during the dry season (Fig. 7). In order for these subsoils to remain saturated for such long periods under incessant ET demand there needs to be a constant supply of water. It is hypothesized that the recharge soils of the upper slopes supply water to the responsive soils via the bedrock flowpath and to a lesser extent through interflow.

CONCEPTUAL MODELS OF HILLSLOPE HYDROLOGICAL BEHAVIOR

Deducing the hydrological behaviour of a soil profile without considering its position in the landscape might lead to false interpretations. It is therefore important to obtain a holistic understanding of how hillslopes behave hydrologically. In the case studies reported here soil properties and their spatial distribution (both vertically and horizontally) were interpreted and related to their hydrological response. From these interpretations 2- dimensional descriptions of the hydrological behaviour were formulated. Two examples of these conceptual models are presented in this chapter. For detailed discussions on these examples, study area descriptions and more conceptual examples see Le Roux et al. (2010) and van Tol et al. (2010a).

Conceptual Model of the Hydrological Behaviour of a Hillslope in the Weatherley Catchment

A conceptual description of a hillslope in the Weatherley catchment is presented in Fig. 8 (van Tol et al., 2010). The soil forms and their associated properties along with their spatial distribution were interpreted to develop the conceptual model of the hydrological behaviour of the hillslope.

The dominant processes (flowpaths and storage mechanisms) in Fig. 8 are indicated by numbers plus a letter in box inserts related in each case to arrows of different sizes; these inserts are included in the text enclosed in a bracket. A discussion of these processes follows in what can be considered as a hypothesis, based on hydropedology, of the hillslope hydrology.

When it rains infiltration dominates in the upper regions of this hillslope (1a). Gentle slopes as well as dense vegetation impede overland flow and facilitate infiltration. The absence of any signs of wetness in the soil of the upperslope indicates that vertical drainage through the profile is dominant. The texture is non-luvic and the clay content is therefore relatively uniform with depth. No, or very little, lateral flow is expected to occur at the A/B horizon interface. These are considered to be true recharge soils since no signs of wetness were recorded in soil profile 240 up to a depth of 1500 mm, indicating that water does not perch in the pedon within this depth. Water draining through 240 therefore either infiltrates the subsurface layers (2a) or flows at the soil/bedrock interface (3a). The latter was not reached with auger observations down to 2400 mm.

Any water which does infiltrate the fractured rock would then either flow vertically and recharge regional aquifers (2b) or, when it encounters a layer with restricted permeability (aquitard), it would flow laterally (3b) and recharge perennial hillslope groundwater downslope.

The presence of interflow soils located where the rock bedding plain surfaces near Uc8 is an indication that the bedding plane (sandstone rock shelf of the Molteno formation) has restricted vertical permeability promoting considerable flow at the soil/bedrock interface (3a). The greater part of the water draining through the soil of the upper slope is therefore expected to flow laterally at this soil/bedrock interface.

Return flow (ex-filtration) to the soil surface (4) is expected as water flowing at the soil/bedrock interface reaches the protruding Molteno shelf. The amount of water exceeds the storage capacity of the soil and returns to the surface contributing to overland flow. It is expected that the overland flow has a relatively short duration as the water will re-infiltrate when it reaches the soils below the rock outcrop (1b).

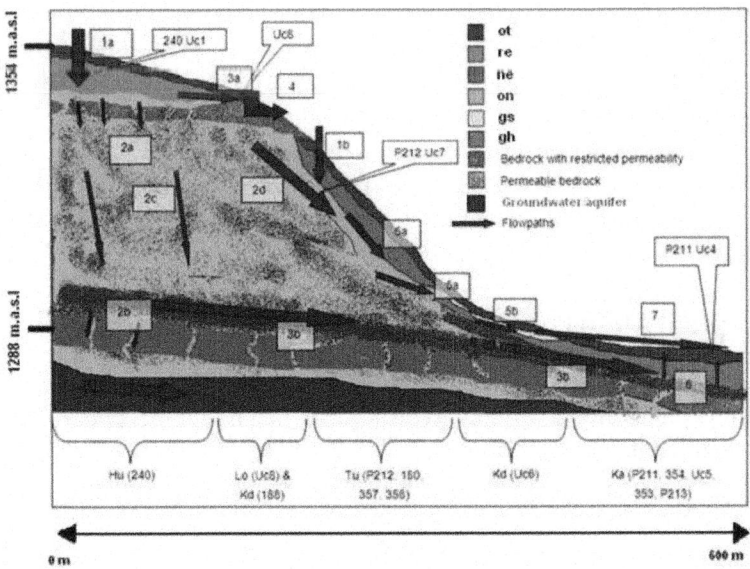

Figure 8: Conceptual hydrological behaviour of the selected hillslope based on soil interpretations. Various processes are indicated by the numbered arrows.

Subsurface lateral flow (5a) in the form of flow at the soil/bedrock interface is indicated by the on horizon (WRB – cambic) present in the deep subsoil of the Tukulu (WRB – (Orthieutric Cambisol) soil of the midslope at P212. This soil body is situated on the Molteno Formation. Groundwater responsible for the redoximorphic features of the on horizon is evidently supplied from the recharge soils as return flow from the bedrock (2d). This return flowpath is expected to result in a fairly constant supply of water during the wet seasons to the on horizon, reflecting its association with perennial groundwater.

The gs horizon (WRB – albic) in the soils of the lower slopes is an indication of the lateral flow of groundwater dominating at the A/B horizon interface (5b). Gleyed soils cover the entire TMU 4 & 5 positions of this hillslope. The gleyed conditions are indications that these profiles are saturated for long periods. The gh horizons (WRB – gley) have a low hydraulic conductivity that impedes infiltration. Precipitation does not infiltrate into these soils due to the saturated state of the gh horizon. The water maintaining saturation in these lower areas must therefore have another origin. It is believed that there is another layer with restricted permeability present in the hillslope (Fig. 8). This layer deflects water which has infiltrated through the recharge soils of the upperslope towards the lower lying areas (3b), resulting in the presence of a perennial aquifer. These very wet soils respond rapidly to precipitation providing overland flow to the stream, the process described as saturation excess overland flow (7). Near surface macropore flow might also play a significant role in this area, as water from the gh horizon pushes up into the more porous ot horizon (WRB – ochric) and then flows laterally. The ot horizons of the soils in the lower slope have Fe and Mn mottling, confirming periodic saturated conditions. Since the gh horizons in the lower footslope and toeslope positions (Fig. 8) are saturated for long periods, the dominant flow direction within the pedon is upwards (6). Evapotranspiration will presumably extract more water from the soil than can infiltrate.

Conceptual Model of Hillslope Hydrological Behaviour in the Two Streams Catchment

The conceptual model of the hydrological behaviour of two hillslopes in the Two Streams catchment (29°12'23"S/30°39'10"E) is presented in Fig. 9 (Le Roux et al., 2010). The number arrows represent dominant hydrological processes. Note that the slope on the right faces south and that on the left hand side of Fig. 9 faces north. This difference in aspect has resulted in distinct morphological characteristics of the soils in the slopes; north facing slopes in the southern hemisphere are generally drier compared to south facing slopes due to more direct sunlight. The soils on the south facing slope are therefore more reduced and exhibit yellow colours compared to the dominant red colours on the north facing slope.

The dominant process in this catchment is infiltration and vertical drainage of precipitation (1 in Fig. 9). The soils in the upper slopes are considered freely drained with no indications of periodic saturation in the A and B horizons. These soils are considered to be recharge soils. Water exits the solum and drains into the highly weathered saprolite (so).

Vertical drainage remains dominant until more impermeable rock is reached, where the flow is deflected in a lateral direction (2). This lateral moving component feeds the stream channel from the side causing prolonged conditions of saturation and gleyed soils (gh horizons) next to the stream channel. Some water can however enter cracks and fissures in the relatively impermeable rock and feed deeper lying water bodies.

The vertical sequence of horizons of the upper slopes; orthic A (ot), yellow brown apedal B (ye) and red apedal B (re) (WRB – ferrilic), might be an indication of lateral movement close to the surface horizons (3). Near surface macropore flow is often observed in areas with topsoil horizons rich in organic matter. Organic matter increases the macroporosity of soils and favours the generation of preferential flow. The accumulation of clay in the on horizon in the midslopes of the south facing slope confirms this lateral flowpath (3). Clay is eluviated from the upperslopes and transported laterally until it accumulates at the break of slope in the on horizons. The importance of studying hydrological processes at hillslope scale is emphasized by the soil distribution pattern of the Two Streams catchment. In both hillslopes the subsoil colour sequence is red then yellow and grey in the valley bottom. Although most of the soils show no indications of lateral flow, the colour sequence suggests that there is significant lateral flow at hillslope scale.

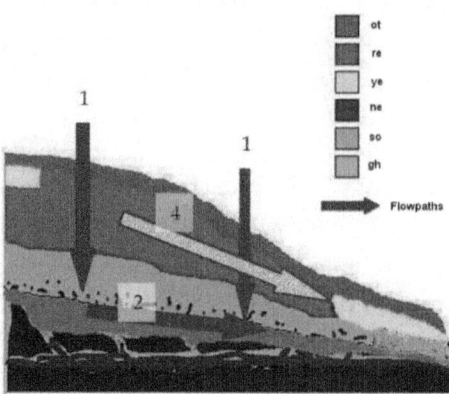

Figure 9: Conceptual model of hillslope hydrological behaviour in the Two Streams catchment (Le Roux et al., 2010).

HYDROPEDOLOGICAL APPLICATIONS IN HYDROLOGICAL MODELLING

Introduction

The hydrological response of catchments is dependent on the combined responses of the individual hillslopes within the catchment (Sivapalan, 2003). The hillslope is generally accepted as a fundamental landscape unit (Lin et al., 2006; Weiler & McDonnell, 2004), and is the smallest scale for holistically understanding and simulating hydrological processes (Tromp van Meerveld & Weiler, 2008). It is therefore not surprising that the hillslope forms the basic building block for a number of hydrological models.

The current dominant paradigm in hydrological modelling involves using an a priori set of small scale theories and process descriptions (e.g. Darcy and Richard's equations) and splitting the catchment into small enough uniform elements for these theories to work. The models arising from this paradigm emphasize the explicit mapping of landscape heterogeneities and process complexities which, according to McDonnell et al. (2007), are an impossible task in even the most intensively studied catchments. Consequently the models based on current theories rely strongly on calibration, mimicking past data, to account and compensate for the lack of understanding of the actual hydrological processes and heterogeneities in the landscape (McDonnell et al., 2007; Sivapalan 2003). This results in models that 'work' but for the wrong process reasons (Weiler et al., 2004) and models highly over parameterized with many combinations of the parameters resulting in the same final result. This leads to a large degree of modelling uncertainty and models unsuitable for predictions in ungauged basins (Beven, 2001 and McDonnell et al., 2007).

Another current weakness in hydrological modelling is the gap between experimentalists and modellers. For example, an experimentalist may propose a conceptual model of hydrological behaviour of a system based on observations, measurements and experience. The appropriateness of the concept can only be verified when a numerical model is built (Bredehoeft, 2005). Unfortunately modellers usually do not incorporate the experimentalist's knowledge into the model structure (McDonnel et al., 2007; Sivapalan 2003; Tromp-van Meerveld et al., 2008; Weiler et al., 2004). When they do, simulations are followed by calibration exercises, which, together with limited understanding of the complex processed involved lead to further confusion by "correcting" with imperfect data (Dunn et al., 2008). On the other hand experimentalists have focussed on the documentation of the unconventional behaviour of new hillslopes instead of the systematic examination of first order controls of hillslope hydrological behaviour, without intercomparisons to obtain common process behaviours (Weiler et al., 2004). The transference or extrapolation value of these hillslope studies is therefore minimal. Some researchers argue that every hillslope is therefore unique (Beven, 2001). This is possibly true to a certain extent, since after hundreds of experiments we appear to be no further towards a common conceptualization of hillslope hydrology, and experimentalists have not yet expressed what the minimal set of measurements are to characterize even a single hillslope (McDonnel et al., 2007; Tromp-van Meerveld et al., 2008; Weiler et al., 2004)! There is a great need for closer collaboration between experimentalist and modellers (Siebert and McDonnell, 2002). Neither the conceptual model of the experimentalist nor the numerical model of the modeller should be considered untouchable, but the common focus should be to, through iteration of concepts and numbers, represent the physical process numerically.

In this section a modelling exercise using hydropedological data for model parameterization and configuration is described. Although model outputs are presented, the main aim of the exercise was to evaluate the capability of a mechanistic model to utilize hydropedological inputs. Simulations were run over two rainfall seasons in the Weatherley research catchment and simulated streamflow and soil moisture contents were compared to measured values. Only one model run was conducted i.e. there were no calibrations with measured data to improve model outputs.

The Hydrological Model and Model Setup

ACRU-Int

ACRU is an agrohydrological, daily time step, multi-layered soil water budgeting model. The standard version, ACRU2000, comprises of two soil layers (A and B- horizon) and a deep groundwater layer. Soil inputs include; the thickness of soil horizons, water contents at the start of simulation (SMAINI and SMBINI), Permanent Wilting Point (PWP), Drained Upper Limit (DUL), saturation (Po), Plant Available Water (PAW), drainage rates (ABRESP and BFRESP) and the erodibility of the soil (K-factor). Except for the latter all inputs are required for both soil horizons (Schulze, 2007).

PWP, DUL and Po are largely determined by the soil texture, organic matter and the bulk density. Typical values for these parameters are proposed in chapter 5 of the ACRU user manual (Smithers and Schulze, 2004) for different textural classes and clay distribution models, i.e. change in clay content with depth. The clay distribution models and typical texture classes were assigned to the 501 soil series of the binomial soil classification of South African soils (MacVicar et al., 1977). Relatively accurate PWP, DUL and Po values are therefore easily accessible for all South African soils. The PAW is the difference between DUL and PWP and is used to calculate the initial water content, expressed as a percentage of PAW (Smithers al., 2004).

ACRU–Int is a revised version of the standard ACRU2000 model. In addition to the 2 soil layers (A&B-horizons) an intermediate layer (saprolite) between the soil and deep groundwater levels was introduced (Lorentz et al., 2007). The intermediate layer has a mechanism whereby lateral release of water can be induced when certain threshold positive pressures at the saprolite/bedrock interface is achieved. The lateral releases from the intermediate zone can be routed to any layer of a downslope land segments. This is ideal for imitating flowpaths at hillslope scale. The model allows redistribution of saturated water (RESP), i.e. between DUL and Po, from the A to the B-horizon (ABRESP) and from the B-horizon to the groundwater (BFRESP). The distribution is expressed as a fraction of the water above DUL draining vertically downwards from the respective horizons on a daily time step. In Schulze (1995) typical RESP values are presented for different textural classes. Low RESP values in a particular horizon will result in the buildup of water above DUL in the soil horizons above it favouring the generation of lateral flow in that horizon. Thus a low ABRESP value will favour lateral flow in the A horizon and a low BFRESP value will favour lateral flow in the B horizon. Reductions in RESP are therefore suggested based on the "Interflow Potential" (IP) of different soil series, high IP = RESP × 0.3 and moderate IP = RESP = 0.6 (Schulze,

1995). The influence of the redistribution fractions on the simulated soil water contents is illustrated in Fig. 10.

Water contents in one land segment were simulated with ACRU-Int for six years with actual climatic data from the Weatherley catchment (Fig. 10). Four different RESP fractions were used a) ABRESP = 0.01 & BFRESP = 0.01, b) ABRESP = 0.99 & BFRESP = 0.99, c) ABRESP = 0.99 & BFRESP = 0.01 and d) ABRESP = 0.01 & BFRESP = 0.99. Simulation a and b show similar trends with a buildup of water in the A-horizon, relatively low water contents in the B-horizon and a general decrease in the water content of the C-horizon due to very little vertical drainage from the A to lower horizons. In simulation b, water is allowed to drain freely to the B-horizon and then to the C-horizon, resulting in relatively low water contents in the A and B-horizons but accumulation in the C-horizon. Simulation c show water freely draining from the A-horizon but due to the impeding C-horizon (BFRESP = 0.01) buildup in the B-horizon, slowly reducing the water content in the C-horizon.

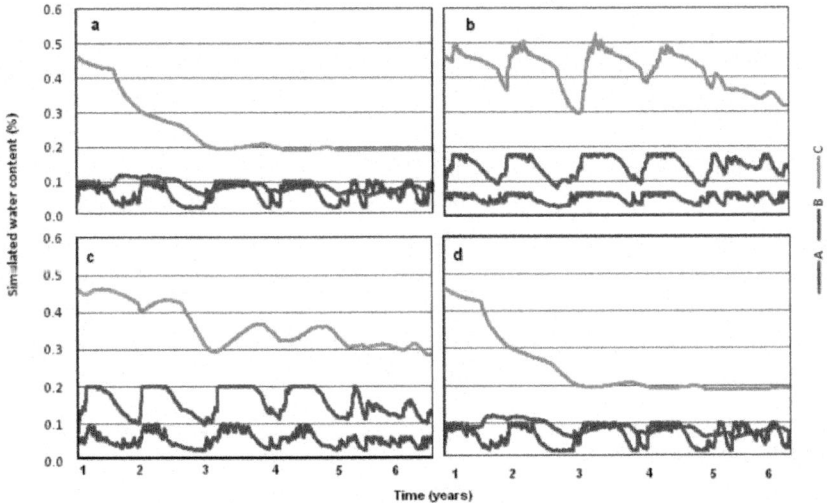

Figure 10: Water contents (%) for three horizons in the Weatherley catchment with different RESP values simulated over 6 years with ACRU-Int.

It is clear from Fig. 10 that the RESP values play an integral part in the simulation of soil water contents and consequently on the outflow of different land segments. Exactly how these values were obtained is however not clear. For example; one would expect a direct relationship between the saturated hydraulic conductivity (K_s) and the RESP value. In Schulze (1995) however a siltyloam horizon with Ks of 6.8 mm.h-1 has a RESP of 0.45 whereas a sandyclayloam with K_s of 4.3 mm.h^{-1} has a RESP of 0.50, similarly, a loam horizon and a sandyslayloam soil were attributed the same RESP value (0.50) although the K_s of the former is triple that of the latter. The heterogeneous horizonation in terms of the textural distribution is the driving force for lateral flow generation in soils and is the basis for assigning "Interflow Potential" values to different soil series. If texture differences are the main reason for differences in RESP values is it not spurious to reduce the RESP value based on the IP? The volume of water draining vertically in a profile is also related to the slope of the land. Steeper slopes generally favour more lateral flow, and less vertical distribution of saturated water. Also, soils with shallow horizons ought to distribute a greater percentage of water in a particular day compared to soils with deep horizons although the texture and hydraulic conductivities are similar.

Another important parameter in ACRU-Int, not considered a soil input but definitely influenced by the soil, is QFRESP. According to definition QFRESP is: Stormflow response fraction for the catchment/subcatchment, i.e. the fraction of the total stormflow (1.0) that will run off from the catchment/subcatchment on the same day as the rainfall event (Smithers et al., 2004). QFRESP is inversely correlated with catchment area and will increase with an increase in slope angle, area covered by impervious material, and rainfall intensity. Soils prone to topsoil crusting as well as very shallow or very wet soils should therefore give high QFRESP values. No physical based methods to obtain suitable QFRESP values are however proposed by the modellers, and thus the appropriateness of the values used depends on calibration. ACRU can also account for unsaturated flow of water (IUNSAT) and flow through cracks or fissures in swelling soils (ICRACK). The latter is divided into three classes based on the clay content. Both IUNSAT and ICRACK can be excluded from simulations.

Model Configuration

For the purpose of the modelling exercise the Weatherley catchment was divided into 7 land segments, each one with distinct hydrological responses (Fig. 11). The division was made derived from several pedological, soil physical, hydrogeological, geophysical and geochemical studies, as well as in-field observations of visible hydrological processes in the selected catchment over the past few years (Lorentz. 2001; Lorentz et al., 2004; Lorentz et al., 2007; van Huysteen et al., 2005; van Tol et al., 2010b). Some soil and landscape attributes, obtained from representative soil profiles of the land segments are presented in Tables 1 and 2. Two representative hillslopes, with diverse hydrological behaviour, were identified based on the properties of the land segments, their sequence from the crest to the river, and the area covered by the individual segments. Conceptual 2-dimensional flow models were then developed and applied to construct flow routings for the two hillslopes (Le Roux et al., 2010). Hillslope 1 includes land segments 1 – 4, and hillslope 2 includes land segments 5 – 7(Fig. 12). Land segments 4 and 7 represent the valley bottom or wetland area; they drain to a separate land segment (9) which represents the stream network. Since the deep groundwater levels are always below the stream channel and do not contribute to low flows, all the drainage out of the intermediate zone into the groundwater layer was routed to another land segment (8). Land segment 8 is therefore not linked to any streamflow generation process. The routing between land segments is presented in Fig. 12.

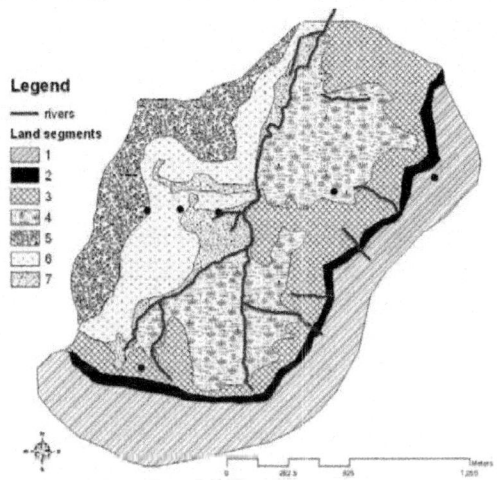

Figure 11: Different land segments in the Weatherley catchment.

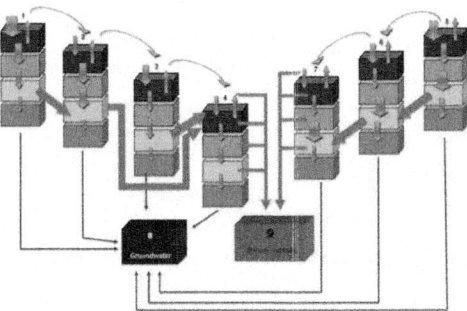

Figure 12: Flow routing for the simulation, the magnitude of the arrows giving an indication of dominant flow directions in various land segments, the number of each one corresponding to the number in Fig. 11.

Model Parameterization

Imitating the dominant hydrological processes was one of the major aims of this simulation exercise. According to Sivapalan (2003), parameters without measured values require calibration with gauged data in order to reflect reality. This leads to uncertainty in predictions when dealing with unguaged basins. Parameter values, not directly available from soil profile descriptions, were therefore physically estimated based on the definition of the parameter and the (limited) understanding of the process influenced by the parameter. These calculations include:

- **ABRESP, BFRESP and INTRESP**: The difference between Ks of the top and lower horizon gives an indication of the vertical distribution from the former to the latter. Ks values were calculated using ROSSETA (Schaap, 2000) for all the horizons of profiles representing the different land segments using texture class distributions and bulk densities (Db). The ratio of the Ks of the lower horizon to the Ks of its overlying horizon was used to estimate the particular RESP value. Ks of the R-horizon were measured. Representative texture class, Db and estimated Ks values are presented in Table 1 and resulting ABRESP, BFRESP and INTRESP values in Table 2.

Table 1: Some attributes of different land segments used in to calculate model parameters.

LandSeg.	Representative soil profiles[*1]	Horizon[*2]	Sand (%)	Silt (%)	Clay (%)	Texture class	Db (Mg m^{-3})	Ks (cm day^{-1})	s >0.7[*3]
1	202	A	79.1	10.9	10.0	Sandy Loam	1.68	46.00	1
		B	71.4	12.6	16.0	Sandy Loam	1.61	27.00	
		INT	57.3	16.7	26.0	Sandy Clay Loam	1.72	6.00	
		R						0.10	
2	204	A	78.0	10.0	12.0	Sandy Loam	1.60	52.48	278
		B	78.0	12.0	10.0	Sandy Loam	1.71	36.61	
		INT	75.0	13.4	11.6	Sandy Loam	1.71	26.87	
		R						0.02	
3	212&205	A	74.0	16.0	10.0	Sandy Loam	1.59	45.84	62
		B	70.0	15.0	15.0	Sandy Loam	1.68	19.25	
		INT	50.0	32.0	18.0	Loam	1.74	5.87	
		R						0.20	
4	206	A	50.0	40.0	10.0	Loam	1.52	22.37	622
		B	46.3	37.8	15.9	Loam	1.67	7.79	
		INT	35.1	26.9	38.0	Clay Loam	1.73	2.09	
		R						0.20	
5	210	A	56.0	28.4	15.7	Sandy Loam	1.59	15.36	99
		B	56.4	23.9	19.7	Sandy Loam	1.71	7.60	
		INT	56.2	23.5	20.3	Sandy Clay Loam	1.66	9.18	
		R						2.00	
6	209	A	50.4	33.9	15.7	Loam	1.54	14.75	437
		B	37.7	35.0	27.3	Loam	1.67	3.11	
		INT	5.0	50.8	44.2	Silty Clay	1.73	1.27	
		R						0.20	
7	208	A	45.8	39.8	14.5	Loam	1.47	17.04	541
		B	35.9	43.2	21.0	Loam	1.62	5.12	
		INT	31.2	42.9	26.0	Loam	1.71	2.58	
		R						0.20	

[*]230 1 Detailed descriptions together with analytical data for these profiles are presented in Van Huyssteen et al. (2005).

*2 A = A horizon, B = B horizon, INT = C horizon, R = Rock

*3 Number of days during overall period of 662 days with rainfall > 1 mm, when the water content of the A-horizon is >0.7 of Po

Table 2: Soil parameters used simulating the Weatherley catchment

LandSeg	Area (km²)	Horizon	Depth (m)	Po	PWP	DUL	ABRESP	BFRESP	INTRESP	QFRESP
				m³ m⁻³						
1	0.315	A	0.4	0.37	0.09	0.19	0.6	0.2	0.02	0.002
		B	0.42	0.39	0.09	0.19				
		INT	1.6	0.35	0.16	0.25				
2	0.072	A	0.1	0.40	0.09	0.19	0.7	0.7	0.00	0.92
		B	0.3	0.36	0.09	0.19				
		INT	1.2	0.35	0.09	0.19				
3	0.270	A	0.3	0.40	0.09	0.19	0.4	0.3	0.03	0.1
		B	1	0.37	0.09	0.19				
		INT	1	0.35	0.13	0.25				
4	0.375	A	0.5	0.43	0.13	0.25	0.3	0.3	0.10	0.94
		B	0.25	0.37	0.13	0.25				
		INT	0.25	0.35	0.20	0.31				
5	0.183	A	0.16	0.40	0.09	0.19	0.5	1.0	0.22	0.15
		B	0.9	0.36	0.09	0.19				
		INT	0.8	0.37	0.16	0.25				
6	0.225	A	0.45	0.42	0.13	0.25	0.2	1.0	0.16	0.66
		B	0.6	0.37	0.13	0.25				
		INT	0.7	0.35	0.25	0.32				
7	0.091	A	0.35	0.45	0.13	0.25	0.3	1.0	0.08	0.82
		B	0.2	0.39	0.13	0.25				
		INT	1.1	0.36	0.13	0.25				

- **QFRESP**: Except for the Molteno outcrop, very few areas in the catchment have only soil related factor influencing the qu impervious top layers. Soil crusting and very shallow soils are also not the norm. The ick flow response is therefore the water content of the topsoil before and during rain events, where saturated A-horizons will impede infiltration, generating overland flow and consequently peak flow. Zere (2007) estimated daily soil water contents, based on weekly neutron probe measurements, over a 6 year period for 28 profiles in the Weatherley catchment. This was used to determine the number of days that A-horizons of the representative profiles were close to saturation (>0.7 of Po) on days with more than 1 mm of rain, during an overall number of 662 days in a 6 year period. The result is presented in Table 1 and resulting QFRESP values in Table 2. Approximately half of land segment 2 is covered by the impervious Molteno rock outcrop and was taken into account for calculation of QFRESP.

The thickness of A and B-horizons was obtained from Van Huysteen et al., 2005 for the profiles representing the different land segments. Where the lower depth of the profile was reached the B2 or C horizon was used as the depth of the intermediate zone, if not, an extra 0.5 m was added to the B2 or C horizon to acquire the intermediate zone depth. PWP, DUL and Po values were estimated based on typical texture class values proposed in Smithers et al., 2004. The simulation period is from 1st Jan 1998 till 31st August 2001. Simulated results are reported for two rain seasons, starting from the 1st of September 1999 to allow the model to 'settle' and incorrect data regarding initial water contents to even out. Trees were planted on parts of the catchment in 2002 and 31st August 2001 was therefore selected as the end of simulation to avoid dissimilarity between vegetative covers.

Rainfall and, when possible, minimum and maximum temperature data were obtained from the BEEH (2003) database. Other climatic data was obtained from the quaternary catchment database. Streamflow was measured at a Crump weir at the catchment outlet and data regarding the streamflow obtained from BEEH (2003) database. Daily simulated soil water contents were compared to daily water contents calculated from weekly neutron water meter readings (Zere, 2005). Soil water contents are expressed as a fraction of porosity (Po).

Modelling Results

Streamflow

Simulated and measured outflow from catchment for the simulation period is presented in Fig. 13. A R^2 value of 0.64 was attained with a linear line deviating almost 100% from the 1:1 line. The divergence is towards measured flow, i.e. a greater volume of flow was measured than simulated. Figures 15 and 16 accentuate the cause of the deviation from the 1:1 line.

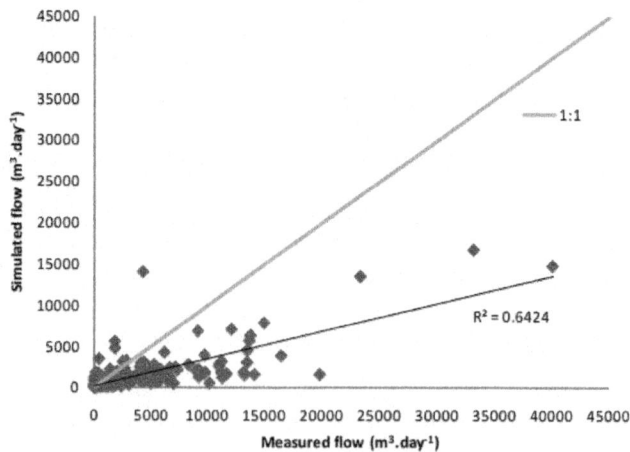

Figure 13: Simulated vs. measured flow during the selected simulation period.

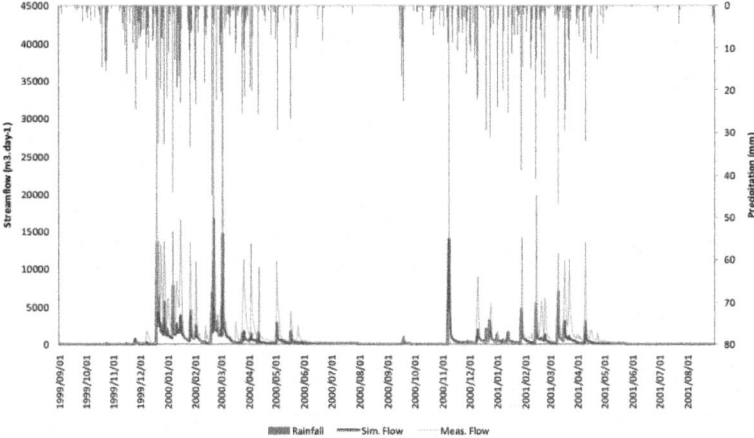

Figure 14: Simulated vs. measured daily streamflow flow plotted against daily rainfall for the simulation period.

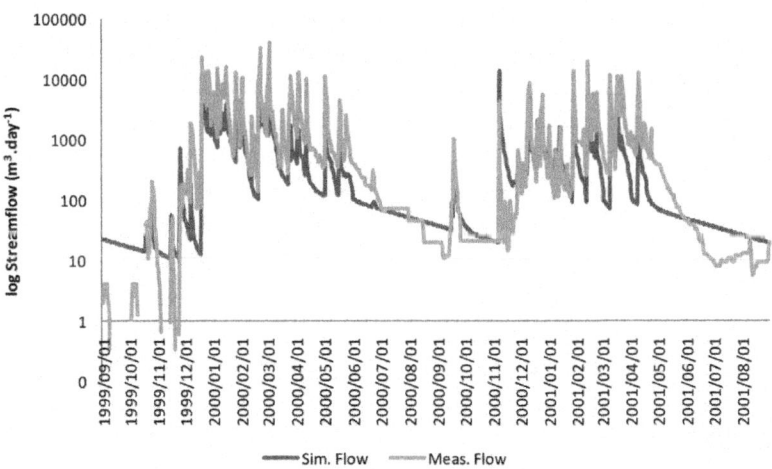

Figure 15: Log of simulated vs. measured flow over the selected simulation period.

Fig. 14 illustrates daily measured flow compared to simulated flow and also the influence of rainfall on flow volumes. It is clear from Fig. 14 that flows are overestimated especially towards the end of the rain seasons. Simulated low flows compared well with measured flows. Rain early in each season resulted in much smaller volumes of streamflow compared to similar size storms just before the end of the rain season. This over estimation of high flows and good representation of low flows is also emphasized in Fig. 15 where comparisons are plotted on a log scale.

Fig. 15 shows high flows being overestimated by an order of magnitude in under some conditions. Low flows and streamflow recession are however simulated reasonable well in drier periods. The average daily difference between simulated and measured results over the simulation period is 854 m^3 day^{-1}. This increases to 1194 m^3 day^{-1} during the rainfall months (October to April) and decreases to 114 m^3 day^{-1} for months normally associated with little or no rainfall. For ten day periods after any recorded rainfall, the difference between simulated and measured streamflows is only 16 m^3 day^{-1}.

Soil Water Contents

The comparisons between simulated and measured soil water contents of land segment 1 are presented in Fig. 16. The figure is divided into four segments starting with daily rainfall data at the top then simulated vs. measured water contents of the A and B-horizon and then simulated vs. measured water contents of the intermediate zone or C-horizon at the bottom. The water contents are expressed as a fraction of the porosity for the different horizons.

Fig. 16 shows a very good correlation between measured and simulated water contents of the A-horizon. Measured water contents for this horizon is slightly higher for most of the simulation period but drainage is quicker than simulated at the end of the rain season. Measured water contents are far higher than those simulated for the B-horizon (Fig. 16). Seasonal variation is evident in the simulated water contents but not as profound in the measured values. Sharp increases and decreases are noted in measured values, but not in simulated water contents. Measured and simulated water contents of the C-horizon compared well, although the measured contents show less seasonal variation than those simulated (Fig. 16). The response of this horizon to rainfall at the beginning of the simulation period shows a lag time of about 2 months.

Discussion

Underestimation of quick flows can be attributed to low QFRESP values. The method used to estimate QFRESP awarded high values for the wetland regions i.e. land segments 4 and 7. Van Tol (2010) showed that approximately 92% of the precipitation on the wetland will arrive at the catchment outlet on the same day. It was therefore surprising that QFRESP values of 0.92 and 0.82 for land segments 4 and 7 respectively were insufficient to generate comparable peak flows (Table 2). There are two possible explanations for this: some portions of land segments 3 and 5 form part of the wetland and should therefore be given higher QFRESP values; upslope land segments 1 and 5 make a larger contribution to daily flows than what their QFRESP values suggest. The latter is supported by findings of Lorentz et al. (2007), as well as our later trench and slotted pipe experiments which show a significant volume of water flowing laterally in the A-horizon of upslope land segments. This near 'surface macropore flow' can contribute approximately 16% of event water (Lorentz et al., 2007) and should be taken into consideration when assigning QFRESP values to land segments. The result for the A horizon of land segment 1 presented in Fig. 16 do not however support that suggestion. The reliable simulation of water content for this horizon indicates that the simulated amount of water infiltrating the profile is more or less in agreement with actual infiltration, and that the QFRESP value i.e. 0.002, for this land segment is satisfactory. One could argue that increasing QFRESP and lowering ABRESP would result in similar water contents in the A-horizon, as there would be less recharge of the B horizon and therefore a larger volume build-up in the A horizon. This will however deprive water from the already underestimated B-horizon and will alter the reasonably wellsimulated water balance of the intermediate zone (Fig. 16). Similarly one cannot simulate less surface water from the A-horizon of land segment 5 by assigning a higher QFRESP value, as water contents of this horizon are underestimated for the greatest part of the simulated period. Lowering the ABRESP would decrease the water content of the already underestimated B-horizon and changing the BFRESP will adversely adjust the relatively well simulated water contents of the intermediate zone.

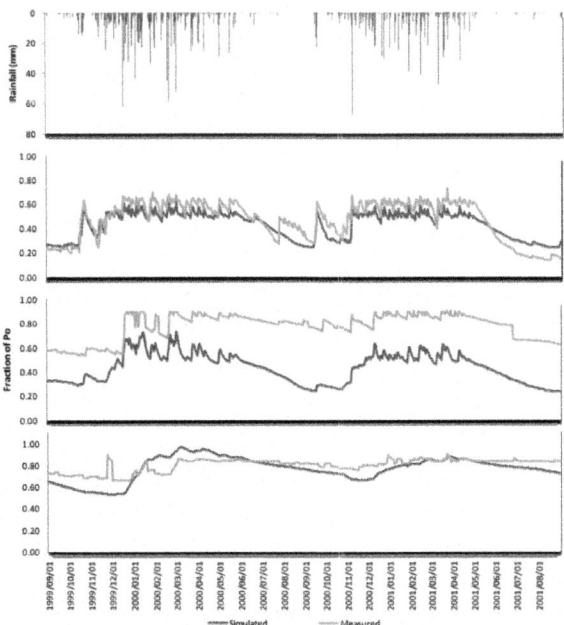

Figure 16: Rainfall and simulated vs. measured water contents of A, B and C-horizons (top to bottom respectively) of land segment 1.

In both land segments 1 and 5 the water contents of the C-horizon are simulated fairly well. Assigning higher QFRESP values to compensate for underestimation of peak flows and then lowering ABRESP, BFRESP and also INTRESP should probably result in comparativelywater content simulations and might increase the accuracy of streamflow predictions. That would however be the antithesis of the aim of this study and will definitely not contribute to predictions in ungauged basins.

It is clear that a method for attaining accurate QFRESP values is needed. This method should encompass soil and landscape properties and should preferable be dynamic in nature, as one of the major driving forces for quick flow generation is the antecedent water content of the topsoil. The latter is in accord with the well known 'variable source area' concept. The influence of the antecedent water on streamflow is further emphasized in Fig. 13. In the 1999/2000 season the first significant increase in streamflow was recorded on the 8th of December after 153 mm of precipitation had been recorded from the beginning of September. For the 2000/2001 season the first significant increase in streamflow was recorded on the 7th of November following 110 mm of precipitation from the beginning of September. Before the 8th December 2000 and the 7th of November, 'peak flows' were overestimated by the model as the storage capacity of the catchment had not yet been filled. Simulated low flows correspond very well with measured flows (Fig. 14). Simulations in Le Roux et al. (2010) using ACRU2000 (in lumped and distributed mode) and ACRU-Int (using 1 and 3 land segments), could simulate high flows moderately well but was unsuccessful in simulating low flows. It is believed that the model configuration used in this study represents the actual processes generating base flow, i.e. drainage from the soil and not from groundwater, and it was therefore encouraging to observe the connection between simulated and measured outflows.

Water contents of the A-horizon were over estimated for the majority of the simulation period in land segments 2, 3, 4, 6 and 7. The reality is that the land segments represent an area, whereas the profiles with the measured data only represent a point in that area. This should be kept in mind. Lateral flow, such as near surface macropore flow, occurs in a downslope direction. When the representative profile is in the lower regions of the land segment, these lateral contributions might have accumulated in the horizons resulting in higher measured compared to simulated water contents. The same applies for the underestimation of water contents in the B-horizons of all land segments except for 3. Routing of A and B-horizons to A, B and C horizons of different land segments might aid in solving this problem and ensure a better imitation of actual hydrological processes.

CONCLUSIONS

The grand challenge for predictions in ungauged basins (and for hydrological modeling as a whole) requires a holistic approach based on an improved understanding of the complex hydrological system and landscape heterogeneity. Soil acts as first order control in catchment hydrology through governing of water flowpaths and thereby influencing residence times and storage mechanisms. Water controls soil formation to such an extent that the soil carries signatures (morphological properties) of the hydrological processes involved in soil genesis. Based on this interactive relationship, soil scientists can make a valuable contribution to the improved understanding of the hydrological behaviour of the system.

The interactive relationship between soil and hydrology was demonstrated through conceptualizing the dominant hydrological response of soil profiles in the Weatherley catchment with six years of measured soil water contents. Results supported the interpretations and emphasized the different hydrological behaviour of different soils. This led to identification of three hydrological soil types namely recharge, interflow and responsive soil types, based on water movement and storage within these soils. The interpretation and conceptualization of soil distribution patterns of hillslopes in the Weatherley and Two Streams catchments reflected the actual hydrological behaviours of these hillslopes. Soil property distribution at hillslope scale proves to be a vehicle for identifying and studying first order controls of hillslope hydrological behaviour.

There is abundant evidence from international and local sources that soil information can improve efficiency of hydrological models. Because hydropedological studies are still in their infancy in South Africa (and in the world) huge gaps exist between the modelers (hydrologists) and experimentalists (soil scientists). Future studies should aim to close the gap by more interaction with all stakeholders. Only through interdisciplinary research can problems like pollution, land use change, floods and climate change be solved. There is need to expand these hydropedological studies in order to improve qualitative and quantitative understanding, conceptualization and characterization of flowpaths, connectivity's, thresholds, non-linearity's and residence times at hillslope scale. This can only be achieved through intensive studies of soils in hillslopes over a wide range of climates and geologies.

REFERENCES

1. Asano, Y., Uchida, T. & Ohte, N., 2002. Residence times and flow paths of water in steep unchannelled catchments, Tanakami, Japan. Journal of

Hydrology 261, pp. (173-192).
2. BEEH, 2003. Weatherley Database V1.0. School of Bio-resources Engineering and Environmental Hydrology, University of Natal, Pietermaritzburg.
3. Beven, K. J., 2001. On fire and rain (or predicting the effects of change). Hydrological processes 15, pp. (1397 – 1399).
4. Bockheim, J. G. & Douglass, D. C., 2006. Origen and significance of calcium carbonate in soils of southwestern Patagonia. Geoderma 136, pp. (751 – 762).
5. Boorman, D. B., Hollis, J. M. & Lilly, A., 1995. Hydrology of soil types: a hydrologically based classification of the soils of the United Kingdom. IH Report No.126. Institute of Hydrology, Oxfordshire, UK.
6. Clothier, B. E., Green, S. R. & Deurer, M., 2008. Preferential flow and transport in soil: progress and prognosis. European Journal of Soil Science 59, pp. (2 – 13).
7. Driessen, P. & Deckers, J., 2001. Lecture notes on the major soils of the world. http://www.fao.org/DOCREP/003/Y1899E/y1899e09.htm. (Retrieved 26/05/2008).
8. Dunn, S. M., Freer, J., Weiler, M., Kirkby, M. J., Seibert, J., Quinn, P.F., Lischied, G., Tetzlaff, D. & Soulsby, C., 2008. Conceptualization in catchment modelling: simply learning? Hydrological processes 22, pp. (2389 – 2393).
9. Fanning, D. S. & Fanning M. C. B., 1989. Soil: Morphology, Genesis and Classification. Wiley & Sons, New York. pp. (360–369).
10. Hillel, D., 1980. Fundamentals of soil physics. Academic Press, New York. IUSS Working Group WRB. 2006. World reference base for soil resources 2006. 2nd edition. World Soil Resources Reports No. 103. FAO, Rome.
11. Karvonen, T., Koivusalo, H., Jauhiainen, M., Palko, J. & Weppling, K., 1999. A hydrological model for predicting runoff from different land use areas. Journal of Hydrology. 217, pp. (253-265).
12. Kutílek, M. & Nielson, D. R., 2007. Interdisciplinary of hydropedology. Geoderma, 138, pp. (252 – 260).
13. Le Roux, P. A. L., 1996, Die aard, verspreiding en genese van geselekteerde redoksmorfe gronde in Suid Afrika. Ph. D. Thesis. University of the Orange Free State, Bloemfontein.
14. Le Roux, P. A. L., van Tol, J. J., Kuenene, B. T., Hensley, M., Lorentz, S. A., Everson, C. S., van Huyssteen, C. W., Kapangaziwirri, E. & Riddell,

E. S., 2010. Hydropedological interpretations of the soils of selected catchments with the aim of improving the efficiency of hydrological models. Report No. K5/1748. Water Research Commission, Pretoria, South Africa.

15. Lilly, A., Boorman, D.B. & Hollis, J.M., 1998. The development of a hydrological classification of UK soils and the inherent scale changes. Nutrient Cycling in Agroecosystems 50, pp. (299 – 302).

16. Lin, H. S., 2003. Hydropedology: bridging disciplines, scale and data. Vadose Zone Journal, 2, pp. (1 – 11).

17. Lin, H. S., Kogelman, W., Walker, C. & Bruns, M. A., 2006. Soil moisture patterns in a forested catchment: A hydropedological perspective. Geoderma 131, pp. (345 – 368).

18. Lin, H. S., Bouma, J., Owens, P. & Verpraskas, M., 2008. Hydropedology: Fundamental issues and practical applications. Catena 73, pp. (151 – 152).

19. Lin, H. S., 2010. Earth's critical zone and hydropedology: concepts, characteristics and advances. Hydrol. Earth. Syst. Sci. 14, pp. (25 – 45).

20. Lorentz, S. A., 2001. Hydrological systems modelling research programme: hydrological processes. Report No. 637/1/01. Water Research Commission, Pretoria.

21. Lorentz, S. A., Thorton-Dibb, S., Pretorius, C. & Goba, P., 2004. Hydrological systems modelling research programme: hydrological processes. Report No. 1061 & 1086/1/04. Water Research Commission, Pretoria.

22. Lorentz, S. A., Bursey, K., Idowu, O., Pretorius, C. & Ngeleka, K., 2007 (a). Definition and upscaling of key hydrological processes for application in models. Report No. K5/1320. Water Research Commission, Pretoria.

23. MacVicar, C. N., De Villiers, D. E., Loxton, R. F., Verster, E., Lambrechts, J. J. N., Merryweather, R. F., Le Roux, J., Van Rooyen, T. H. & Von Harmse, H. J., 1977. Soil Classification – a binomial system for South Africa. Dept. Agric. Tech. Serv. Pretoria.

24. McDonnell, J. J., Sivapalan, M., Vaché, K., Dunn, S., Grant, G., Haggerty, R., Hinz, C., Hooper, R., Kirchner, J., Roderick, M. L., Selker, J. & Weiler, M., 2007. Moving beyond heterogeneity and process complexity: A new vision for watershed hydrology. Water Resources Research 43, pp. (1–6).

25. McGlynn, B. L., McDonnell, J. J. & Brammer, D. D., 2002. A review of the evolving perceptual model of hillslope flowpaths at the Maimai catchments, New Zealand. Journal of Hydrology 257, pp. (1-26).

26. Mosley, M. P., 1982. Surface flow velocities through selected forest soils South Island, New Zealand. Journal of Hydrology 55, pp. (65-92).
27. Nahar, N., Govindaraju, R. S., Corradini, C. & Morbidelli, R., 2004. Role of run-on for describing field-scale infiltration and overland flow over spatially variable soils. Journal of Hydrology. 286, pp. (36-51).
28. Netterberg, F., 1978. Dating and correlation of calcretes and other pedocretes. Trans. Geol. Soc. S. Afr., 81, 379 – 391.
29. Nieber, J. L., Bauters, T. W. J. Steenhuis, T. S. & Parlange, J. Y., 2000 Numerical simulation of experimental gravity-driven unstable flow in water repellent sand. Journal of Hydrology, 231, 295-307Park, S. J., Mcsweeney, K. & Lowery, B., 2001. Identification of the spatial distribution of soils using a process-based terrain characterization. Geoderma. 103, pp. (249-272).
30. Park, S. J. & Van De Giesen, N., 2004. Soil-landscape delineation to define spatial sampling domains for hillslope hydrology. Journal of Hydrology. 295, pp. (28 – 46).
31. Schaap, M. G., 2000. ROSETTA version 1.2. U. S. Salinity Laboratory. ARS-USDA. Schulze, R. E. 1995. Hydrology and agrohydrology: A text to accompany the ACRU 3.00 agrohydrological modelling system. Water Research Commission, Report No 63/2/84. WRC, Pretoria.
32. Schulze, R. E., 2007. Soils: Agrohydrological information needs, information sources and decision support. In Schulze, R.E (Ed). 2007. South African atlas of climatology and agrohydrology. Report No. 1489/1/06. Water Research Commission, Pretoria.
33. Seibert, J. & McDonnell. J. J., 2002. On the dialog between experimentalist and modeller in catchment hydrology: Use of soft data for multicriteria model calibration. Water Resour. Res. 38 (11), 1241, doi:10.1029/2001WR000978.
34. Shankar, N. & Achyuthan, H., 2007. Genesis of calcic and petrocalcic horizons from Coimbatore, Tamil Nadu: Micromorphology and geochemical studies. Quaternary International 175, pp. (140 – 154).
35. Siddle, R. C., Noguchi, S., Tsuboyama, Y. & Laursen, K., 2001. A conceptual model of preferential flow systems in forested hillslopes: evidence of seld-organization. Hydrol. Process. 15, pp. (1675-1692).
36. Sivapalan, M. 2003a. Prediction in ungauged basins: a grand challenge for theoretical hydrology. Hydrol. Process., 17, pp. (3163 – 3170).
37. Sivapalan, M., Takeuchi, K., Franks, S.W., Gupta, V.K., Karambiri, H., Lakshmi, V., Liang, X., McDonnell, J.J., Mendiono, E. M., O'Connell,

P.E., Oki, T., Pomeroy, J.W., Schertzer, D., Uhlenbrook, S., & Zehe, E., 2003. IAHS decade on prediction in ungauged basins (PUB), 2003 – 2012: Shaping an exciting future for the hydrological sciences. Hydrol. Sci. J. 48 (6) pp. (857 – 880).

38. Smithers, J. & Schulze, R. E., 2004. ACRU Agrohydrological modelling system: user manual v4.00. School of Boiresources Engineering and Enviromental Hydrology, University of Natal, Pietermaritzburg.

39. Soulsby, C. & Tetzlaff, D., 2008. Towards simple approaches for mean residence time estimation in ungauged basins using tracers and soil distributions. Journal of Hydrology 363, pp. (60 – 74).

40. Soil Classification Working Group, 1991. Soil Classification - A taxonomic system for South Africa. Mem, agric. Nat . Resour. S. Afr. No. 15. Dept. Agric. Dev., Pretoria.

41. Soil Survey Staff, 1992. Keys to Soil Taxonomy, 5th edn. Pocahontas Press Inc., Blacksburg, Virginia.

42. Ticehurst, J. L., Cresswell, H. P., McKenzie, N. J. & Clover, M. R., 2007. Interpreting soil and topographic properties to conceptualise hillslope hydrology. Geoderma 137, pp. (279 – 292).

43. Tromp-van Meerveld, I. & Weiler, M., 2008. Hillslope dynamics modelled with increasing complexity. Journal of Hydrology 361, pp. (24 – 40).

44. Uchida, T., McDonnell, J. J. & Asano, Y., 2006. Functional intercomparison of hillslope and small catchments by examining water source, flowpath and mean residence time. Journal of Hysrology 327, pp. (627-642).

45. Uhlenbrook, S., Wenninger, J. & Lorentz, S., 2005. What happens after the catchment caught storm? Hydrological processes at the small, semi-arid Weatherley catchment, South-Africa. Advances in Geosciences 2, pp. (237 – 241).

46. Van Breedeman, N. & Brinkman, R., 1976. Chemical equilibria and soil formation. In G. H. Bolt & M. G. M. Bruggrnwert (eds.). Soil chemistry. A. Basic elements. Elsevier, Amsterdam.

47. Van Huyssteen, C. W. & Ellis, F., 1997. The relationship between subsoil colour and degree of wetness in a suite of soils in the Grabouw district, Western Cape I. Characterization of colour-defined horizons. S. Afr. J. Plant Soil, 14, pp. (149 – 153).

48. Van Huyssteen, C. W., Hensley, M., Le Roux, P. A. L., Zere, T. B. & Du Preez, C. C., 2005. The relationship between soil water regime and soil profile morphology in the Weatherley atchment, an afforestation area in the Eastern Cape. Report no. 1317/1/05. Water Research Commission,

Pretoria.

49. Van Huyssteen, C. W., 2008. A review of the advances in hydropedology for application in South Africa. S. Afr. J. Plant Soil, 25, pp. (245 – 254).

50. Van Tol, J. J., Le Roux, P. A. L., Hensley, M. & Lorentz, S. A., 2010. Soil as indicator of hillslope hydrological behaviour in the Weatherley Catchment, Eastern Cape, South Africa. Water SA. 36, pp. (513 – 520).

51. Van Tol, J. J., Le Roux, P. A. L. & Hensley, M., 2010a. Soil properties as indicators of hillslope hydrology in the Bedford catchments. S. Afr. J. Plant & Soil 27, pp. (242 – 251).

52. Verpraskas, M. J. & Bouma, J., 1976. Model experiments on mottle formation simulating field conditions. Geoderma 15, pp. (217-230).

53. Webster, R. 2000. Is soil variation random? Geoderma 97, pp. (149 – 163).

54. Weiler, M. & McDonnell, J., 2004. Virtual experiments: a new approach for improving process conceptualization in hillslope hydrology. Journal of Hydrology 285, pp. (3 – 18).

55. Wenninger, J., Uhlenbrook, S., Lorentz, S. & Leidbungut, C., 2008. Identification of runoff generation processes using combined hydrometric, tracer and geophysical methods in a headwater catchment in South Africa. Hydrological Sciences-Journal-des Sciences Hydrologiques 53, pp. (65 – 80).

56. Zere, T. B., 2005. The hydropedology of selected soils in the Weatherley catchment in the Eastern Cape of South Africa. Ph.D. Dissertation. University of the Free State, Bloemfontein.

Chapter 9

UPDATED BRAZILIAN'S GEOREFERENCED SOIL DATABASE – AN IMPROVEMENT FOR INTERNATIONAL SCIENTIFIC INFORMATION EXCHANGING

Marcelo Muniz Benedetti[1], Nilton Curi[2], Gerd Sparovek[3], Amaury de Carvalho Filho[4] and Sérgio Henrique Godinho Silva[2]

[1]Centro de Tecnologia Canavieira
[2]Universidade Federal de Lavras
[3]Escola Superior de Agricultura "Luiz de Queiroz" /USP
[4]Empresa Brasileira de Pesquisa Agropecuária Brazil

INTRODUCTION

The Brazilian soils represented by this database are denominated accordingly the available classification system which follows a hierarchy based on attributes and diagnostic horizons constituting the frame of the system. All the classes of soil present properties and characteristics determining a differential behavior due to the action of formation agents on the originating material. This heterogeneity has been studied for many years in order to permit a major understanding about Brazilian soils.

The classification systems allow the information organization tracking knowledge development. The classification represents the art of designing systems with some intuitive ideas about their subdivisions and priorities. In this way, it can be understood why and how soil classification systems prepared by people with varied knowledge, techniques and practical expertise differ from each other.

The soil classification is still away from the improved level of development reached by the botanical or zoological classification, but relevant progress has been achieved (Resende et al., 2007). Although there isn't a worldwide unified system available yet, attempts have been made in that direction. Among the many systems of pedological classification, the Soil Taxonomy

(United States, 1975, 1999) developed in the United States must be mentioned. It constitutes the most elaborated and comprehensive system; even though, it presents some problems regarding to tropical soils (Resende et al., 2007). To allow the classification of all kinds of soil in the planet, a system with special characteristics has been continuously developed by the FAO to elaborate the map of soils of the world (FOOD AND AGRICULTURE ORGANIZATION OF THE UNITED NATIONS, 1974). In this way, there is a tendency in many countries, including Brazil, to develop their own soil classification systems keeping a coherent relation with the FAO system and the Soil Taxonomy (Resende et al., 2007).

During the most intensive period of soil surveys in Brazil, there was not a consolidated soil classification system available yet. In this way, maps and pedological surveys reports present a considerable range of differences in applied nomenclature, soil profiles classifications and legend compositions.

The Brazilian system of soil classification (Empresa Brasileira de Pesquisa Agropecuária [EMBRAPA], 1999, 2006) introduced significant modifications both in concepts and nomenclature. Because of that, an actualization and standardization is necessary to allow comparative analysis with the classifications applied in the previous surveys, which constitute the main source of data and relevant information for the system's development itself.

Updates of legends and pedological surveys have been done often with many objectives: a) just to update the legend; b) to unify legends of different maps with a specific objective, as the evaluation of agricultural suitability of lands; c) to allow fitting in other kinds of soil classifications; and d) to make easier the exchange of expertise among researchers from different countries. The update is also essential for the construction and use of a database to allow the treatment and use of information in a quick, adequate way firmed over the actualized bases. In the last decades, the quick evolution of the methods and instrumentations used for the acquisition, storage, recovering, manipulation, analysis, access and distribution of data has facilitated the treatment of a great amount of data by the soil scientists. These techniques are indispensable for the monitoring and evaluation of the soil systems, their components and processes (Baumgardner, 1999). The way as the data are stored in a database helps the organization, searching and updating of information. With this, the same data can be used for different applications reducing the space and effort (Assad & Sano, 1998). As examples it can be mentioned: (a) Digital Soil Map of the World (FAO, 1996); (b) SOTER - The World Soils and Terrain Database (Van Engelen, 1999); (c) CANSIS - Canadian Soil Information System (Coote & Macdonald, 1999); (d) NASIS - National Soil Information System (Soil Survey Staff, 1991); (e) Hydraulic Properties of European Soils (HYPRES) database (Nemes et al., 1999); and (f) Unsatured Soil Hydraulic Database (UNSODA), in its second version already (Nemes et al., 2001).

In this way, a database of Brazilian soils was developed starting from soil surveys done by the former "Soil's Conservation and Survey National Service" (SNLCS) of Embrapa (actual Embrapa Solos), previous institutions and the RADAM Brazil Project (Cooper et al., 2005). This basis allows just quantitative evaluation because of the lack of an updated soil's profile classification related to nomenclature changes and time based taxonomic distinction criteria. Included in the described context, this work makes an analysis of the structure and an update of the soil's database profile classification to make available a more friendly and practical database for future research and analysis. Through a consistent and compatible identification of the database registers relatively to the soil's profiles, it is possible to update the basis comparing similar soils in distinct regions, defining central- and dispersion-values of the attributes, or generating studies focusing on the update of the classification system under analysis. Besides, this study furnishes information to help in the identification of the different classes of soil applying an approximated correspondence involving the Brazilian, American Soil Taxonomy and International Classification WRB/FAO systems.

SOIL DATABASE

This work has used information originated from soil surveys in regional scale, covering a great portion of the Brazilian territory (Figure 1), contained in a Brazilian database soil (Cooper et al., 2005).

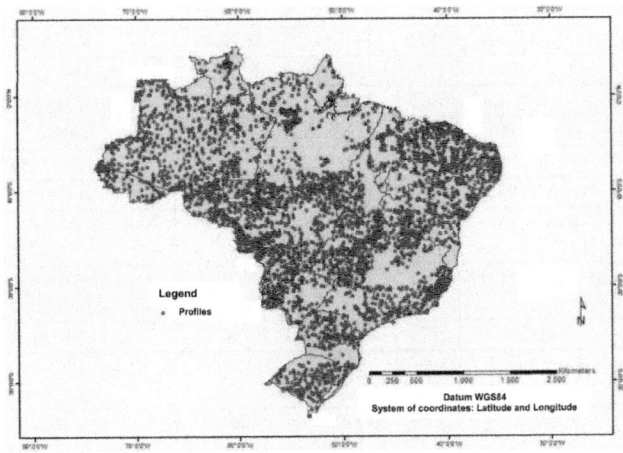

Figure 1: Brazil map showing the political division and locations of the soil profiles used in the study.

The works which originated the database came from official soil surveys chosen to assure the total coverage of the Brazilian territory. They were published between 1960 and 1986 by the former Soil's Conservation and Survey National Service (SNLCS-Embrapa), previous institutions, and the RADAM Brazil Project as described in table 1.

Table 1: Source and year of the surveys used in the database

RADAM Volumes	Region	Year	Bulletin	Region	Year
1	São Francisco River and Aracaju	1973	34 SNLCS - BT N°60	North of Minas Gerais	1979
2	Teresina and Jaguaribe	1973	35 SNLCS - BP N°27 Vol. I	Paraná	1984
3	São Luís and Fortaleza	1973	36 SNLCS - BP N°27 Vol. II	Paraná	1984
4	Araguaia and Tocantins	1974	37 SNLCS - BP N°28	Pole Trombetas –PA	1984
5	Belém	1974	38 SNLCS - BP N°29	Pole Carajás – PA	1984
6	Macapá	1974	39 SNLCS - BP N°30	Urucará - AM	1984
7	Tapajós	1975	40 SNLCS - BP N°35 Vol. II	Maranhão	1986
8	Boa Vista / Tumucumaque	1975	41 SNLCS - BP N°36 Vol. II	Piauí	1986

RADAM Volumes	Region	Year	Bulletin	Region	Year
9	Tumucumaque /Tocantins	1975	42 SNLCS - BP N° 32	Barreirinha - AM	1984
10	Santarém	1976	43 SNLCS - BP N°18	Pole Roraima - RR	1983
11	Neblina Peak	1976	44 SNLCS - BP N°19	Tefé - AM	1983
12	Rio Branco	1976	45 SNLCS - BP N°20	Espírito Santo	1977
13	Javari/Contamana	1977	46 SNLCS - BP N°23	Ceará	1972
14	Içá – AM	1977	47 SNLCS - BP N°24	Bahia	1972
15	Juruá – AM	1977	48 SNLCS - BP N°31	Carneiro - AM	1984
16	Porto Velho	1977	49 SNPA - BT N° 12	São Paulo	1960
17	Purus – AM	1978	50 SNLCS - BP N°35 Vol. I	Maranhão and Piauí	1986
18	Manaus – AM	1978	51 SNLCS - BP N°15	Pole Pré-Amazônia Maranhense	1982
19	Guaporé	1979	52 SNLCS - BP N°16	Ariquemes - RO	1982
20	Juruema	1980	53 EPFS - BT N°13	Espírito Santo and Minas Gerais	1970
21	Fortaleza	1981	54 DPP – BT N°26 Vol. II	Pernambuco	1972
22	Tocantins	1981	55 SNLCS - BP N°26	Maranhão and Piauí	1984
23	Jaguaribe/Natal	1981	56 DPP – BT N°18	South of Mato Grosso	1971
24	Salvador	1981	57 EPFS - BT N°15	Paraíba	1972

25	Goiás	1981	58 SNLCS - BT N°45	Espírito Santo	1978
26	Cuiabá	1982	59 SNLCS - BT N°38	Left Margin of São Francisco River - BA	1976
27	Corumbá	1982	60 SNLCS - BT N°52 Vol. II	Right Margin of São Francisco River - BA	1977-1979
28	Campo Grande	1982	61 DPP - BT N°21	Rio Grande do Norte	1971
29	Brasília	1982	62 DPP - BT N°28	Ceará	1973
30	Aracaju	1983	63 SNLCS - BP N°14	Apiaú - RO	1982
31	Goiânia	1983	64 SNLCS - BP N°17	Mato Grosso	1983
32	Rio de Janeiro / Vitória / Goiânia	1983	65 SNLCS - BP N°9	Barreirinha - AM	1982
33	Porto Alegre/ Uruguaiana / Lagoa Mirim	1975	66 SNLCS - BP N°36 Vol. I	Piauí	1986

* RADAM - Exploratory Soil Surveys (Radam Brazil Project); DPP - Division of Pedologic Research; EPFS - Staff of Pedology and Soil Fertility; SNPA - National Service of Agronomic Research (Committee on Soil); SNLCS - Soil's Conservation and Survey National Service (now Embrapa Solos); BP - Research Bulletin; BT - Technical Bulletin.

The database was implemented using the Microsoft Excel® format, including a set of data relative to the characteristics and analytical results of a superficial and a subsuperficial soil horizon of each profile (Table 2). The profiles are identified by codes referent to the publication source together with their original numeration, while for the horizons it was maintained the same designations used in the original works. The variables for which the data were compiled and placed in a distinct column are constituted by: code from the publication source, number of the profile, year of the survey, latitude, longitude, original soil classification, slope, drainage, symbol and depth of the horizon, moist color, macroclastic composition (pebbles and cobbles), particle size distribution (coarse sand, fine sand, silt and clay), content of oxides obtained by sulfuric acid digestion (SiO_2, Al_2O_3, Fe_2O_3), pH in H2O and in KCl, organic C, organic matter, total N, exchangeable cations (Ca^{2+}, Mg^{2+}, $Ca^{2+} + Mg^{2+}$, K^+, Na^+, Al^{3+}, H^+, $H^+ + Al^{3+}$), CEC, base sum, base saturation and aluminum saturation (Cooper et al., 2005).

Starting from this initial basis, and after proceeding with several adjustments and distortion corrections, in some cases through search in the original works, it was done the actualization of the soil profiles classification accordingly to the new version of the Brazilian system of soil classification (EMBRAPA, 2006) followed by the exportation of the data to a database management system, the Microsoft Access®. In that way, it was constituted the database of Brazilian soils which presents qualitative characteristics described in a numerical format containing 10,950 horizons referring to 5,479 soil profiles and 57 columns with the above mentioned variables and the actualized classification and respective degree of reliability (Benedetti et al., 2008). Each profile has a spatial indication related to its location on the earth surface, which means, has its coordinates (latitude and longitude) in the system of coordinates DATUM: WGS 84. The access to the database with the updated classification of the soil profiles can be done in the following web address: www.esalq.usp.br/gerd.

For the update of soil profiles database classification, compatible with the new version of the Brazilian soil classification system (EMBRAPA, 2006), it were considered the characteristics of the two horizons of each profile (superficial and subsuperficial) contained in the database, considering the criteria established for the attributes and diagnostic horizons, and the concept and definition of the known classes of soil. The taxonomic fitting of the soil profiles was done up to the fourth categorical level, and for some soils it was included the distinction in fifth level referred to the occurrence of alic character (Al saturation $\geq 50\%$ and $Al^{3+} \geq 0.5$ $cmol_c/kg$) when pertinent.

Degrees of confidence, represented by numbers from 1 to 4, were adopted to express the adequacy of the adjustment obtained in the taxonomic fitting, in one of each four categorical levels, mentioning that many profiles would need complementary data for more precisely updating the classification. The value 1 indicates taxonomic fitting with elevated degree of confidence, which reduces progressively with the increasing of the values.

Table 2: Code, description and units of representation of the database variables

Code	Variable	Description	Units
ClassfSoil	Soil Classification	Soil name according to the original document	
DecVal	Slope	Inclination of the soil surface in relation to a flat reference	%
Drain	Drainage	Velocity at which water is removed from the soil profile	
SoilDepth	Soil Depth	Lower limit of the horizon	cm
HzSimb	Soil Horizon Symbol	According to the pedological convention	
HzDepth	Horizon Depth	Depth range registering the beginning and end of the horizon	cm
Munsell Color	Munsell Color	Description according to Munsell notation	
CG	Cobble	Material> 20 mm in diameter	(%)
FG	Pebble	Material between 20-2 mm in diameter	(%)
CS	Coarse Sand	Material between 2-0.2 mm in diameter	(%)
FS	Fine Sand	Material between 0,2-0.002 mm in diameter	(%)
Sand	Sand	Material between 2-0.005 mm in diameter	(%)
Silt	Silt	Material between 0.05-0.002 mm in diameter	(%)

Clay	Total Clay	Material < 0.002 mm in diameter	(%)
SiO_2	SiO_2	Silicon oxide content	(%)
Al_2O_3	Al_2O_3	Aluminum oxide content	(%)
Fe_2O_3	Fe_2O_3	Iron oxide content	(%)
pH_H_2O	pH measured in water	pH value determined in a mixture with a specified relationship between soil and water	
pH_KCL	pH measured in KCL solution	pH value determined in a mixture with a specified relationship between soil and 1mol L^{-1} KCl	
C	Organic Carbon	Content of organic carbon in soil	(%)
MO	Organic Matter	Content of organic matter in soil	(%)
N	Total Nitrogen	Content of total nitrogen in soil	(%)
ExCa	Ca^{2+}	Content of exchangeable calcium	($cmol_c$/kg)
ExMg	Mg^{2+}	Content of exchangeable magnesium	($cmol_c$/kg)
ExK	K^+	Content of exchangeable potassium	($cmol_c$/kg)
ExNa	Na^+	Content of exchangeable sodium	($cmol_c$/kg)
ExAl	Al^{3+}	Content of exchangeable aluminum	($cmol_c$/kg)
ExH	H^+	Content of exchangeable hydrogen	($cmol_c$/kg)
Ex_H_Al	$H^+ + Al^{3+}$	Content of exchangeable hydrogen and aluminum	($cmol_c$/kg)
CTC_pH7	Cation Exchange Capacity at pH 7.0	Sum of exchangeable Ca, Mg, K, Na, H and Al (T value)	($cmol_c$/kg)
SumB	Base sum	Sum of exchangeable Ca, Mg, K and Na (S value)	($cmol_c$/kg)
SatB	Base Saturation	Relationship between sum of bases and cation exchange capacity (V value)	%
SatAl	Aluminum Saturation	Relationship between exchangeable aluminum and sum of bases more exchangeable aluminum (m value)	%

The value 2 expresses some doubt in relation to the classification at the considered categorical level, as a function of the unavailability of some analytical data of minor relevance, and the previous profile classification suggests the present fitting as the most probable. The value 3 follows a similar line of reasoning, but in this case the doubt is increased due to some missing data, being the present classification based exclusively on the previous denomination. The value 4 was applied for situations when it was not possible to complete the update.

BRAZILIAN SOILS: A PANORAMIC VISION

The set of data, referring to physical, chemical and morphological attributes, which includes also information about location, depth and drainage, made possible the updating of the classification of almost all the soil profiles compounding the database. Figure 2 presents the representativeness of the classes of soil in elevated categorical level when compared with their total area of occupation in the Brazilian territory.

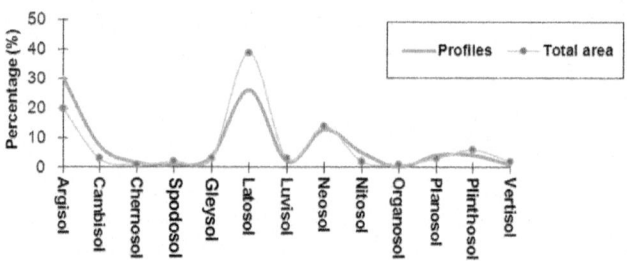

Figure 2: Percentage of profiles in the database compared with the percentage of each class of soil in total area of Brazil.

The soils were classified up to the fourth categorical level, being the alic character used when pertinent for distinction of soils in the fifth categorical level. With all classes in the most elevated categorical level (order) represented in the database, from the present actualization resulted 281 classes of soil in the subgroup level (besides three classes in suborder level referring to profiles that couldn't be classified in lower levels), for which the relation with the original classification and respective profile number are available for downloading in the following address: www.esalq.usp.br/gerd.

Through the comparison between the previous classification and the present fitting of the soil profiles in their various hierarchic levels many relevant aspects could be observed. From this evaluation emerged some characteristics expressed in the previous denomination that couldn't find a correspondence among the classes envisioned by the present SiBCS (Brazilian System of Soil Classification), both in the previous and actual versions, especially in the fourth categorical level, which resulted in loss of information. Keeping in mind to release these information to the international community and also to establish an approximated correlation between the soil classes in the Brazilian classification with the American classification "Soil Taxonomy" and with the International WRB/FAO systems, each soil class is commented individually as follows:

Argisols

With expressive participation in the database, referent to 1,660 soil profiles representing 30.29% of the total, the Argisols comprehend the majority of the soils previously classified as Podzolic soils, that present textural B horizon with low activity of clay, or with high activity of clay and low base saturation.

In this class it were also included three soil profiles previously classified as Structured Dusky Earth, that present medium texture and could not be fitted as Nitosols (class that put together almost all the former Structured Dusky Earth), and a Rubrozem profile, which by the first edition of the SiBCS would be classified as Alisol (EMBRAPA, 1999), class excluded of the present version. With the extinction of this last class and pertinent updates of the definitions of Argisols, which by now comprehend soils with elevated clay activity conjugated with dystrophic or alic character (EMBRAPA, 2006), some distortions were corrected as the case of the profile 83 in the soil survey from the "Uruguaiana/Lagoa Mirim report" (Brasil, 1975d). Originally classified as Ta Alic Gray-Brown Podzolic soil, this soil would not have taxonomic fitting accordingly to the previous edition of the system (EMBRAPA, 1999), stated that the high activity of the clay fraction would exclude it from the Argisol class, while the concentration of exchangeable Al inferior to 4 $cmol_c$/kg made impossible its fitting as Alisol, also not fitting the requirements of any other class in the order level. The introduced modifications allowed the classification of this soil as Argisol, at the same time that the reformulation of the concepts of this class in the suborder level (2nd categorical level) leaded to its fitting as Gray-Brown soil, recently included class, and in third level as Alitic, class now incorporated to the system and that corresponds to part of the Alisols in the previous version.

The inclusion of the Gray-Brown Argisols, as mentioned above, allowed the distinction between soils which have the darkening of the superior portion of the textural B horizon as remarking characteristic, referent to the former Gray-Brown Podzolic soil. Because of the strong correspondence between the previous- and the currently-employed criteria on these soils reconnaissance, the taxonomic fitting of the soil profiles without specification of the B horizon color is based on the previous denomination. However, for some soils in this class it was not possible the fitting in lower taxonomic levels, because it is recognized only one class in the third categorical level, referring to soils with alitic character.

A good correlation was also observed between the actual classes of the second categorical level, referent to Gray-, Yellow- and Red-Argisols and the former Gray-, Yellow- and Dark Red- Podzolic soils, respectively. As a function of the current concepts, this last class presents some soils in which the color hue a little bit yellower than 2.5YR (in general 3.5YR) are now fitted as Red-Yellow Argisols. Different from the others, the former Red-Yellow Podzolic soils present relevant variation in the taxonomic fitting in the second categorical level, distincted in Yellow Argisols (with dominance of colors having hue 7.5YR or yellower) and Red-Yellow Argisols (B horizon color redder than 7.5YR and yellower than 2.5YR, except when hue is 5YR with value and chroma < 4), and also in Red Argisols (hue 2.5YR or redder). Therefore, in the absence of information about the B horizon color, it was considered for these soils a more elevated degree of uncertainty in the actualization of the classification in this categorical level relatively to the others.

In the great group level (third categorical level), the available data allow for the majority of the profiles a taxonomic fitting with increased confidence. However, for the Yellow Argisols it is notable the impossibility of distinction based on the activity of clay fraction, differently from the other classes in the same level. In this way, the profile 23 from the soil survey of the Fortaleza region (Brasil, 1981a), previously classified as Ta Alic Red-Yellow Podzolic soil, for example, fits as Yellow Argisol, due to its yellow colors, and as Dystrophic in the third level, although it presents high activity of clay. Also in the fourth categorical level some differential characteristics expressed in the former classification are not considered in the soils distinction by the new system. The profile 32 of the soil survey of Fortaleza region (Brasil, 1981a), formerly classified as plinthic abruptic Alic Red-Yellow Podzolic soil, for example, fits as abruptic Dystrophic Yellow Argisol, because it is not considered in the SiBCS the possibility of fitting soils having simultaneous occurrence of such characteristics. Besides, the occurrence of transitional character indicated by the latosolic and cambic denomination are not distinguished in some cases, as example of the profile 79 of the soil survey of Rio Branco region (Brasil, 1976c), previously classified as latosolic Alic Red-Yellow Podzolic soil, and the profile AE2 of the soil survey of the pilot area at Barreirinha-AM (EMBRAPA, 1982) denominated moderate A cambic Tb Alic Red-Yellow Podzolic soil, which following the current system fits as typic Aluminic Red-Yellow Argisol.

Establishing a high categorical level correspondence with the system proposed by the FAO, the Argisols could be approximated to the classes Acrisols, Lixisols and Alisols, while for the American classification Soil Taxonomy these soils would be approximated to the classes Ultisols and Oxisols (Kandic). These classes present as main diagnostic features the clay illuviation to the subsuperficial horizons generating a textural gradient.

Cambisols

This class covers the vast majority of soils previously identified by the same name, used in order level to designate soils having restricted pedogenetic development, characterized by the presence of incipient B horizon, excepting those of the eutrophic character, with high activity clay and chernozemic A horizon, now included in Chernosol class.

Among profiles that make up the database, 397 were classified as Cambisols, which represents 7.25% of the total. While keeping the former name in the first categorical level, even without major changes in the criteria for identification, the soils of this class have suffered a systematic structuration in all subsequent levels. The distinctions between them, represented in a low systematic way by the former nomenclature, are now distributed and hierarchically organized in the second, third and fourth categorical levels.

In the suborder level, three classes are recognized: Humic Cambisols, characterized by the presence of humic A horizon; Fluvic Cambisols, referring to the soils of alluvial origin; and Haplic Cambisols, which cover the other soils of the class, whose characteristics do not meetthe requirements of the earlier ones. Of these, only the Fluvic Cambisols, class now incorporated into the classification system (EMBRAPA, 2006) and whose distinction requires a comprehensive characterization of the entire depth of the profile, or even the environment of occurrence, making its recognition difficult from the data available here, are not listed in the database. It was also observed for some profiles, a certain difficulty for the fitting as Humic, due to lack of further information about the superficial horizon.

The absence of more complete analytical data also hampered the taxonomic fitting in lower categorical levels, which increases the uncertainty about the classification. The impossibility to confirm the carbonatic nature, for example (results related to the concentration of $CaCO_3$ are not present in the database), led to the fitting as Carbonatic in the third categorical level, based on the previous name, or some characteristic that would suggest the occurrence of that character within the specified control section. Thus, the classification of the profile 244 of the soil survey of the right margin of the São Francisco River (EMBRAPA, 1977-1979) as typic Carbonatic Haplic Cambisol is based on the supposed occurrence of the carbonatic character as indicated by the previous classification (moderate A carbonatic C Tb Eutrophic Cambisol) within the first 120 cm depth.

In general, the current structuration of the classification system provided a more accurate discrimination, better representativity and a more adequate basis of comparison for Cambisols. Such result is not always observed for other classes of SiBCS. Certain distinctions afforded by the former nomenclature, however, are not represented in the current system, for example soils previously identified as Brown Cambisols, typical of the subtropical environment of southern Brazil, which currently are not being discriminated. Some characteristics are also not included in the fourth categorical level, referring to intermediate or special characteristics. Thus, for the profile 29 of the soil survey of Mato Grosso (EMBRAPA, 1983), previously classified as chernozemic A podzolic Tb Eutrophic Cambisol, the transitional characteristic expressed in the original classification was not considered, having been framed as typic Eutrophic Tb Haplic Cambisol. The plinthic character used to distinguish soils with not enough plinthite to be characterized as plinthic horizon (EMBRAPA, 2006), is not included in classes distinction of the 4th categorical level for the Eutrophic Ta Haplic Cambisols, since the plinthic class is not considered in the fourth categorical level. Thus, soils originally classified as moderate A plinthic Ta Eutrophic Cambisols like the profile 128 of the soil survey of the State of Maranhão (EMBRAPA, 1986a), or the profile 13 of SNLCS-BP-26 (EMBRAPA, 1984), lost the reference indicative of plinthite presence (expressed both in the previous classification as in the designation of the subsurface horizon), being classified as typic in the fourth level. The Brazilian Cambisols have close correlation with the American and International systems due mainly to their transitional feature in the formation process, characterized by a not very expressive B horizon in the soil profile. Their names are derived from the classification of the soil map of the world by FAO, where they are known as Cambisols. The American classification reports the origin of their names in the conditions of pedogenesis, calling them Inceptisols.

Chernosols

A total of 87 soil profiles, representing approximately 1.59% of the database, were classified in this class, which groups soils previously named as Brunizens, Reddish Brunizens and Rendzinas, plus some eutrophic Cambisols, with high clay activity and chernozemic A horizon. The reunion of these soils in a class on level of order reflects the importance attached to chemical and mineralogical characteristics together with the presence of chernozemic a horizon, with important implications for the sustainable use and management.

In the second categorical level, we observed a strong correspondence between Rendzinas and Reddish Brunizens with the current classes of Rendzic Chernosols and Argilluvic Chernosols, respectively, while the former Brunizens and Cambisols were classified as Haplic. However, for some profiles, the taxonomic classification was a little prejudiced because there was not enough available complementary information particularly with regard to colors of all the extension of the B horizon needed to distinguish the Ebanic Chernosol (dark color, almost black), or even because this horizon was not represented in the database. Thus, only the profile 57 of the soil survey of the south of the former State of Mato Grosso (Brasil, 1971a), formerly called Reddish Brunizem, could be classified as Ebanic Chernosol.

In the level of great group (third categorical level) the main limitation for a better taxonomical classification was related to the lack of data referent to the carbonatic character. Thus, the classification as Carbonatics was based on the previous name or even on the indication of the carbonatic character given by the subordinated designation of the subsurface horizon (EMBRAPA, 1988b).

The insufficiency of data also made it difficult to classify in the fourth categorical level (subgroup). Thus, several profiles were classified as typical as an attempt, taking into account the impossibility to confirm the occurrence of distinctive characteristics of other classes. In addition, some properties expressed in the previous name don't correspond to the classes provided by the actual system, both in its previous version (EMBRAPA, 1999) as in the current one (EMBRAPA, 2006), for not being included in the key, a class that can distinguish it, which results in loss of information. This situation is exemplified by the profile 34 of the RADAM Brazil Project (Brasil, 1981a), originally classified as abruptic solodic Reddish Brunizem, now classified as abruptic Orthic Argilluvic Chernosol. Also for profiles 48 (Brasil, 1972a) and 34 (Brasil, 1981a) of the RADAM Brazil Project, both named as vertic abruptic Reddish Brunizem, the differential characteristic due to the abrupt textural change is not addressed in the current classification.

The Chernosols occupy small areas in the Brazilian territory, but they have great expressiveness in temperate regions. They can be correlated with the Chernozems, Kastanozems, Phaeozems and Greyzems classes in the legend of the WRB / FAO. In the American system these soils correspond to Mollisols that exhibit high activity of clay.

Spodosols

Corresponding to soils that are distinguished by the presence of the spodic B horizon, this class includes the former Podzols and Hydromorphic Podzols, represented in the database by 53 soil profiles, which equals to 0.97% of the total.

In the second categorical level (suborder), these soils were classified as Humiluvic and Ferriluvic. However, the taxonomic fitting in this level was somewhat difficult due to the lack of complementary morphological information and, in some cases, the lack of data relative to the B horizon. For these profiles, the value 3 was used to represent the degree of confidence in order to provide the continuity of classification, since the distinction of classes in the subsequent categorical levels bases on equivalent concepts and denominations, especially in the great group level. In general, the classification in this hierarchical level also has a certain amount of uncertainty (degree of confidence 2), due to the lack of information about environmental conditions, once the presence of water table or saturation by water within 100 cm from the soil surface are distinctive criteria of classes. Therefore, the classification of the profiles, as Hydromorphics or Orthics, was based on the previous classification that, in general, distinguished soils subjected to hydromorphic conditions or not. Similarly, the fourth level of classification (subgroup) was also a bit impaired by the absence of complementary data of profiles that would allow a more precise taxonomic fitting . Therefore, the actualization of classification of soil profiles that belong to the class of Spodosols, in level of suborder, great group or subgroup, is about a kind of approximation, in correlation with the previous classification.

The Spodosols of American Soil Taxonomy include diagnostic characteristics similar to the spodic B horizon of the Brazilian Espodosols. The legend of the WRB / FAO retains the name formerly used by the mapping of soils in Brazil, as can be seen in the database the designation of Podzols.

Gleysols

Class of soils consisting of mineral material, whose genesis is related to strong hydromorphic conditions, expressed in the presence of gley horizon, and absence of distinctive features of Plinthosols or Planosols. This class is represented in the database by 214 soil profiles, which correspond to 3.91% of the total. They include soils formerly and generically named as Gleysols or Gley soils, which often were separated in Humic Gley, Low Humic Gley, Tiomorphic Gley and Saline Gley. It includes, in addition to those, the Gray Hydromorphic soils.

As observed for other classes, the lack of data concerning the full extension of the profiles resulted in difficulty for some taxonomic framework, inclusive in the level of order, due to their subsurface horizons contained in the database do not show enough coloration to distinguish the gley horizon . In these cases, the classification in this class was based on the previous name, in the assumption that other horizons of the profiles in question meet the requirements of this diagnostic horizon.

All classes of the second categorical level defined by the current system are contained in the database, although for some profiles the taxonomic classification at this level also provides some degree of uncertainty, especially for Salic Gleysols, since no data are presented relative to the electrical conductivity. Also for some soil profiles classified as Tiomorphic Gleysols, in which the pH of the horizons contained in the database are slightly above 3.5, therefore insufficient to characterize sulfuric horizon or sulfidric materials, requiring evaluation of other horizons for a more precise definition, the classification in the second level was established based on the former designation, with value 2 for the degree of confidence. The distinction of these soils in the third categorical level was satisfactory, without significant limitations. Already the fitting in the fourth level, highlights the impossibility of differentiation based on some characteristics formerly included in the distinction of these soils, such as the great capacity of swelling and shrinkage of the soil material, expressed in the vertic designation, such as the profile 146 of the soil survey of the State of Maranhão (EMBRAPA, 1986a), previously classified as moderate A solodic with tiomorphism vertic Ta Eutrophic Gleysol, which was classified as solodic Orthic Tiomorphic Gleysol.

It is also important to emphasize the question of the current fitting of the former Gray Hydromorphic, whose distinction related to the presence of textural B horizon, is addressed only in the fourth categorical level, referring to the class of argisolics. However, in the event of sufficient characteristics for classification as plinthic, such as the profile 132 of the RADAM Brazil Project (Brasil, 1981b), that distinction is impossible, since these ones, as well as the leptics, take precedence in the sequence of the identification key. Moreover, it is not possible to differentiate soils with other special characteristics, such as the presence of fragipan, as occurs with the profile 37 of the soil survey of the left margin of the São Francisco River (EMBRAPA, 1976c).

The Gleysols are developed over unconsolidated materials (sediments and saprolites) and are greatly influenced by the occurrence of prolonged flooding due to the position of the water table near the surface. They are recognized in the legend of the WRB / FAO as Gleysols, except for classes of Salic Gleysols, that correlate to Solonchaks. Unlike what occurs in the two systems mentioned above, the American classification Soil Taxonomy identify the inherent properties of these soils in lower levels of classification. The Gleysols can be correlated with the Entisols (Aqu-alf-and-ent-ept), while Gleysols having saline character can best approach classes of the Aridisols and Entisols (Aqusulfa-hydra-salic).

Latosols

Characterized by a very advanced stage of development, as evidenced by the insignificant occurrence or even absence of weatherable primary minerals, and the presence of latosolic B horizon, the Latosols constitute the largest class of territorial expression and agricultural potential of the country, being explored with various crops, reforestation and pasture (Ker, 1997). They represent 25.79% of the database, with a total of 1,413 soil profiles.

For this class, which name was maintained in the level of order in the current system, some subdivisions were already established since the 60's, as exemplified the denominations of Red-Yellow Latosol, Dark Red Latosol and Dusky Red Latosol, to which, over time, it were added Yellow Latosol, Brown Latosol, Ferriferous Latosol and Una Latosol. These designations were, partially, maintained in the distinction of classes in the second categorical level, although the differentiation criteria have changed. In the current system, the distinction of these soils in this level is based only on the color of the B horizon, while previously competed in addition to this, other characteristics such as the iron oxide content obtained by sulfuric acid digestion, which in low contents (< 8 % of Fe_2O_3) is a sure indicator of strong cohesion ("hard-setting soils"), condition very far from the granular structure (high friability), typical of most of the Brazilian Latosols. In this context, the previous classification system was more informative. Thus, the former Dark Red-, Dusky Red-, and Ferriferous- Latosols, generally refer to the currently Red Latosols, with the distinction of the iron oxide content used in the third categorical level. The former denominations RedYellow-, Yellow-, and Brown- Latosols still remain, but while for this last class there is a strong correspondence between past and present distinctive criteria, for the other two classes there are not, beyond nomenclature, a direct relationship with the classes of the current system. The Yellow Latosols tend to stay in the homonymous class, while the former Red-Yellow Latosols are distributed in the classes of Yellow-, Red-, and Red-Yellow Latosols; otherwise the Una Latosols from the database were all classified as Yellow Latosols. In view of this, the lack of information about the color of the B horizon constitutedone of the main difficulties for updating the classification of these soils in the second categorical level. In this case, the taxonomic fitting was oriented by the previous name, with differentiation concerning the degree of confidence, according to the following pattern: Dusky Red- and Dark Red-Latosols were classified as Reds, and Yellow- and BrownLatosols in the homonymous classes, all with degree 2 of confidence; similarly, the RedYellow Latosols were fitted in the class of the same denomination, but with degree 3 of confidence, due to the increased uncertainty of this adjustment.

In the third categorical level distinction, that is based on the characteristics of the exchange complex, in association with the contents of Fe_2O_3 from the sulfuric acid digestion, the taxonomic framework was very satisfactory as well as the subsequent level (subgroup), which seems to indicate major concern in the establishment of possible distinction of these soils, perhaps because of their high potential for agricultural use, as well as their large territorial expression in the country. The word Latosol implies soils formed under conditions of significant weathering-leaching, resulting in materials with a residual concentration of secondary minerals, including kaolinite, gibbsite, and iron- and aluminum- oxides, in different proportions. The understanding of their genesis facilitates the identification of their corresponding names in the American Classification Soil Taxonomy and legend of WRB/FAO, as Oxisols and Ferralsols, respectively.

Luvisols

Grouping of soils characterized by the presence of textural B horizon or nitic B horizon, with high clay activity and high base saturation, excluding those with chernozemic A horizon. This class is represented in the database by 118 soil profiles (2.15% of the total). Encompasses soils originally classified as Non-Calcic Brown soils and Red-Yellow Podzolic soils, with little participation of Gray-Brown Podzolic soils and Dark Red Podzolic soils. It also includes some profiles formerly called Reddish Brunizens, whose A horizons do not have enough base saturation ($\geq 65\%$) to characterize the chernozemic A horizon, according to the current criteria.

Although having achieved a very satisfactory taxonomic fitting in the level of order, some loss of accuracy in the update was already observed from the second categorical level, as a function of the available data do not allow an precise evaluation of the dominance of colors of the B horizon, necessary to distinguish between Chromic- and Haplic- Luvisols. This fact resulted in a higher degree of uncertainty regarding the taxonomic fitting, also reflected in lower categorical levels. Another aspect that difficulted the classification of some soils, due to the unavailability of data for the entire extension of the profiles, is relative to the depth of the solum (horizons A + B), characteristic used to distinguish, in the third categorical level (great group), and the class of Palic Chromic Luvisols.

For the same reason, the distinction of some soils in the fourth categorical level was also slightly affected, considering the utilized criteria to identify the classes of saprolithics and lithics. Moreover, some differential characteristics indicated in the previous denomination are not considered by the current classification, as the profile 58 of the soil survey of the State of Pernambuco (Brasil, 1972e), currently classified as planosolic Orthic Haplic Luvisol, in which the presence of solodic character and vertic characteristic expressed in the previous classification (solodic vertic planosolic Non-Calcic Brown soil), are not considered in the distinction of this type of soil in the current system.

The Luvisols can be framed in the class of Luvisols of the WRB/FAO legend that presents the B horizon with accumulation of high activity clays. In the American classification the best correspondent classes of soils would be the Alfisols and Aridisols (Argids).

Neosols

Soils generally young, with little development and in process of formation, the Neosols are characterized by the absence of any diagnostic B horizon. With 684 soil profiles, that represent 12.48% of the database, are included in this class four major soil groups recognized by the classification scheme previously adopted in Brazil, that are the Litholic Soils, Regosols, Alluvial Soils and Quartzous Sands. In close correspondence with the above groups four classes are recognized in the second categorical level: Litholic Neosols, Regolithic Neosols, Fluvic Neosols and Quartzarenic Neosols, all represented in the database. Because of this, as well as the close correlation of distinctive criteria adopted by the current classification system and the previous scheme, there were few profiles that presented some problems for taxonomic fitting in the suborder level.

In this sense, should be highlighted some soils previously classified as Litholic Soils, currently framed in the class of Regolithic Neosols, due to present the C horizon extending below 50 cm depth. However, some profiles of Quartzous Sands, might not even be fitted as Neosols, due to the change of the criteria adopted in distinguishing sandy soils, that from the 80's replaced the lower limit of 15% clay by the requirement of the granulometric compositions of the sandy and loamy sand textural classes. Consequently, the classification of these profiles was not updated.

The fitting of these soils in the third categorical level was also done without difficulty, in some cases guided by the previous classification, like some profiles of Hydromorphic Quartzous Sands for which the color of the subsuperficial horizon is not presented. However, the profiles AE9 and AE10 of the soil survey of the pilot area of Barreirinha-AM (EMBRAPA, 1982d), previously classified as solodic Ta Alic Alluvial Soils, that currently fit as Fluvic Neosols, could only be classified up to the second categorical level, because it is not included in the current system the distinction of soils of this class that present dystrophic character together with high activity clay. In this way, by correlation with similar soils, it is suggested the inclusion of the class of the Dystrophic Ta to the fitting of Fluvic Neosols in the third categorical level.

In relation to the fourth categorical level, in general, the distinction of the soils presented high agreement with the differentiations expressed in the previous denomination, although with certain degree of uncertainty in the taxonomic fitting, due to the absence of complete data of the profiles. For the profile P06 of the RADAM Brazil Project (Brasil, 1981c), it was not possible, however, the distinction regarding the occurrence of concretions, as indicated in the previous classification (concretionary Eutrophic Litholic Soil), characteristic that deserves to be evaluated as a differential characteristic of classes in the subgroup level.

The Quartzarenic Neosols updated in this database can be correlated with the Arenosols of the WRB/FAO legend, while in the American Soil

Taxonomy this class would fit in the third categorical level (great group) as Quartzipsamments. Regolithic Neosols have been recognized as Regosols by WRB/FAO. Regarding their correlation with the American system this can be done only in the second categorical level (suborder) as Psamments. Soils that present lithic contact within 50 cm depth (Litholic Neosols), can be correlated toLeptosols of the WRB/FAO legend. The American classification uses this property in the fourth categorical level (subgroup) and can be related to Lithic of Orthents suborder and also with Lithic of great group Psamments. The Fluvic Neosols are soils originated from alluvial sediments with close correlation with the Fluvisols of the WRB/FAO legend, while in the American Soil Taxonomy they would be fitted in the second categorical level as Fluvents.

Nitosols

In the database of soils of Brazil, this class is represented by 260 profiles, which equals to 4.75% of the total. Encompasses soils previously referred as Structured Dusky Red Earth, Structured Brown Earth, Similar Structured Dusky Red Earth and Similar Structured Brown Earth, beyond some profiles recognized as Red-Yellow Podzolic soils which present low textural gradient. More specifically for these last soils, there is a higher degree of uncertainty regarding the classification as Nitosols, because the database do not include information on morphological characteristics necessary for unambiguous identification of the nitic B horizon, distinctive of this class, whose requirements include structural development and significant presence of clay skins.

With the re-structuration promoted by the new version of SiBCS, regarding the inclusion of the class of Brown Nitosols in the second categorical level, many of the soils previously classified as Haplics, according to the first edition of the system (EMBRAPA, 1999), are now fitted in this new class.

In this work we option for the fitting as Brown Nitosols only the soils formerly called Structured Brown Earth or Similar Structured Brown Earth, provided they meet the requirement of specified color, or, in the absence of this information, assigning higher value for the degree of confidence (greater degree of uncertainty). This situation highlights the need for a more precise definition of the distinctive criteria of the class, where the comparison of related soils, members of the database under study could contribute. It should be registered also the impossibility of classification in lower categorical levels of the profile C48 of soil survey of the State of Paraná (EMBRAPA, 1984), originally identified as chernozemic A Eutrophic Structured Brown Earth, because in the current system classes that allow taxonomic fitting as Brown Nitosols with eutrophic character are not considered. Relatively to other classes in the second categorical level, we observed a strong correspondence between the former Structured Dusky Red Earth and Red Nitosols, that beyond these soils also comprises a small proportion of Red-Yellow Podzolic soils. However, some soils previously recognized as Structured Dusky Red Earth, many of which having high iron oxide content, were classified as Haplic Nitosols, due to present 3.5YR or 4YR hues in the B horizon.

Regarding the fourth categorical level, as observed for other classes, some differences shown in the previous classification do not found correspondence which classes of the current system, such as the latosolic Eutrophic Similar Structured Dusky Red Earth (Brasil, 1982d), which fits as typic Eutrophic Haplic Nitosol. Also the profile C63 of the soil survey of the State of Paraná (EMBRAPA, 1984), previously classified as cambic Alic Similar Structured Brown Earth was fitted as typic Aluminic Brown Nitosol, for not being included possibilities of distinction relative to intermediate characteristics represented in the previous designation.

In this class are classified clayey to very clayey soils with slight increase of clay in the subsurface horizon, not establishing enough textural gradient for fitting as textural B horizon, but being possible to observe clear clay skins on soil aggregates. These observations can be found in the classes of the Nitisols, Lixisols and Alisols of the world soils legend of the FAO, while for the American classification Soil Taxonomy, it can be observed the correlation encompassing Ultisols, Oxisols (Kandic) and Alfisols.

Organosols

Due to the limited expressiveness of these soils in the country, generally restricted to small areas, as well as more difficult access, they tend to be contemplated with a few profiles sampled in more generalized pedological surveys as the ones considered here. Thus, the presence of the class of Organosols in the database was minimal, covering only 11 profiles, which equals to 0.20% of the total.

Characterized by its essentially organic constitution, the distinctive criteria of this class in order level present an adequate correlation with the set of soils previously included under the general denomination of Organic Soils. Thus, we obtained a very satisfactory taxonomic fitting in the highest categorical level, although for some profiles, considering the unavailability of complete data, the fitting in the current classification system has been performed by direct correlation with the previous classification.

For the subsequent categorical level, all profiles were classified as Haplics. Only for the profile 37 of RADAM Brazil Project (Brasil, 1981d), which presents very low values of pH in water in the subsurface horizon, there is about the possibility of occurrence of sulfuric horizon or sulfidric material (distinctive characteristics of Tiomorphic class), thus requiring more data for confirmation.

The main difficulty in upgrading the classification of these soils was observed in the third categorical level (great group), because the distinctive criteria involve determinations not performed in epochs before the publication of the first edition of the Brazilian System of Soil Classification (EMBRAPA, 1999), which makes very difficult a more accurate evaluation, even from complete morphological and analytical characterization of soils profiles. Classification at this level was therefore undertaken in an attempt form, fitting all the profiles in the class of Saprics, which groups soils constituted of organic material in an advanced stage of decomposition.

Also in the fourth categorical level, the absence of analytical data, especially concerning the electrical conductivity, reduces the confidence of taxonomic fitting. Only the profile C137 of the soil survey of the State of Paraná (EMBRAPA, 1984), referred in SiBCS (EMBRAPA, 2006) as representative of the terric Sapric Haplic Organosols, was classified with higher degree of confidence in all categorical levels.

Most organic constituents, in various stages of decomposition, in relation to the mineral ones are the basic criteria for the distinction of Organosols in the Brazilian system of classification and Histosols by the American system and FAO.

Planosols

In relation to the grouping of soils with planic B horizon, which is characterized by abrupt transition in association with sharp increase of clay in relation to the overlying horizon, the Planosols tend to occur in environments of restricted drainage, susceptible to hydromorphism at least temporarily. This class represents 4.22% of the updated of profilesof the database, which is equivalent to 231 profiles. Refer to soils formerly called Planosols and Solodized-Solonetz, now included in the same class in the order level, distinguished in the second categorical level as Haplic Planosols and Natric Planosols , respectively. A very satisfactory adjustment between the classes previously recognized and the current ones was achieved.

However, despite the strong correlation between the current criteria for distinguishing these soils and the previously criteria used, because there are not available morphological data and analytical results of the entire profiles, some of them were classified based on the original name, such as the profile 66 of the technical bulletin 16, of the former Division of Research Pedologic (Brasil, 1972e), called weak A Ta Eutrophic Planosol, for which the subsurface horizon of the database refers to the C horizon, with no indication of the occurrence of colors having low chroma. Due to this, this soil was tentatively classified as typic Eutrophic Haplic Planosol.

The non-availability of analytical data also contributed for reducing the confidence of the update of the classification in the subsequent categorical levels. The distinction of Salic Soils, at the subgroup level, for example, was not possible, due to lack of data on electrical conductivity. Major limitations for the taxonomic fitting were observed, however, in the fourth categorical level. For example, for the profile 89 of soil survey of the State of Ceará (Brasil, 1973a), originally called weak A vertic solodic Planosol, it was not possible with the available data, the confirmation of the solodic character, although it may be present in some horizon within the first 120 cm depth. Thus, this soil was classified as vertic Eutrophic Haplic Planosol. Similarly to observed for the other classes, some differential characteristics represented in the previous name are not included in the distinction of soils by the current system. As a result, the soil represented by the profile 200 of the RADAM Brazil Project (Brasil, 1981b), formerly designated plinthic Tb endoeutrophic Planosol, was classified as typic Eutrophic Haplic Planosol, because it is not included the plinthic class.

The soils that in the first edition of the Brazilian classification system (EMBRAPA, 1999) were differentiated in the second categorical level, such as Hydromorphic Planosols, were excluded from the current version (EMBRAPA, 2006). Actually they are classified as gleissolics in the fourth categorical level, but by the ordering of the classes in the identification key, in some cases they can not be distinguished as such. The profile 81 of the RADAM Brazil Project (Brasil, 1981a), for example, even though there are enough characteristics for its fitting as gleisolic, by the current criteria it is fitted as solodic Eutrophic Haplic Planosol, due to the solodic character take precedence in the distinction of soils at this level.

As for the Dystrophic Haplic Planosols, the gleisolic class take precedence in relation to plinthic and solodic in the classification key at the fourth level, not allowing the distinction in the case of simultaneous occurrence of soils with these characteristics, as occurs for the profile 97 of the RADAM Brazil Project (Brasil, 1982b), originally classified as moderate A plinthic Tb Alic Planosol, by the current system fits as gleisolic Dystrophic Haplic Planosol. Considering the above comments, and considering the strong influence of hydromorphic conditions on sustainable use and management of these soils, as well as environmental and pedogenesis aspects, it is strongly suggested the distinction of Planosols with these characteristics in higher categorical level. This has practical application and considerable geographical expression mainly in the state of RioGrande do Sul (Costa et al., 2009). It seems appropriate to consider the possibility of keeping the distinction in the second categorical level, as adopted by the previous version of SiBCS (EMBRAPA, 1999).

The suborder of Natric Planosols can be identified in the American classification Soil Taxonomy and WRB/FAO as Natr-(ust-ud)-alf and Solonetz, respectively. Classes that do not fall into this suborder are denominated at second categorical level as Haplics and can be correlated with the classes of Planosols (WRB/FAO), and Albaquults, Albaqualfs and Plinthaqu (alf-ept-ox-ult)s of Soil Taxonomy.

Plinthosols

In this class are included soils characterized by expressive plinthization, presenting significant occurrence of plinthite, either in its mild or hard (petroplinthite) form, resulting from changes in the formation environment. In the database of soils of Brazil this class has a participation corresponding to 5.40%, which is equal to a total of 296 soil profiles. It covers the vast majority of the former Hydromorphic Laterites, which were later recognized as Plinthosols, a designation that was kept in the current system, and, with the modifications introduced in its latest version, referring to the concept of concretionary and lithoplintic horizon, now include those soils previously listed under the generic name of Concretionary Soils, including the Lateritic Concretionary Soils.

The main difficulty of current fitting of the soils in this class precisely refers to the last soils, which, because of the lack of more precise criteria for their identification, constituted a large and highly heterogeneous group. Thus, they were classified as Plinthosols according to the reference of the occurrence of concretions (petroplinthite) expressed in its name, giving high degree of confidence only for those soils with at least one horizon with 30 cm or more thickness and volume of pebbles and cobbles equals to 50% or more, enough for characterizing the concretionary horizon. For the others, considering that often the coarse material was discarded during sampling, or it was not measured, and the available data do not cover the entire extension of the profiles, it was considered greater uncertainty in the taxonomic fitting, with a moderate degree of confidence (value 2) in case there is some indication of the occurrence of petroplinthite (given by the horizon designation, for example), or lower degree of confidence (value 3), otherwise. All these soils, also including a smaller number of profiles originally called Concretionary Plinthosols or Concretionary Laterites, were grouped in a single class at second (Petric) and third categorical level (Concretionary), being distinguished only in the suborder level, due to the occurrence of some diagnostic horizon (incipient B or cambic B, latosolic B or textural B), or other differential characteristic, as mentioned in the previous name.

Other Plinthosols of the database were classified, in the second categorical level, as Argilluvics and Haplics, with great domain of the first in relation to the second ones. The change in the latest version of the system, referring to the argilluvic character (defined as having a texture ratio superior to 1.4) instead of the presence of textural B horizon to distinguish the Argilluvics (EMBRAPA, 2006), seems to be mainly responsible for this fact. In strong contraposition to the criteria adopted for Petric Plinthosols, the distinction of both classes in the subsequent categorical level is based on the saturation of exchange complex, and subsidiarily on clay activity for soils with very high aluminum content, differentiating soils with Alitic-, Aluminic-, Dystrophic- or Eutrophic-character, all of these classes represented in the database.

Also in the fourth level it is clearly demonstrated the great difference in the criteria used in distinguishing between Argilluvic Plinthosols and Haplic Plinthosols, and the Petric Plinthosols, which constitute a very distinct group within this class. The occurrence of appreciable amounts of plinthite in the soil profile is also identified in the class of the Plinthosols in the WRB/FAO international classification. The American system of classification adopts this character in lower categorical level (subgroup) and can be found correspondence of Plinthosols with different classes of the Soil Taxonomy, which present the plinthic character. These classes comprise the Oxisols, Ultisols, Alfisols, Entisols and Inceptisols.

Vertisols

The soils of this class, due to the dominance of very high activity clays, are characterized by a great capacity of swelling and shrinkage of soil material with wetting and drying cycles, a factor that restricts their pedogenetic development. Having little occurrence in Brazilian territory, Vertisols represent only 1.00% of the profiles in the database, comprising a total of 55 profiles.

As the identification of soils in this class includes criteria based on morphological characteristics evaluated in the field, the first categorical level was established by direct correlation with the original classification, whose name was maintained by the current system, just as the essence of the distinctive criteria, which allowed a satisfactory fit. For some profiles, however, there was some difficulty in taxonomic fitting, due to the presence, sometimes in the surface horizon, other times in subsurface horizon, of clay content below 300 g / kg, established as thresholds to distinguish the class of Vertisols (up to 20 cm depth, after mixing), but also for the vertic horizon (EMBRAPA, 2006). Once in the database are not included all the horizons of the profile, thereby preventing a more adequate evaluation, it was maintained their classification as Vertisols, assigning the value 2 for the degree of confidence in case of one of the horizons do not meet the specified requirement, and value 3 when both had amounts of clay below the minimum limit required. This situation, however, shows some discrepancy between the definitions of vertic horizon and of the class of Vertisols, as well as the need to adjust them. It may be appropriate to assess, from the characteristics of the profiles contained in the database, the adequacy of a minimum amount of clay established for distinction of these soils.

In the second categorical level, it were recognized classes of Hydromorphic Vertisols, Ebanic Vertisols and Haplic Vertisols. Some difficulty in classification was observed mainly for fitting of some soils as Hydromorphics, due to the subsurface horizons refer to depths greater than 50 cm, being necessary information on colors of the overlying horizons to confirm the classification (in this case it was assigned the value 2 to express the degree of confidence). Other profiles, on the other hand, did not present information on moist color, being classified as Haplics, with value 3 for the confidence degree.

Also in the great group level it was observed some difficulty in taxonomic fitting of the profiles, mainly due to the lack of relevant data, such as the contents of equivalent $CaCO_3$. This form, the fitting in the class of the Carbonatics was based on previous classification, corroborated by the subordinated designation of horizons of the profile (EMBRAPA, 1988b), being the other profiles classified as Orthics, since the previous name did not present indication of the occurrence of salic character.

The absence of analytical data has hampered the taxonomic fitting in the fourth categorical level, as example of the electrical conductivity, necessary for the identification of saline character. The characteristics of vertic horizon identify typical pedological features due to shrinkage and swelling of the minerals of this class. These features are called friction surfaces ("slickensides"), and are common both in the U.S. classification for the Vertisols as well as in the international legend of the FAO for those soils also called Vertisols.

CONCLUSIONS

The database used in this work reveals an adequate representativeness of the soil classes distributed over the Brazilian territory. This database can function as an advance to stimulate the exchanging of information and experience among researchers at international level.

REFERENCES

1. Assad, E. & Sano, E. (1998). Sistema de informações geográficas: Aplicações na Agricultura. EMBRAPA, Brasília, Brasil.
2. Baumgardner, M. (1999). Soil databases. In: SUMNER, M.E., ed. Handbook of soil science. Boca Raton, CRC Press, pp. H1-H4.
3. Benedetti, M. ; Sparovek, G.; Cooper, M.; Curi, N. & Carvalho Filho, A. (2008). Representatividade e potencial de utilização de um banco de dados de solos do Brasil. Revista Brasileira de Ciência do Solo. , v.32, No.6, pp. 2591-2600.
4. Brasil (1972a). Estudo expedito de solos nas partes central e sul do Estado da Bahia, para fins de classificação e correlação. Ministério da Agricultura. Departamento Nacional de Pesquisa Agropecuária. Divisão de Pesquisa Pedológica. (DNPEA. Boletim Técnico, 24 Sudene. Série Pedologia, 17), Recife, Brasil.
5. Brasil (1958). Levantamento de reconhecimento dos solos do Estado do Rio de Janeiro e Distrito Federal: contribuição à carta de solos do Brasil. Ministério da Agricultura. Serviço Nacional de Pesquisas Agronômicas. Comissão de Solos. (Boletim, 11), Rio de Janeiro, Brasil.
6. Brasil (1960). Levantamento de reconhecimento dos solos do Estado de São Paulo: contribuição à carta de solos do Brasil. Ministério da Agricultura. Serviço Nacional de Pesquisas Agronômicas. Comissão de Solos. (Boletim, 12), Rio de Janeiro, Brasil.
7. Brasil (1971a). Levantamento de reconhecimento dos solos do sul do Mato Grosso. Ministério da Agricultura. Departamento Nacional de Pesquisa Agropecuária. Divisão de Pesquisa Pedológica. (Boletim Técnico, 18), Rio de Janeiro, Brasil.
8. Brasil (1973h). Levantamento exploratório de solos do Estado do Ceará. Ministério da Agricultura. Departamento Nacional de Pesquisa Agropecuária. Divisão de Pesquisa Pedológica. 2v. (Boletim técnico, 26; SUDENE. DRN. Série Pedologia, 14), Recife, Brasil.
9. Brasil (1972e). Levantamento exploratório-reconhecimento de solos do Estado de Pernambuco. Ministério da Agricultura. Departamento Nacional de Pesquisa Agropecuária. Divisão de Pesquisa Pedológica. 2v. (DNPEA. Boletim Técnico, 16; SUDENE. DRN. Série Pedologia, 14), Recife, Brasil.
10. Brasil (1971b). Levantamento exploratório-reconhecimento de solos do Estado do Rio Grande do Norte. Ministério da Agricultura. Departamento Nacional de Pesquisa Agropecuária. Divisão de Pesquisa Pedológica.

(DNPEA. Boletim Técnico, 21; SUDENE. Série Pedologia, 9), Rio de Janeiro, Brasil.

11. Brasil (1970c). Levantamento exploratório dos solos da região sob influência da Companhia Vale do Rio Doce. Ministério da Agricultura. Escritório de Pesquisas e Experimentação. Equipe de Pedologia e Fertilidade do Solo. (Boletim Técnico, 13), Rio de Janeiro, Brasil.

12. Brasil (1972f). Levantamento exploratório-reconhecimento de solos do Estado da Paraíba. II. Interpretação para uso agrícola dos solos do Estado da Paraíba. Ministério da Agricultura. Escritório de Pesquisas e Experimentação. Equipe de Pedologia e Fertilidade do Solo. (Boletim Técnico, 15; SUDENE, Série Pedologia, 8), Rio de Janeiro, Brasil.

13. Brasil (1979c). Levantamento exploratório-reconhecimento de solos do norte de Minas Gerais. Ministério da Agricultura. EMBRAPA-SNLCS/SUDENE DRN (Boletim Técnico, 60), Recife, Brasil.

14. Brasil (1973-1983). Levantamentos Exploratórios do Projeto Radam Brasil – Volumes 1 a 33. Ministério das Minas e Energia. Departamento Nacional de Produção Mineral. 1973a, b, c, 1974a, b, c, 1975a, b, c, d, 1976a, b, c, 1977a, b, c, d, 1978a, b, 1979, 1980, 1981a, b, c, d, e, 1982a, b, c, d, 1983a, b, c; (Levantamento de Recursos Naturais), Rio de Janeiro, Brasil.

15. Cooper, M.; Mendes, L.; Silva W. & Sparovek, G. (2005). A national soil profile database for Brazil available to international scientists. Soil Science Society of America Journal, Madison, v. 69, pp. 649-652.

16. Coote, D. & Macdonald, K. (1999). The Canadian soil database. In: SUMNER, M.E., (Ed.). Handbook of soil science. Boca Raton: CRC Press, pp. H41-H51.

17. Costa, A.; Araújo, E.; Marques, J. & Menezes, M. (2009). Avaliação do risco de anoxia para o cultivo de eucalipto no Rio Grande do Sul utilizando-se levantamento de solos. Piracicaba: Scientia Forestalis, 37 pp. 367–375.

18. Embrapa (1988b). Definição e notação de horizontes e camadas do solo. Serviço Nacional de Levantamento e Conservação de Solos. 2. ed. (Embrapa-SNLCS. Documentos, 3), Rio de Janeiro, Brasil.

19. Embrapa (1999). Sistema Brasileiro de Classificação de Solos. Centro Nacional de Pesquisa de Solos. Rio de Janeiro, Brasil.

20. Embrapa (2006). Sistema Brasileiro de Classificação de Solos. Centro Nacional de Pesquisa de Solos. 2 ed, Rio de Janeiro, Brasil.

21. Embrapa (1984). Levantamento de reconhecimento de média intensidade

dos solos e avaliação da aptidão agrícola das terras da área do Pólo Trombetas, Pará. Serviço Nacional de Levantamento e Conservação de Solo. (EMBRAPA-SNLCS. Boletim de Pesquisa, 28), Rio de Janeiro, Brasil.

22. Embrapa (1984), Levantamento de reconhecimento dos solos e avaliação da aptidão agrícola das terras de uma área de colonização no Município de Careiro, Estado do Amazonas. Serviço Nacional de Levantamento e Conservação do Solo. (Boletim de Pesquisa, 31), Rio de Janeiro, Brasil.

23. Embrapa (1983), Levantamento de reconhecimento de media intensidade dos solos e avaliação da aptidão agrícola das terras de 21.000 hectares no município de Tefé, Amazonas. ServiçoNacional de Levantamento e Conservação do Solo. (Boletim de Pesquisa, 19), Rio de Janeiro, Brasil.

24. Embrapa (1983), Levantamento de reconhecimento de média intensidade dos Solos e avaliação da aptidão agrícola das Terras do Pólo Roraima. Serviço Nacional de Levantamento e Conservação do Solo. (Boletim de Pesquisa, 18), Rio de Janeiro, Brasil.

25. Embrapa (1984). Levantamento de reconhecimento dos solos e avaliação da aptidão agrícola das terras de uma área de colonização do Município de Barreirinha, Estado do Amazonas. Serviço Nacional de Levantamento e Conservação de Solo. (Boletim de Pesquisa, 32), Rio de Janeiro, Brasil.

26. Embrapa (1984). Levantamento de reconhecimento dos solos e avaliação da aptidão agrícola das terras do Municipio de Urucará, Estado do Amazonas. Serviço Nacional de Levantamento e Conservação de Solo. (Boletim de Pesquisa, 30), Rio de Janeiro, Brasil.

27. Embrapa (1983). Levantamento de Reconhecimento de Solos do Mato Grosso. Serviço Nacional de Levantamento e Conservação de Solos. (Boletim de Pesquisa, 17), Rio de Janeiro, Brasil.

28. Embrapa (1982). Levantamento de reconhecimento dos solos do Município de Ariquemes, Estado de Rondônia. Serviço Nacional de Levantamento e Conservação do Solo. (Boletim de Pesquisa, 16), Rio de Janeiro, Brasil.

29. Embrapa (1982), Levantamento de reconhecimento dos solos do Pólo Pré-Amazonia-Marenhense, Estado do Maranhão. Serviço Nacional de Levantamento e Conservação do Solo. (Boletim de Pesquisa, 15), Rio de Janeiro, Brasil.

30. Embrapa (1984). Levantamento de Reconhecimento de Solos e Aptidão Agrícola das Terras da Área do Pólo Carajás Estado do Pará. Serviço Nacional de Levantamento e Conservação de Solo. (EMBRAPA-SNLCS - Boletim de Pesquisa, 29), Rio de Janeiro, Brasil.

31. Embrapa (1977). Levantamento de reconhecimento dos solos do Estado do Espírito Santo. Serviço Nacional de Levantamento e Conservação de Solos. (Boletim de Pesquisa, 20), Rio de Janeiro, Brasil.
32. Embrapa (1978b). Levantamento de reconhecimento dos solos do Estado do Espírito Santo. Serviço Nacional de Levantamento e Conservação de Solos. (Boletim técnico, 45), Rio de Janeiro, Brasil.
33. Embrapa (1984). Levantamento de reconhecimento de solos do Estado do Paraná. Serviço Nacional de Levantamento e Conservação de Solos. (EMBRAPA-SNLCS - Boletim de Pesquisa, 27; IAPAR. Boletim Técnico, 16), Londrina, Brasil.
34. Embrapa (1979b). Levantamento de reconhecimento dos solos do centro-sul do Estado do Paraná (área 9): informe preliminar. Serviço Nacional de Levantamento e Conservação de Solos. (Boletim Técnico, 11), Curitiba, Brasil.
35. Embrapa (1977-1979). Levantamento exploratório-reconhecimento de solos da margem direita do rio São Francisco, Estado da Bahia. Serviço Nacional de Levantamento e Conservação de Solos. 2v. (Embrapa-SNLCS. Boletim Técnico, 52; SUDENE. Série Recursos de Solos, 10), Recife, Brasil.
36. Embrapa (1976c). Levantamento exploratório-reconhecimento de solos da margem esquerda do rio São Francisco, Estado da Bahia. Serviço Nacional de Levantamento e Conservação de Solos. (Embrapa-SNLCS. Boletim Técnico, 38; SUDENE. Série Recursos de Solos, 7), Recife, Brasil.
37. Embrapa (1982). Levantamento e reconhecimento de baixa intesidade dos solos e avaliação da aptidão agrícola das terras do projeto de colonização Apiaú - Território Federal de Roraima. Serviço Nacional de Levantamento e Conservação de Solo. (EMBRAPA-SNLCS. Boletim de Pesquisa, 14), Rio de Janeiro, Brasil.
38. Embrapa (1982). Levantamento de reconhecimento de baixa intensidade dos solos e avaliação da aptidão agrícola das terras de área piloto do município de Barreirinha - Estado do Amazonas. Serviço Nacional de Levantamento e Conservação de Solo. (EMBRAPASNLCS. Boletim de Pesquisa, 9), Rio de Janeiro, Brasil.
39. Embrapa (1986a). Levantamento exploratório - reconhecimento de solos do Estado do Maranhão. Serviço Nacional de Levantamento e Conservação de Solos. 2v. (EMBRAPASNLCS. Boletim de Pesquisa, 35. Brasil. SUDENE-DRN. Série Recursos de Solos, 17), Rio de Janeiro, Brasil.

40. Embrapa (1986b). Levantamento exploratório - reconhecimento de solos do Estado do Piauí. Serviço Nacional de Levantamento e Conservação de Solos. 2v. (EMBRAPA-SNLCS. Boletim de Pesquisa, 36. Brasil. SUDENE-DRN Série Recursos de Solos, 18), Rio de Janeiro, Brasil.
41. FAO (1996). The digitized soil map of the world including derived soil properties. Rome, FAO, 1 CD-ROM.
42. FAO (1994). Revised legend with corrections. FAO, Rome: 137p. (World Soil Resources Report, 60).
43. FAO (1974). Soil map of the World: 1:5.000.000 legend. Paris: Unesco, v. 1, Roma, Itália.
44. Ker, J. (1997). Latossolos do Brasil: uma revisão. Geonomos, Belo Horizonte, v. 5, n° 1, pp. 17- 40.
45. Nemes, A.; Schaap, M.; Leij, F. & Wosten, J. (2001). Description of the unsaturated soil hydraulic database. UNSODA version 2.0, Journal of Hydrology. Amsterdam, n. 251, pp. 151- 162.
46. Nemes, A.; Wosten, J.; Lilly, A. & Oude Vos Harr, J. (1999). Evolution of different procedures to interpolate particle – size distributions to achieve compatibility within soil databases. Geoderma. Amsterdam, n. 90, pp. 187-202.
47. Resende, M.; Curi, N.; Rezende, S. & Corrêa, G.(2007). Pedologia: base para distinção de ambientes. 5. ed. Editora UFLA, Lavras, Brasil.
48. Soil Survey Staff – SSS (1991). National soil information system (NASIS): soil interpretation and information dissemination sub-system. Draft requirements statement. Lincoln, USDA, Natural Resources Conservation Service, National Soil Survey Center. Washington, USDA.
49. United States (1975). Soil Taxonomy: a basic system of soil classification for making and interpreting soil surveys. Department of Agriculture. Soil Survey Division. Soil Conservation Service. Soil Survey Staff. Agriculture Handbook. Washington, USDA.
50. United States (1999). Soil taxonomy: a basic system of soil classification for making and interpreting soil surveys. Department of Agriculture. 2nd ed. (Agriculture Handbook, 436), Washington, USDA.
51. Van Engelen, V.W.P. (1999). Soter: the world soils and terrain database. In: SUMNER, M.E., (Ed.). Handbook of soil science, Boca Raton, CRC Press, pp. H19-H28.

Chapter 10

AN APPLICATION APPROACH TO KALMAN FILTER AND CT SCANNERS FOR SOIL SCIENCE

Marcos A. M. Laia[1,2] and Paulo E. Cruvinel[1,2]

[1]Embrapa Instrumentation

[2]Physics Institute of São Carlos, University of São Paulo Brazil

INTRODUCTION

Over the years, experts on soil science have brought together researchers from various fields with the aim of pooling efforts in order to characterize the properties of soils. The use of these results in agriculture through various activities can be directed towards a development whose natural resource base can be maintained in a long term. Conservation, minimization of soil pollution together with the development of irrigation and more efficient and more cost effective drainage systems optimize the efficiency of water use and nutrients in agricultural production.

The soil can be preserved through management methods, which seek to prevent deterioration, induced erosion and deposition of sediments into rivers. Environment degradation can be caused naturally and by humans. The way in how to use the soil can cause degragation and excessive compression, salinisation and acidification. Several conventional techniques have been used to find answers to the various physical mechanisms and also the chemical and biological processes that occur in soils. Among those techniques are included: neutron probe, gravimetry, direct transmission of rays, plotters, microscopy and mercury intrusion, however these mechanisms have limitations.

New techniques that enable an accurate prediction aim at managing the flow of contaminants through unsaturated soil zone. With the goals of acquiring non-invasive samples and reaching higher resolutions on γ and x-rays computerized tomography (CT) and on Nuclear Magnetic Ressonance (NMR), which provide cross-section images from the analysed objects. NMR, however, presents strong restrictions for its use in porous media that contain paramagnetic materials (Crestana & Nielsen, 1990) and, besides this, it is difficult or even impossible to quantify the results by correlating the NMR signal and the content of water. However, by means of CT it is possible to figure out a good correlation between the x-ray linear attenuation coefficients and the water content in soils.

The quality of an image is one of the key requirements for its analysis and it is desirable that the reconstructed object is very close to the tested sample. The use of algorithms developed for the areas of human knowledge has grown considerably and has improved image processing for visual information analysis and human interpretation or the automatic perception of machines. In human interpretation, x-ray images are used not only in medicine, but also in geology, in archaeology, and in soil science for agricultural purposes.

On the other hand, the perception of automatic machines for present automatic recognition of faces, characters, fingerprints, computer vision, control of robots for surveillance, automatic processing of satellite images for fire recognition, climate change and identification of storms and hurricanes. Image processing aims at modeling the characteristic of the human eye, study processed images, such as Fourier transform and other separable image transforms. It also allows designing filters used to retrieve an image and use masks so that the processed image is more applicable than the original (real).

This study aims at improving the quality of tomographic images by filtering the signals before their reconstruction and reaching an image quality in the reconstructed slice obtained by the projections close to the real one. Digital processing algorithms can be used to work with the image by using computer vision techniques such as segmentation (for feature extraction) and Hough transform (used in order to detect geometry, ie, detection of pores), which allow the count of pores in the soil. Classification algorithms can also be used to characterize the type of soil as well as its chemical components based on the values of pixels and density.

Both kinds of algorithms can be an auxiliary tool for a reliable analysis on the impacts of the use of agricultural machinery on the soil, which result in the increase of soil density due to the compression caused by the machinery axis. This increase is directly linked to the reduction of larger diameter pores that cause a decrease in water and in nutrients and also to the gas diffusion and root penetration.

Besides, this chapter aims at presenting the use of unscented Kalman filter (UKF) and the algorithm used to separate a noise from a signal. This will be done by showing that the unscented Kalman filter together with artificial neural networks is the best option for filtering. There will be an overview and a specification after each equation of the algorithm. Moreover, this study complemented previous works that aimed at creating a better algorithm by modeling and testing results with different types of Kalman filter and with different approaches in the physical model and kinds of noise (Laia et al., 2007; Laia & Cruvinel, 2008a; Laia et al., 2008b; Laia & Cruvinel, 2009; Laia & Cruvinel, 2011), which also present a comparison with an artificial neural network (ANN) solution.

COMPUTERIZED TOMOGRAPHY SCANNER

In radiology, computed tomography (CT) consists of an image that is derived from computerized processing of data obtained from a series of x-ray angled projections which reproduce a cross-section (a "slice") of the object under study. CT, such as conventional radiology, is based on the fact that x-rays are partially absorbed by various materials. While materials like plastic and water are easily traversed by x-rays, others, like metals, are not.

This technique had already been widely applied to medical areas, however its use in soil science was introduced by Petrovic (Petrovic et al., 1982), Hainswoth and Aylmore (Hainswoth & Aylmore 1983) and Crestana (Crestana, 1985). Petrovic made it possible to use x-ray CT to measure the density of soil volumes, while Crestana demonstrated that CT can solve problems related to studies of the water physics in the soil. From these studies, a project that developed a scanner for soil science has been created (Cruvinel, 1987; Cruvinel et al., 1990; Cruvinel et al., 2009).

Computerized Tomography in Soil Science

The application of CT in soil science to investigate the same physical properties plays an important role in studying the transportation of water and solutes within this environment. The direct transmission of γ or x-rays provides a major contribution in order to solve the various problems in the field of soils with results on a scale of the order of millimeters, however many responses are still expected on the level of particles, macropores and micropores.

Figure 1 illustrates different CT scanners that are dedicated to soil science and were developed and installed at Embrapa Instrumentation (São Carlos, SP - Brazil), which are based on sources of γ and x-rays for the study of soils and plants and also allow the use of various sources of radiation and intensities of energy. Several studies have been developed to improve the visualization of the images acquired and the reconstruction algorithm, as well as the equipment developed (Venturini, 1995; Minatel, 1997; Granato 1998; Mascarenhas et al, 1999).

250   Mechanisms of Soil Stabilization

a

b

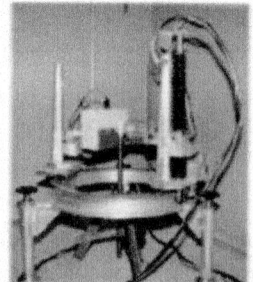
c

d

Figure 1: Scanners developed at Embrapa Instrumentation: a. mini CT scanner developed for agricultural applications (Cruvinel, 1987); b. portable CT scanner for field use (Naime, 1994); c. tomography with micrometric precision (Macedo, 1997) and d. Compton scattering scanner (Cruvinel & Balogun, 2006).

Compared to classical methods, such as the direct transmission of γ-ray and gravimetric tests, the CT scanner presents the advantage of measuring a) heterogeneities in the soil b) the density of the soil and c) the moisture content pixel by pixel, and also gets a two-dimensional picture or three-dimensional soil samples from non-invasive and independent of the geometry and shape, even by using different energies and radioactive sources (Cruvinel et al. 1990).

Basically, a CT indicates the amount of radiation absorbed by each portion of the analyzed section, translates these variations into gray scale and also produces an image. As the x-ray absorption capacity of a material is closely related to its density, different density areas will present different intensities, which can be seen in monochromatic color and if you apply a mask of pseudo-color, it allows their distinction more clearly. Thus, each signal value corresponds to the average absorption of the tissue in the area, expressed in Hounsfield units (named after the creator of CT machines). Each projection represents an average and each set is stored in a projection matrix. Based on the intensity emitted by the x-ray source and the intensity captured by the detector at the other end of the propagation line, one can determine the attenuation weight due to the object that is located between the source and the detector. The data on the attenuation weight is crucial to the reconstruction process and enables a mapping of the linear attenuation coefficient of the object cross section.

This coefficient mapping is represented by pixels whose values are given by the so-called CT Numbers. These numbers are normalized according to the water attenuation coefficient μ_{water}. In other words, the CT numbers are defined in equation 1:

$$CT\ Number = \frac{\mu - \mu_{water}}{\mu_{water}} \times 1000 \quad (1)$$

where μ is the linear attenuation coefficient of the analyzed body.

From this number, it is possible to obtain a map of the attenuation coefficients, which allows a more detailed analysis of the body being studied. In medicine, it was agreed that the water CT number is equivalent to 0 (zero). The integration function of the object along the ray is represented by a line integral. Each set of line integrals of parallel rays form a parallel projection (Figure 2), which can be treated having the signal theory as its base.

The main advantage of CT is to allow the study of cross sections without intrusion, that is, an unparalleled improvement in relation to techniques of soil analysis, which, in general, are invasive and can destroy important features that could be preserved. It is important to obtain a perfect image quality that prevents materials found in the soil from being erroneously interpreted. This is achieved when the material in focus is the same found by an intrusive manual search.

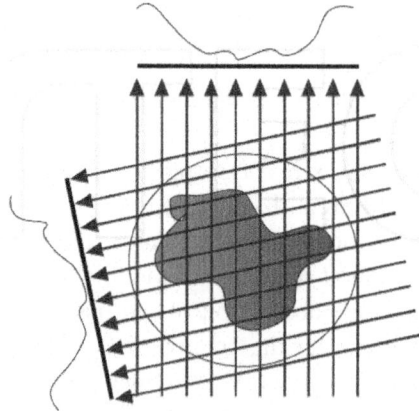

Figure 2: Two parallel projections of an object expressed by a two-dimensional function.

Noise that can occur in CT

In CT there are three main processes related with the interaction of the radiation with the matter, which are the photoelectric effect, the Compton effect and pair production effect (Cruvinel et al. 1990). Besides the issues that are related to the effects arising from the energy range used in the source, others also influence the measurement in computed tomography, like the statistics of photon counting. The probability of detecting photons in a range of exposure time t can be estimated by the distribution function of Poisson (Deremack & Crowe, 1984), (Cruvinel, 1987).

$$P_s(\lambda, t) = \frac{(\bar{\lambda})^n}{\lambda! e^n} \qquad (2)$$

where λ is the number of photons and $\bar{\lambda}$ is the average number of photons emitted in the time interval t and n indicates the efficiency of the photomultiplier, as shown in the following expression:

$$\bar{\lambda} = \xi M t \qquad (3)$$

where M is the average ratio of photons (photons / second) and ξ is the quantum efficiency of the photomultiplier. The uncertainty or noise is given by the standard deviation

$$\sigma = \sqrt{\bar{\lambda}} = \sqrt{\xi M t}. \qquad (4)$$

Therefore, the signal to noise ratio (SNR), presented by the incident signal, is:

$$SNR = \frac{\bar{\lambda}}{\sigma} = \sqrt{\bar{\lambda}}. \qquad (5)$$

In this ratio, it is estimated that, for a small number of photons, there can be considerable noise, however, as $\bar{\lambda}$ increases, the noise might be negligible. The thermionic emission of electrons in the photo multiplier cathode can cause an increase in the noise. Considering that the photo-cathode emits electrons randomly due to the cathode current, a new signal to noise ratio equation is given by equation 7:

$$\sqrt{(\xi M + M)t} \qquad (6)$$

$$SNR = \frac{\sqrt{t}\xi M}{\sqrt{\xi M + M}} \qquad (7)$$

In the display of a tomographic image, there is granularity, which is significant for viewing low contrast objects. During the analysis of a CT image, there is presence of granularity, which is significant for viewing low contrast objects. The term noise in tomography images refers to the variation of the attenuation coefficient at the average value and it is obtained when an image of a uniform object is used. (Hender, 1983). The image noise can be based on the calculus of the standard deviation and also on the Weiner power spectrum of the noise, which is seen as a function of spatial frequency. In other words, it allows observing the intensity and the type of noise that involves the system also influences the image obtained.

The noise in CT images involves rounding errors in the reconstruction program (algorithm noise), electronic noise and noise caused by the display system. The main source of noise inCT images is the change in quantum (quantus mottle), defined as the statistical spatial and the temporal variation in the number of x-ray photons absorbed by the detector.

The noise of the algorithm depends on the pixel size of the display device and also influences the image noise, since it leads to larger pixels to reduce the image noise, however with less quality resolution. Reconstruction algorithms typically make use of anti-aliasing filters to minimize the visual effect of the noise but also present some loss in the spatial resolution.

The electronic noise can be generated by non-ideal electronic devices such as non pure resistors and capacitors, non ideal contact terminals, leakage current of transistors, Joule, and it may also be independent of the signal, due to external interference (electrical or mechanical) (Ziel, 1976).

Besides the noise, the CT images are subject to various artifacts and distortions as polychromatic sources (non-ideal or monochromatic), caused by the following effects: beam hardening, aliasing, different materials in the same voxel (partial volume), displacement of the sample or of the equipment (Duerinckx & Macocski, 1978; Joseph & Spital, 1978; Ibbott, 1980; Granato, 1998).

The low-pass and the median filters can be used to solve problems of signal noise, however they can cause a crucial losses of informations. Systems having different sources of noise do not present an optimal solution when using these types of filters.

THE KALMAN FILTER

The Kalman Filter is a mathematical tool, created more than 30 years ago, which is widely used to solve statistical problems. It is considered a good estimator for a large class of problems and an effective and useful estimator for other classes. It has recently been used in computer graphics for applications involving the simulation of musical instruments in virtual reality (VR), for the reading of speaker's lips in video sequences, among other applications (Pereira, 2000). In 1960, Rudolf Emil Kalman published a paper describing a recursive solution to the problem of linear filtering for discrete data (Kalman, 1960).

Stanley F. Schmidt, who worked on the Apollo project at NASA, has been the first to apply it in a practical way, since he aimed at taking a spacecraft to the moon and bring it back. For this, he had to solve problems in the estimation of trajectories and control. Schmidt worked with what would be the first full implementation of the Kalman filter and it had the same integral control system of Apollo and from this experience, it has been used in most on board estimation systems and also in trajectory control of aircrafts.

The Kalman filter is considered to be an advance concerning the estimation theory. It is used, for example, to estimate a linear-quadratic Gaussian problem, which addresses the difficulty of state estimation of a dynamic linear system disturbed by white Gaussian noise, by using measurements that are linearly related to these states and also corrupted by this noise. This filter consists of a set of mathematical equations that provide a significant and efficient computational solution (recursively) to estimate the state of a process, so that the mean square error can be minimized. The filter allows that states (past, present and even future) are estimated and may do so even when the precise nature of the modeled system is unknown (Welch & Bishop, 2004).

Several models can be found, such as: increased, extended, dual estimation, joint estimation, unscented, among others.

Discrete Kalman Filter

The process to be estimated solves the general problem of estimating a state x, which is a process controlled in a discrete time and generated by a linear stochastic differential equation, ie:

$$x_k = Ax_{k-1} + Bu_k + n_{k-1}, \qquad (8)$$

with a measurement z, which is

$$z_k = Hx_k + v_k. \qquad (9)$$

The random variables n_{k-1} and v_k represent the process noise and the measurement (respectively). They are assumed to be independent (between them) and are considered white noise with the normal probability distributions:

$$P(n) \sim N(0, Q) \qquad (10)$$

$$P(v) \sim N(0, R) \tag{11}$$

In practice, the covariance matrixes of the process noise (covariance Q) and of the measurement noise (covariance R) can change at each measurement.

The discrete Kalman filter (DKF) estimates a process by using a kind of feedback, in which it estimates the process state at some time and then obtains feedback in the form of (noisy) measurement. Thus, the equations can be divided into two stages: the ones to update the time and the ones to update the measures.

The equations that update the time are responsible for projecting an a priori (in time) estimate of the current state and of the error covariance to obtain an a priori estimate for the next step.

The equations that update or correct measures are responsible for the feedback - for example, the incorporation of a new measure into the a priori estimate for an a posteriori estimate.

The equations that update or predict the time can also be considered prediction equations, while the equations that update the measures can be treated as correction equations. Thus, the estimation algorithm is a predictor-corrector algorithm for solving numerical problems. Time update equations (predictor)

$$\hat{x}_k^- = A\hat{x}_{k-1} + Bu_k \tag{12}$$

$$P_k^- = AP_{k-1}A^T + Q. \tag{13}$$

Note that these equations refer to time and covariance estimation over time from step k-1 to step k.

Measurement update equations (corrector)

$$K_k = P_k^- H^T (H P_k^- H^T + R)^{-1} \tag{14}$$

$$\hat{x}_k = \hat{x}_k^- + K_k(z_k - H\hat{x}_k^-) \tag{15}$$

$$P_k = (I - K_k H) P_k^-. \tag{16}$$

The first task during the update of the measure is to compute the Kalman gain, K_k. The next one is to measure the update process and find the z_k value and, then, generate an aposteriori state estimate by incorporating it as in equation 15. The final step is to obtain an estimated covariance error by using equation 16.

Extended Kalman Filter

A solution to nonlinear systems is the extended Kalman filter (EKF) (Welch and Bishop 2004). By analyzing the prediction function of the Kalman filter, illustrated in equation 12, it is possible to observe that the filter behaves linearly. By applying a nonlinear function, one can obtain a good prediction for the next states, the extended Kalman filter. This algorithm applies the Kalman filter to nonlinear systems simply by linearizing all nonlinear models, so that the traditional equations of the filter can be applied. The nonlinear system can be rewritten as:

$$x_k = f(x_k, n_k, u_k), \qquad (17)$$

$$z_k = h(x_k, v_k). \qquad (18)$$

The algorithm is also adapted to solve nonlinear problems through the use of functions, as seen below:

Time update equations (predictor)

$$\hat{x}_k^- = f(\hat{x}_k, 0, u_k) \qquad (19)$$

$$P_k^- = F_k P_{k-1} F_k^T + Q. \qquad (20)$$

Measure update equations (corrector)

$$K_k = P_k^- H_k^T (H_k P_k^- H_k^T + R)^{-1} \qquad (21)$$

$$\hat{x}_k = \hat{x}_k^- + K_k(z_k - h(x_k^-, 0)) \qquad (22)$$

$$P_k = (I - K_k H_k) P_k^- \qquad (23)$$

For the propagation of variances, one must know the Jacobian or the Hessian transition matrices and observe the state functions given by Equations 17 and 18, respectively.

Unscented Kalman Filter

The unscented Kalman filter (UKF) is similar to the extended version (Julier & Uhlmann, 1997). The distribution of states is represented by a Gaussian random variable, however it is specified by using a minimum set of sampling points, which are chosen carefully. The sampled points capture the true mean and the covariance of a random variable and when it propagates through a truly non-linear system, it captures the mean and the covariance accurately to promote a third order estimation for any nonlinearity. Thus, this is done through the use of unscented processing.

The unscented transformation is a method used to calculate the statistics of a random variable that undergoes a nonlinear transformation (Wan & Merwe, 2000). Considering a spread of a random variable (with dimension L) through a nonlinear function y=g(x), one assumes that it must present a mean x and a covariance P_x to calculate the statistics y. It should form a matrix X of 2L+1 sigma vectors (with corresponding weights W_i), according to the following:

$$X_i = \bar{x} \tag{24}$$

$$X_i = \bar{x} + (\sqrt{(L+\lambda)P_x})_i, \text{ for } i = 1,\ldots,L \tag{25}$$

$$X_i = \bar{x} - (\sqrt{(L+\lambda)P_x})_i, \text{ for } i = L+1,\ldots,2L \tag{26}$$

$$W_0^{(m)} = \lambda/(L+\lambda) \tag{27}$$

$$W_0^{(c)} = \lambda/(L+\lambda) + (1-\alpha^2+\beta) \tag{28}$$

$$W_i^{(m)} = W_i^{(c)} = 1/\{2(L+\lambda)\}, \text{ for } i = 1,\ldots,2L \tag{29}$$

where $\lambda = \alpha^2(L+k) - L$ is a scalar parameter. The variable $\lrcorner$ determines the sigma scattering point around the mean $\bar{x}$ and is always a minimum positive value. k is a secondary scalar parameter defined as 0 and b is used to incorporate the a priori knowledge of the distribution of x (for Gaussian distributions, b = 2 is optimal). $(\sqrt{(L+\lambda)P_x})_i$ is the i-ism line of the square root matrix. The sigma vectors are propagated through the nonlinear function:

$$y_i = g(X_i) \text{ for } i = 0,\ldots,2L \tag{30}$$

The mean and the covariance y_i are approximated by using the mean and the covariance of the sample (Figure 3) of the posterior sigma points

$$\bar{y} \approx \sum_{i=0}^{2L} W_i^{(m)} y_i \tag{31}$$

$$P_y \approx \sum_{i=0}^{2L} W_i^{(c)} \{y_i - \bar{y}\}\{y_i - \bar{y}\}^T. \tag{32}$$

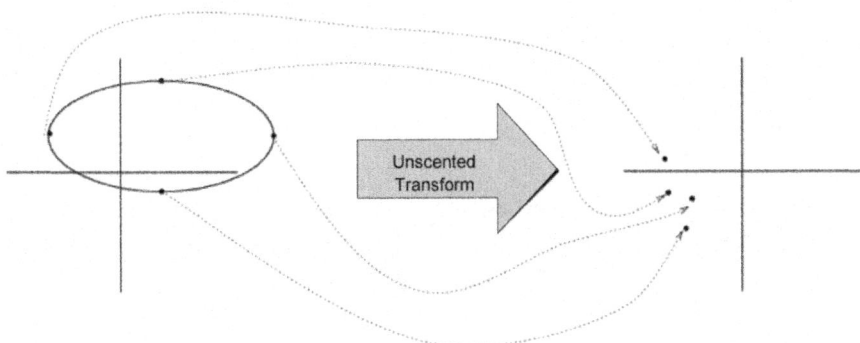

Figure 3: The unscented transform.

This method differs from the general methods of sampling (the Monte-Carlo method such as particle filters), which requires orders of magnitude with more sampling points as an attempt to define and propagate the state (possibly non-Gaussian) distributions. The unscented approaches result in more accuracy for the third order of Gaussian inputs for all nonlinearities.

The unscented Kalman filter is a direct extension of the unscented transform for the equation recursive estimation

$$\hat{x}_k = \hat{x}_k^- + K_k \cdot [y_k - \hat{y}_k^-)] \tag{33}$$

where the state of the random variable is redefined with the concatenation of the original states and the noise:

$$x_k^a = [x_k^T v_k^T n_k^T]^T. \tag{34}$$

The selection of sigma points is applied to a new random variable state to select and calculate the corresponding sigma matrix x_k^a. The unscented Kalman filter is pointed out by equations 35, 36, 37, 38, 39, 40, 41, 42, 43, 44, and 45.

The Jacobian or the Hessians matrices do not need to be calculated and the calculation total numbers are the same of the extended filters related to the nonlinear controls that require feedback from the states. In these applications, the dynamic model is a physically-based parametric model, which is assumed as known.

Unscented Kalman filter algorithm

Sigma points calculation:

$$X_k^n = [\hat{x}_{k-1}^a \quad \hat{x}_{k-1}^a \pm \sqrt{(L+\lambda)} P_{k-1}^a] \tag{35}$$

where X is the set of points with unscented transformation based on the mean and on the a priori covariance.

Prediction equations:

$$\hat{x}_k^- = \sum_{i=0}^{2L} W_i^{(m)} x_{i,k|k-1}^x \tag{36}$$

where $W^{(m)}$ represents the set of sigma point weights used for true mean reconstruction.

$$P_k^- = \sum_{i=0}^{2L} W_i^{(c)} [x_{i,k|k-1}^x - \hat{x}_k^-][x_{i,k|k-1}^x - \hat{x}_k^-]^T \tag{37}$$

where $W^{(c)}$ represents the set of sigma point weights used for true covariance reconstruction.

$$X_{k|k-1}^x = F(X_{k-1}^x, X_{k-1}^v) \tag{38}$$

where F is the function of sigma propagation for state transitions.

Correction equations:

$$Y_{k|k-1} = H(X_{k-1}^x, X_{k-1}^n) \tag{39}$$

where H is the system function of sigma point generation for observation states Y.

$$\hat{y}_k^- = \sum_{i=0}^{2L} W_i^{(m)} Y_{i,k|k-1}^x \tag{40}$$

where y is the observed state estimation reconstructed for the sigma points.

$$P_{\hat{y}_k \hat{y}_k}^- = \sum_{i=0}^{2L} W_i^{(c)} [Y_{i,k|k-1} - \hat{y}_k^-][Y_{i,k|k-1} - \hat{y}_k^-]^T \tag{41}$$

$$P_{x_k \hat{y}_k} = \sum_{i=0}^{2L} W_i^{(c)} [X_{i,k|k-1} - \hat{x}_k^-][Y_{i,k|k-1} - \hat{y}_k^-]^T \tag{42}$$

$$K = P_{x_k y_k}(P_{\hat{y}_k \hat{y}_k})^{-1} \tag{43}$$

where K is the Kalman gain obtained through the noisy covariance

$$\hat{x}_k = \hat{x}_k^- + K(y_k - \hat{y}_k^-) \tag{44}$$

this equation represents the correction of a priori and

$$P_k = P_k^- - K(P_{\hat{y}_k \hat{y}_k}) K^T \tag{45}$$

represents the correction of the a priori covariance.

The Kalman filter was originally designed to solve a problem of state estimation and has been used in many applications. This newer filter also provides a better performance. Thus, the unscented Kalman filter and the extended filter present the same order of complexity. Due to the numerical instability of the noise and the use of the Cholesky factorization to determine the square root of the probability matrix, van der Merwe and Wan have developed the Square-Root Unscented Kalman Filter (SRUKF) (Merwe & Wan, 2001), which allows a better control of the variance matrix values and bypasses the problem of becoming a negative or indefinite matrix. As the original unscented Kalman filter, the square root filter is initialized by calculating the square root of the covariance matrix states by the Cholesky factorization:

$$S_0 = chol\{E[(x_0 - \hat{x}_0)(x_0 - \hat{x}_0)^T]\} \tag{46}$$

However, the spread factor and the Cholesky update are then done in subsequent iterations to directly form the sigma points. In the equation below, the update time of the Cholesky factor is calculated by using a QR decomposition of the matrix composed of the propagated sigma point weights and the square root of the covariance matrix of the additive noise case:

$$S_k^- = qr\{[\sqrt{W_1^{(c)}}(X_{1:2L,k|k-1}^* - \hat{x}_k^-) \sqrt{R^v}\;]\} \tag{47}$$

A subsequent Cholesky update (or regression) in the equation below is needed since the weight zero is perhaps negative:

$$S_k^- = cholupdate\{S_k^-, X_{0,k}^* - \hat{x}_k^-, W_0^{(c)}\} \tag{48}$$

These two steps replace the time update. They are also used to calculate the Cholesky factorization and the error covariance of the observation:

$$S_{\tilde{y}\,k} = qr\{[\sqrt{W_1^{(c)}}(Y_{1:2L,k} - \hat{y}_k^-) \sqrt{R^n}\;]\} \tag{49}$$

$$S_{\tilde{y}\,k} = cholupdate\{S_{\tilde{y}\,k}, Y_{0,k} - \hat{y}_k^-, W_0^{(c)}\}. \tag{50}$$

Unlike the way in which the gain of Kalman filter is calculated in standard unscented, it is calculated here by using two inversions:

$$K_k(S_{\tilde{y}\,k} S_{\tilde{y}\,k}^T) = P_{x_k y_k} \tag{51}$$

Since it is square and triangular, efficient replacements can be used to solve it directly, without having to invert the matrix. Finally, the Cholesky factoration updates the state covariance in the equation below and it is calculated by applying sequential Cholesky regressions:

$$S_k = cholupdate\{S_k^-, U, -1\} \tag{52}$$

The vectors are represented by the columns of the equation regression 53. This update replaces the previous equation 45:

$$U = K_k S_{\tilde{y}\,k} \tag{53}$$

By knowing the process function and with a Kalman filter that supports non linear functions, it is possible to have a significant improvement in the signal. Another solution is to use a neural network to promote a better function of the mapping process by reducing the projections noise. For an estimation of the neural network weights together with the state estimates, two methods of filtering can be used: the estimation and the dual estimation. These arrangements are used to determine the filtering when the initial weights are known and the next state is obtained through a linear mapping as the previous one.

Joint Estimation

Since the transfer function is not known and with the aim of increasing the filter order, a new estimation method has been used, in which it is possible to estimate new parameters from the states of the hidden chains of the Markov model. The main problem involves the identification of functions required to estimate states and parameters. The prediction equations can be described as:

$$x_k = f(x_{k-1}, n_k) \tag{54}$$

$$z_k = h(x_k, v_k) \tag{55}$$

The parameter estimation involves the determination of a nonlinear mapping

$$y_k = g(x_k, W) \tag{56}$$

where x_k is the input, W is the weight and y_k is the output. The nonlinear mapping g is parameterized by the vector W. The nonlinear mapping can be done by an artificial neural network. The learning corresponds to estimating the parameters of W. Training can be done with pairs of samples, consisting of a known input and a desired output (x_k, d_k). The machine error is defined by equation 57. The learning objective is to minimize the expected squared error.

$$e_k = d_k - g(x_k, W) \tag{57}$$

The UKF can be used to estimate the parameters by using a model for network training that writes a new state-space representation:

$$w_k = w_{k-1} + v_k \tag{58}$$

$$y_k = g(x_k, w_k) + e_k \tag{59}$$

where the parameter w_k correspond to a stationary process with an identity matrix of state transition, governed by a procedural noise v_k (the choice of variances determines the filtering performance). The output y_k corresponds to a nonlinear observation w_k. The extended Kalman filter can be applied directly as an efficient technique for the correction of second order parameters.

As the problem in focus is to work with unobserved input x_k and requires an estimation of states and parameters, one should consider a dual estimation problem, taking a dynamic and discrete-time nonlinear system into account

$$x_{k+1} = f(x_k, v_k, W) \tag{60}$$

$$z_k = h(x_k, n_k) \tag{61}$$

where states are x_k and the set of parameters W of the dynamic system should be estimated only from the noisy signal y_k.

The dynamic system can be seen as a neural network, where W is the set of weights and function f corresponds to a function of neural network that uses an input x_k. Thus, by applying these equations to the unscented Kalman filter, one can have a new function for estimating and observing new states. One approach to neural networks can be seen in (Laia & Cruvinel, 2008b; Laia & Cruvinel, 2009).

SYSTEM MODELING

The physical model of photon counting is defined by the equation:

$$I_k = I_0 e^{-\mu_k d} \tag{62}$$

where I_0 is the number of photons leaving the source, μ is the material degree of absorption, d is the distance between the source and the detector and I is the number of photons that cross the material and reach the detector. The counting of photon is affected by a Poissontype noise. For a closer model to the physical one, each projection is analyzed individually, as if they were in time-varying positions. This classical approach enabled to develop a dynamic estimation of noise-free projections (Figure 4):

$$I_{p_k} = f(I_{p_{k-1}}, q_{k-1}) \tag{63}$$

$$I_{s_k} = h(I_{p_k}, r_k) \tag{64}$$

where I_p is a noise-free projection and I_s, a projection disturbed by noise. The variables q and r represent the white noise, ie, present distribution $q \sim N(0, Q)$ and $r \sim N(0, R)$, respectively. Function f can be used as a mapping neural network or as a transfer state matrix. Function h hides the unobserved states and can also be an array. It can also be adjusted by using the Poisson noise and the Anscombe transform. Some of the previous studies that focused on this approach and obtained good results are (Laia et al, 2007; Laia & Cruvinel, 2008b; Laia & Cruvinel, 2009).

A filtering proposal is needed for a new model (based on the physical one) to determine the process variables and how the observation is carried out. The equation of a process defines the previous state x_{k-1}, and through a transformation influenced by a function f and white noise q_k, one can reach a new state x_{k-1}. These states can be hidden from the system output. Thus, it is possible to define a new function g that transforms this variable according to what is observed.

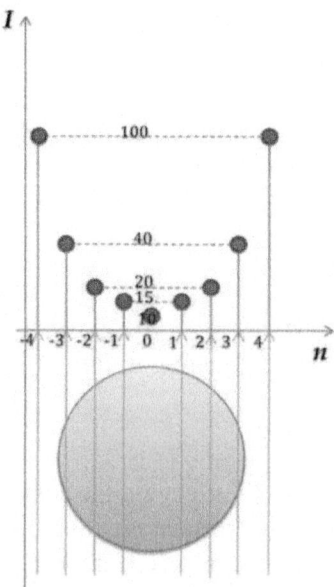

Figure 4: The process for projections acquisition considering an object based in previous studies.

The uncertainty estimation process of the current value has a confidence interval that consists of photon counting and is given by Poisson noise. This uncertainty can be filtered with an estimation of other measures of the equation that are independent on the confidence interval. Thus what ends up being filtered is the noise from the detector, ie, mechanical and electronic noise.

In order to increase confidence and get more reliable values, one can change the focus on what is being estimated. The different distributions can be mapped with specific non-linear transformations that alter the Gaussian distributions, such as the Anscombe (Poisson) or the inverse BoxMuller (Uniform), however they are linear approximations that present cumulative errors. The Kalman filter is limited to solving stochastic problems, given by the following equations:

$$\mu_k = f(\mu_{k-1}, q_k) \tag{65}$$

$$I_k = h(I_0, \mu_k, d, r_k) \tag{66}$$

Equation 65 is the transition function that updates the measure of the attenuation coefficient or the material degree of absorption while variable q is the uncertainty of this process equation. Equation 66 is the observation function that generates the final projection from the degree of absorption and d the distance between the source and the photon detector. Variable r is the noise that is transferred to the projection. This model can be approximated by a second order filter in order to estimate the variables of the process equation and to use the Anscombe transform to the observation equation:

$$\begin{bmatrix} \mu_k \\ \Delta\mu_k \end{bmatrix} = \begin{bmatrix} 1 & 1 \\ 0 & 1 \end{bmatrix} \begin{bmatrix} \mu_{k-1} \\ \Delta\mu_{k-1} \end{bmatrix} + \begin{bmatrix} 0 \\ q_k \end{bmatrix} \tag{67}$$

$$I_k = h(I_0 e^{-\mu_k d}, r_k) \tag{68}$$

Function h is related to the Anscombe transform (Anscombe, 1948), which transforms a Poisson noise into a Gaussian noise, with variance value close to 1. As the noise variation presents a Poisson distribution, the mean and the variance of values are equivalent to the counting of photons. A new system can be defined based on an equation, in which the ray sum μ_k is the variable used in the process (Equation 76) and it also allows the observation of the projection, as it is presented in the array of projections:

$$\mu_k = f(\mu_{k-1}, u_k, q_k) \tag{69}$$

$$I_k = h(I_0 e^{-\mu_k d}, r_k) \tag{70}$$

Variable u_k consists of an external input that is related to the prediction of new states. In order to promote a better estimation of the states without noise, one can use neural network to determine the behavior of the equation of the process that uses the Kalman filter for a dual estimation:

$$\mu_k = f(\mu_k, w_k, q_k) \tag{71}$$

$$I_k = A(I_0, r_k) e^{-\mu_i d} \tag{72}$$

$$w_k = w_k + a_k \tag{73}$$

$$e_k = g(\mu_{k-1}, w_k) - \mu_k \tag{74}$$

where a_k is used to update the weight and e_k indicates the error between inputs and the desirable outputs.

To promote a better estimation of the transition states of the process, an artificial neural network can be used:

$$\begin{bmatrix} \mu_k \\ W_k \end{bmatrix} = \begin{bmatrix} g(\mu_{k-1}, u_k, W_{k-1}) + q_k \\ W_{k-1} + a_k \end{bmatrix} \tag{75}$$

$$I_k = h(I_0 e^{-\mu_k d}, r_k). \tag{76}$$

Now, by focusing on the observation equation, the observation noise variance can be treated with the Anscombe transform through the use combined with its inverse.

$$I_k = A^{-1}(A(I_0 e^{-\mu_k d}) + r_k) \tag{77}$$

where A represents the transformation and its inverse A^{-1}. This makes it possible to work with a Gaussian noise with distribution $r \sim N(0,1)$.

In this model, the input has been used as the current observation of the system (ie, the attenuation coefficient of the noise) so that one could take advantage of the neural network functionality which provides a nonlinear mapping. The neural network itself consists of the interaction between the value measured before and the noise presented now, in order to better predict the data.

An equation based on equation 5 can be developed in order to determine the Poisson noise given by the following expression:

$$I \pm \sqrt{I}.\tag{78}$$

Thus, the observation equation can be written as follows:

$$I_k = I_0 e^{-\mu_k} + \sqrt{I_0 e^{-\mu_k d}} r_k.\tag{79}$$

where r, since it presents a Gaussian distribution, can assume a negative or zero value. This alternative is allowed to deviate what is considered to be only an approximation from the use of the Anscombe transform.

Another model for the equation could be simplified as:

$$I_k = I_0 e^{-\mu_k d} + r_k.\tag{80}$$

where r is replaced by a variance given by the signal noise ratio of

$$\frac{I_k}{\sqrt{I_k}}.\tag{81}$$

This approach allows the inclusion of other noises in the system of photon counting through the propagation of errors. In order to set the noise variance R being treated, one has to take the number of primary quantities into account. It is measured from the observation of the system variables {Io, μ}. The value of I depend on the relationship between these variables. In formal language, that is:

$$I = G(I_0, \mu, d).\tag{82}$$

In case the errors with the measured magnitudes I_o, μ and d are ΔI_o, $\Delta \mu$ and Δd, the photon counting error ΔI is given by the expression:

$$\Delta I = \left|\frac{\partial I}{\partial I_0}\right| \Delta I_0 + \left|\frac{\partial I}{\partial \mu}\right| \Delta \mu.\tag{83}$$

The values of ΔI_o and $\Delta \mu$ are given by mean standard deviation or by their estimator as there are many or few magnitude measures Io and μ.

When the sample size is adequate, it can determine the statistical error of independent magnitudes and these can become the dependent magnitude variance for calculating the statistical error.

$$\sigma_I^2 = \left(\frac{\partial I}{\partial I_0}\right)^2 \sigma_{I_0}^2 + \left(\frac{\partial I}{\partial \mu}\right)^2 \sigma_\mu^2 + \left(\frac{\partial I}{\partial d\mu}\right)^2 \sigma_d^2.\tag{84}$$

Besides the uncertainty of the system variables, the detector presents a characteristic noise. This noise variation is known when the detector closes, in other words, as no photons are released to it, it still gives the score. As it is associated directly to an additive noise, you can add the noise variation into the equation as an error for variable I:

$$\sigma_I^2 = \left(\frac{\partial I}{\partial I_0}\right)^2 \sigma_{I_0}^2 + \left(\frac{\partial I}{\partial \mu}\right)^2 \sigma_\mu^2 + \left(\frac{\partial I}{\partial d}\right)^2 \sigma_d^2 + \left(\frac{\partial I}{\partial I}\right)^2 \sigma_e^2. \qquad (85)$$

σ_I^2 is used to define the noise variance R of the observation equation of the filter. Then, it is necessary to define the variance of the process. In the literature, Haykins (Haykins, 2001) has defined several ways to infer on the process variance Q. Since the signal is observed, the noisy signal variance can be used. As the process variance is directly linked at the µ vector, the equation can be changed for a noise µ :

$$\mu = \ln\left(\frac{I_0}{I}\right)/d \qquad (86)$$

From this vector, a variance Q needs to enter the filter. It can be obtained as follows:

$$Q = \sigma_\mu^2 \qquad (87)$$

Another important step is the definition of the control constants: α, β and κ.

As the process variable µ magnitude differs from the observation of variable I, α = 1 has been chosen. In case this value is greater or lower than ideal, there is no possibility of filtering or causing numerical instabilities.

As suggested in the literature, for a parameter or a joint estimation filters, the variable remained β = 2, while the κ value was given 3 - the number of neurons. With the aims of comparing the efficiency of the filter and setting up a filter to ensure a desirable image quality, both of the algorithms have been applied to various soil samples. The first sample consists of sand grains in a Plexiglas envelope. The second and third samples are portion of natural soil. The fourth and the fifth are present in degraded soil bulks and the sixth shows a portion of naturally cemented soil. In an artificial neural network there are two neurons in the input layer, two in the intermediate and one in the output layer.

For the inputs of the filter, the same uncertainty was used, as all of the samples have been generated in the same CT scanner. The variance $\sigma_{I_0}^2$ was obtained by the maximum of vector projections. After the filtering process, it has been applied to the maximum of the projection matrix. The variance σ_μ^2 was the same used in the process variance Q. For the variance σ_d^2, the value used was 0.05, corresponding to the uncertainty of measurement in millimeters. For the errors in the photon counting detector, the value used was 100 for the variance σ_e^2. The results obtained from this new modeling system are presented below. In Figures 5, 7, 9, 11 and 13, it is possible to visualize the comparison between the signals of a set of projections by using the samples filtered with the SRUKF: Original projections (red), Linear estimation (blue) and non linear estimation with ANN (green). In Figures 6, 8, 10, 12, 14 and 16, it is possible to visualize the reconstructed images by using the filtered back-projection algorithm.

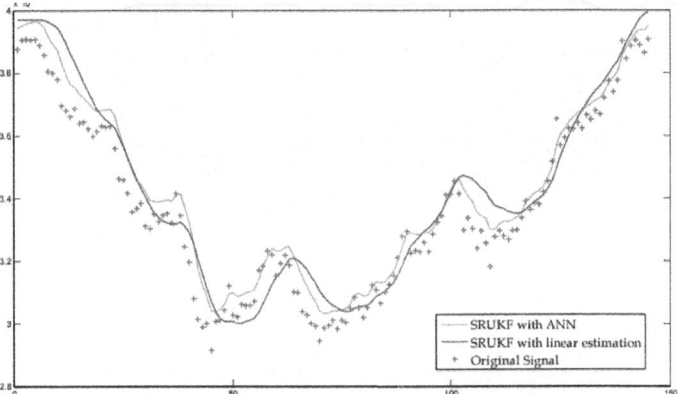

Figure 5: Comparison between the projections of sand grains.

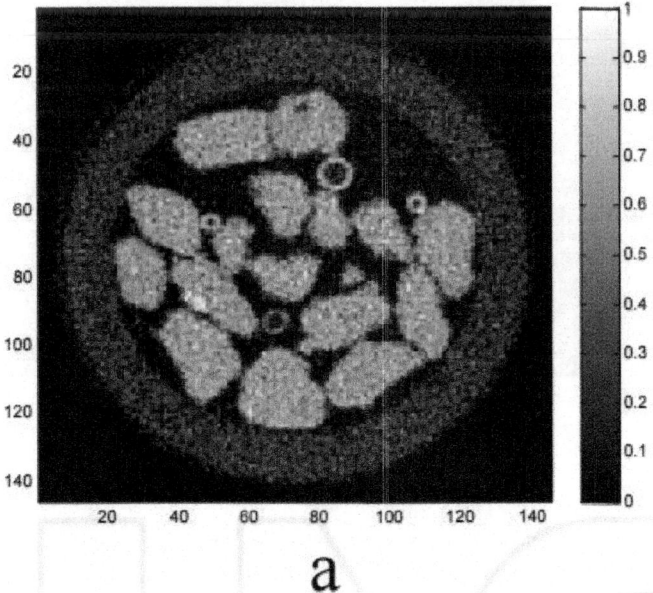

a

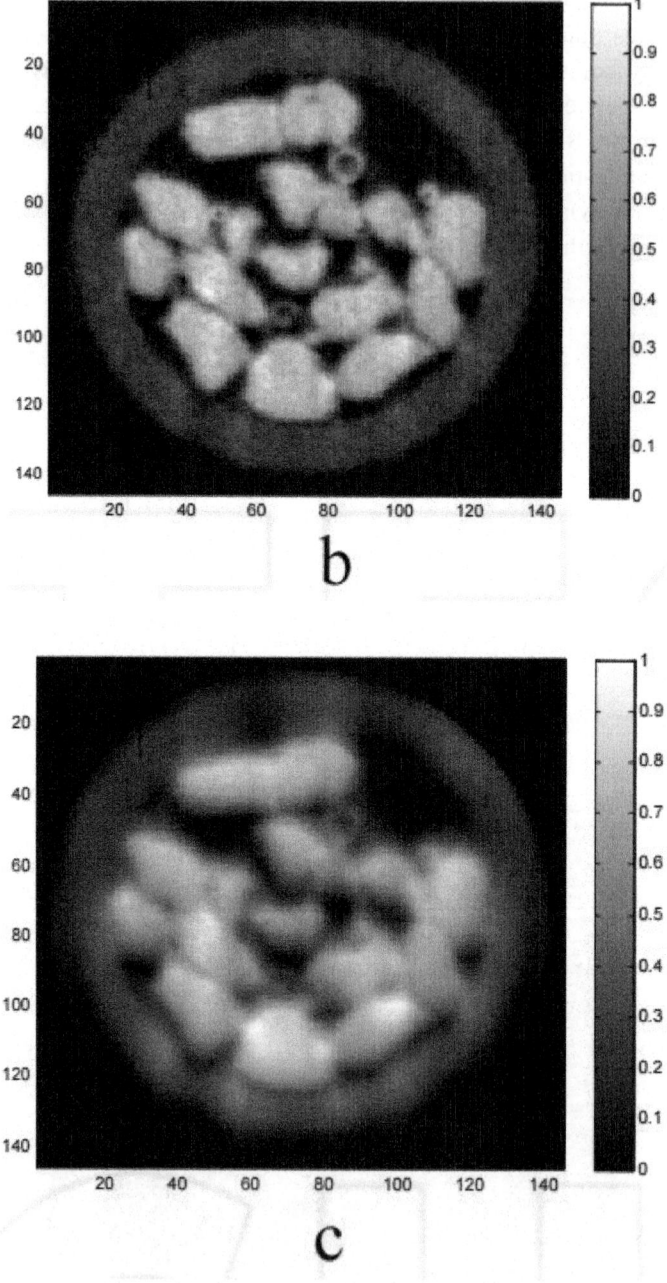

Figure 6: Comparison between the reconstructed images of sand grains. a. Original projections, b. Projections filtered with the SRUKF and the ANN and c. Projections filtered with linear estimation.

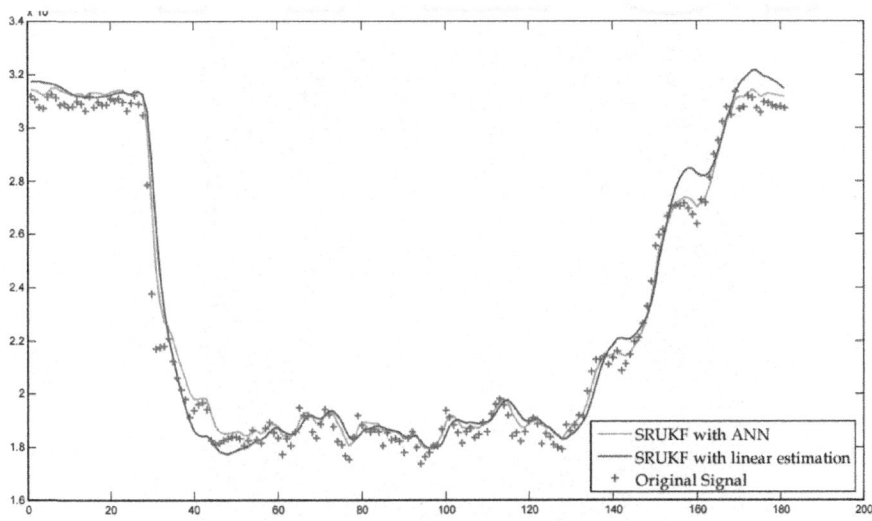

Figure 7: Comparison between the projections of soil bulk samples.

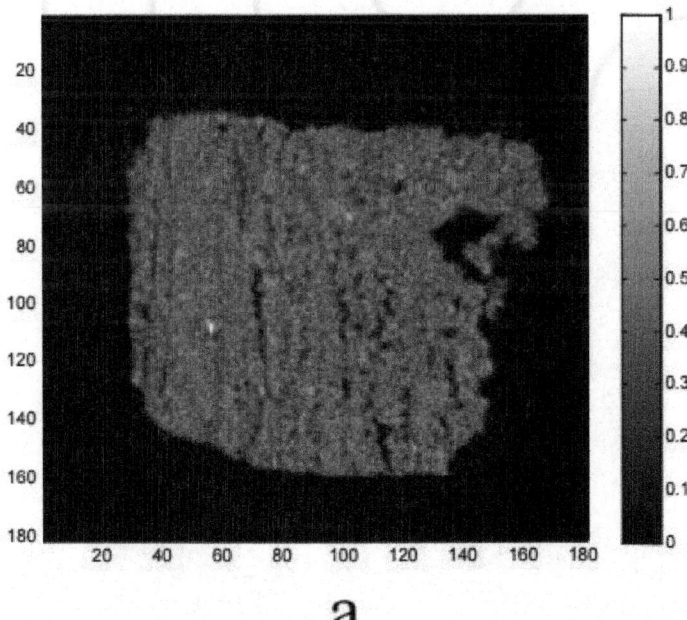

a

272   Mechanisms of Soil Stabilization

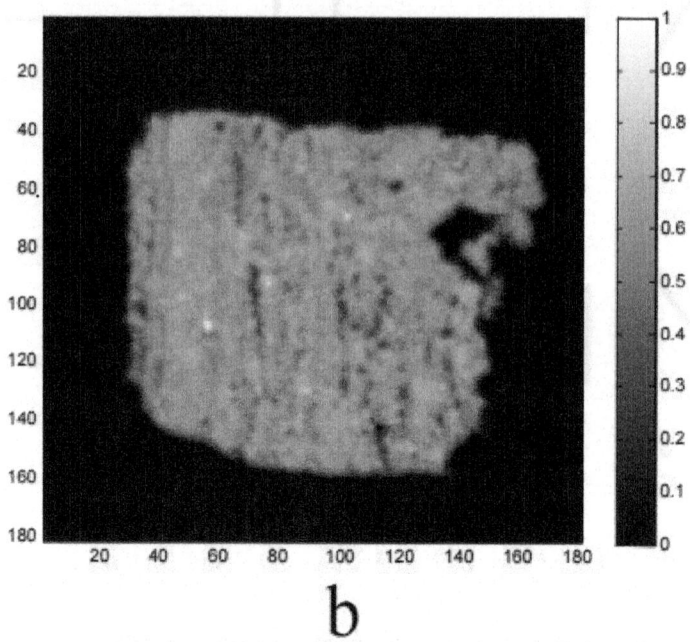

b

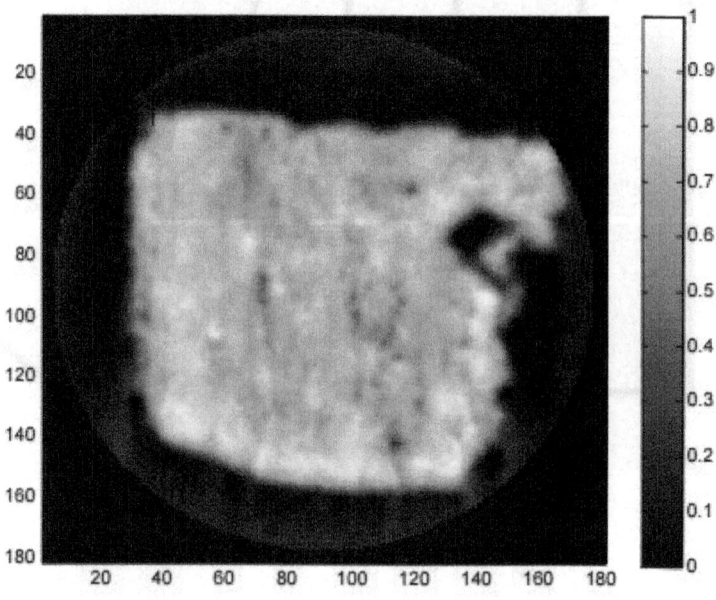

c

Figure 8: Comparison between the agricultural soil reconstructed images for the visualization of bulk information. a. Original projections, b. Projections filtered with the SRUKF and the ANN and c. Projections filtered with linear estimation.

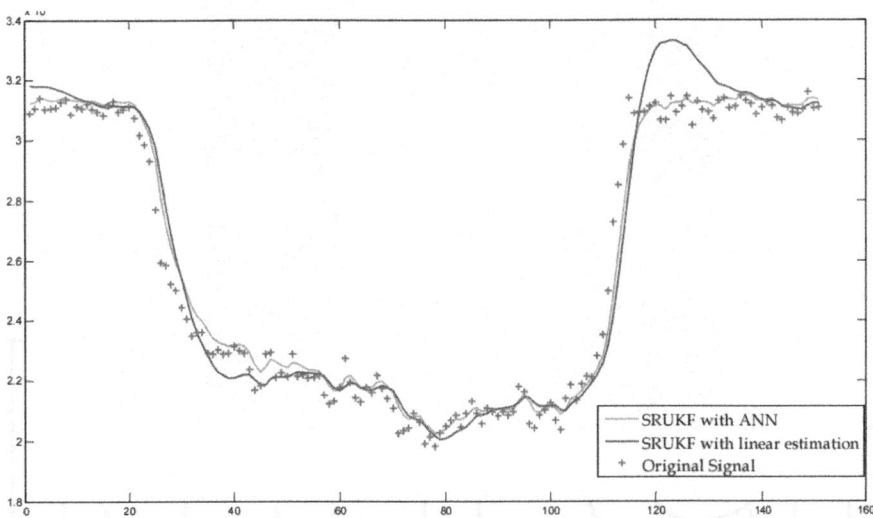

Figure 9: Comparison between projections of the soil bulk sample.

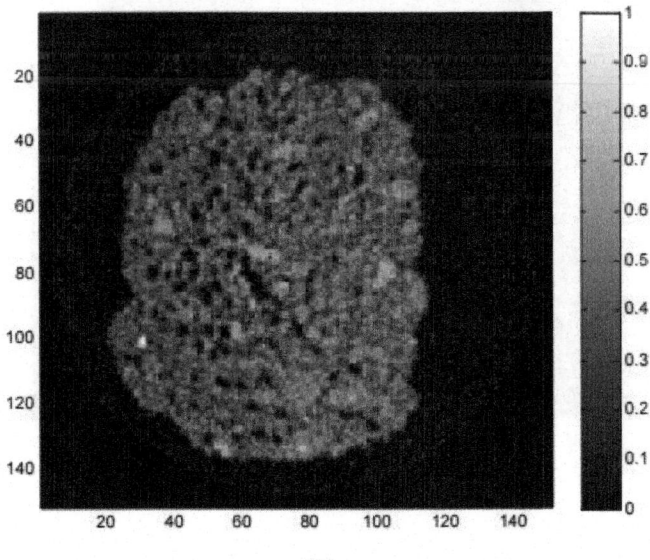

a

274 Mechanisms of Soil Stabilization

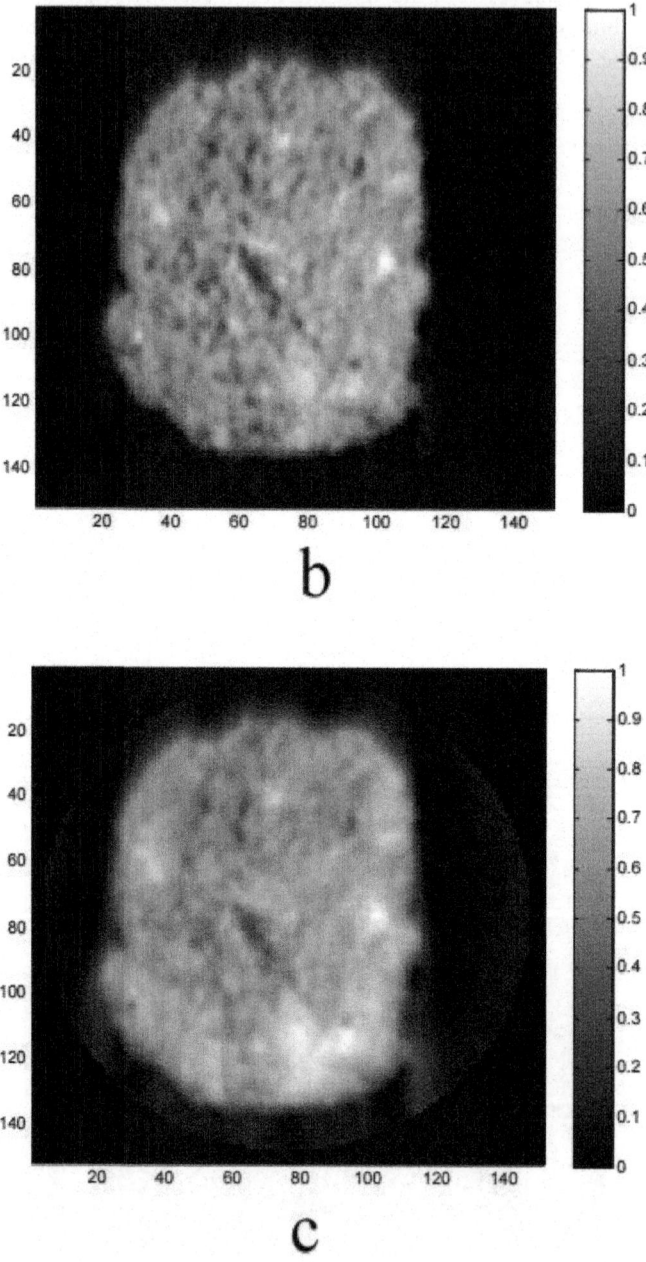

Figure 10: Comparison between the agricultural soil reconstructed images for the visualization of bulk information. a. Original projections, b. Projected filtered projections with the SRUKF and the ANN and c. Projections filtered with linear estimation.

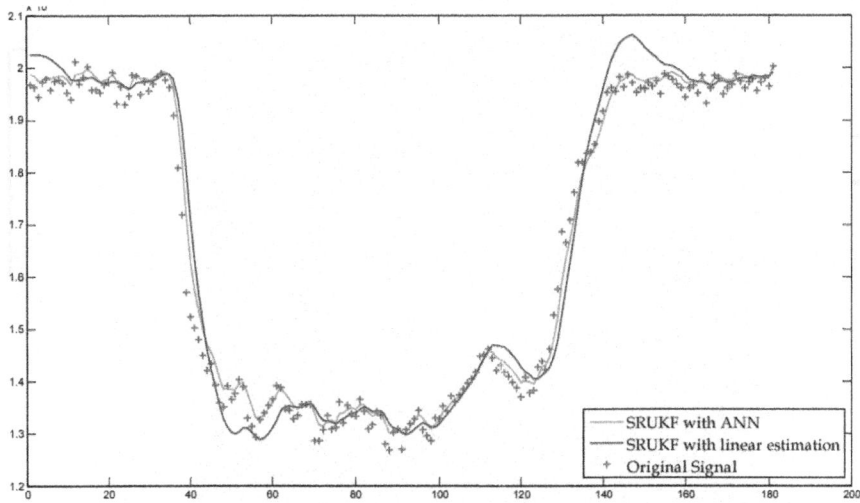

Figure 11: Comparison between the projections of degraded soil samples.

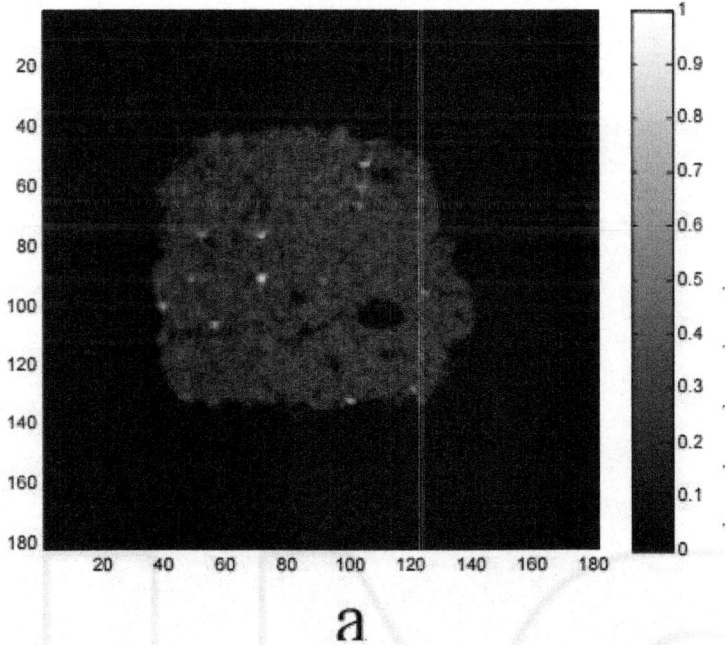

a

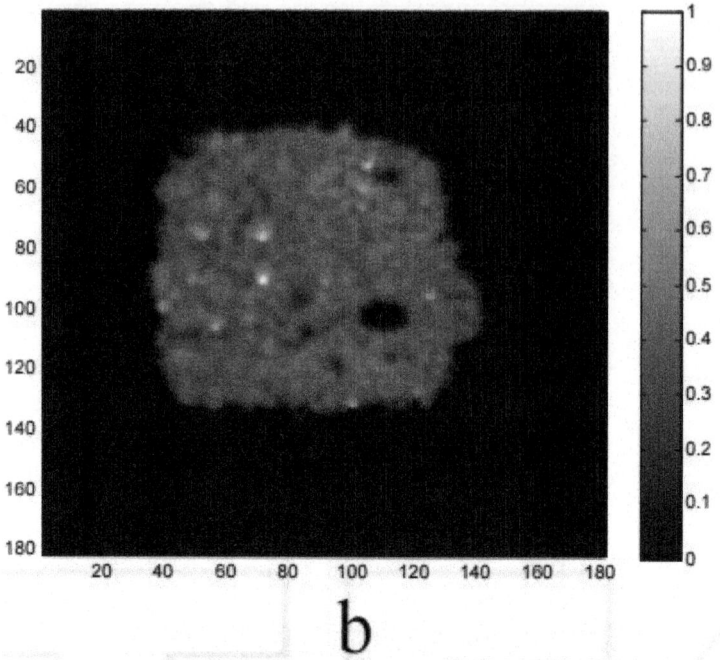

b

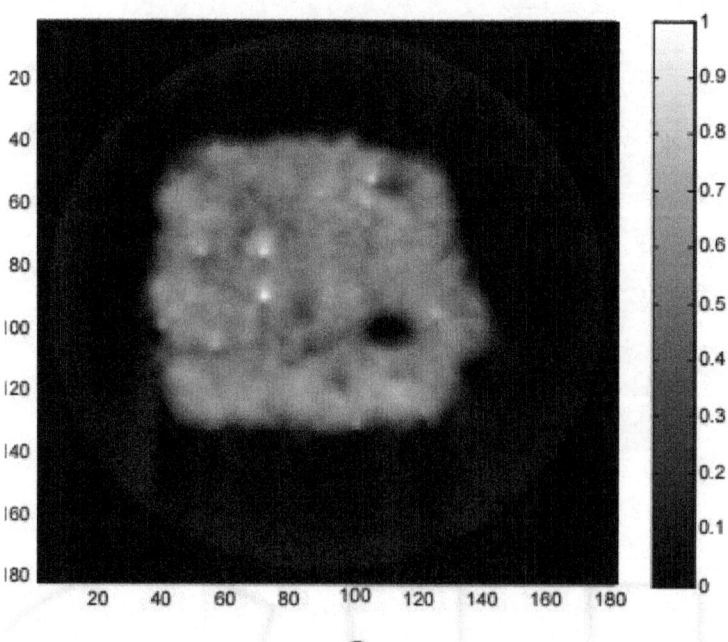

c

Figure 12: Comparison between the reconstructed images of degraded soil sample. a. Original projections, b. Projections filtered with the SRUKF and the ANN and c. Projections filtered with linear estimation.

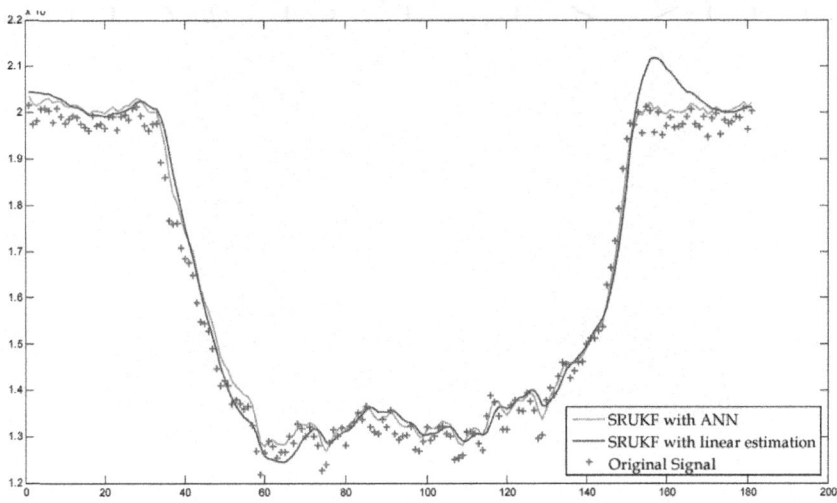

Figure 13: Comparison between the degraded projections of a soil sample.

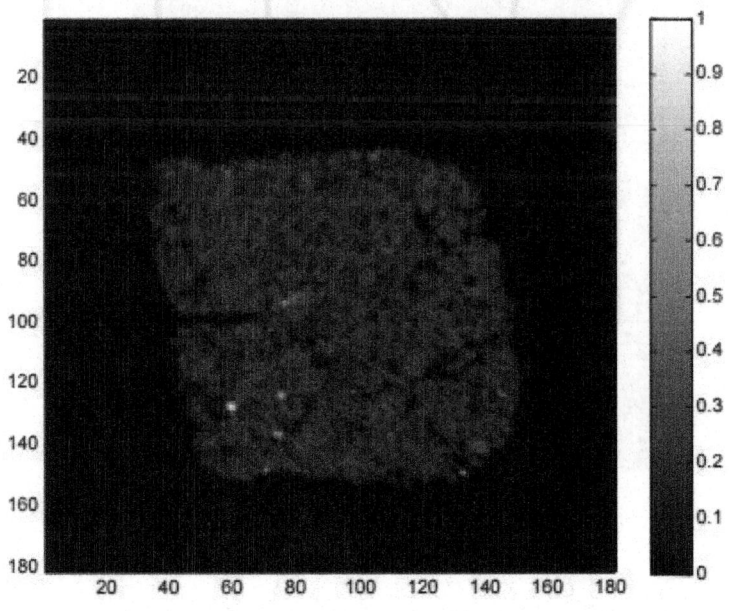

a

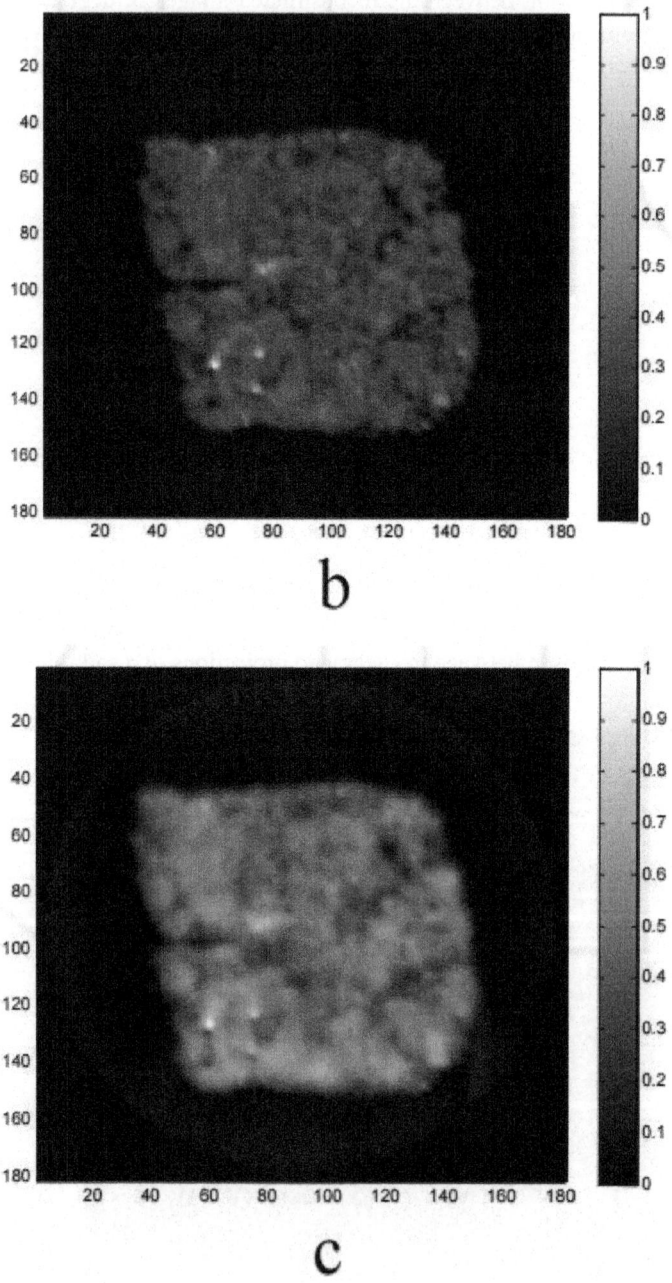

Figure 14: Comparison between the reconstructed images: a. Original projections, b. Projections filtered with the SRUKF and the ANN and c. Projections filtered with linear estimation.

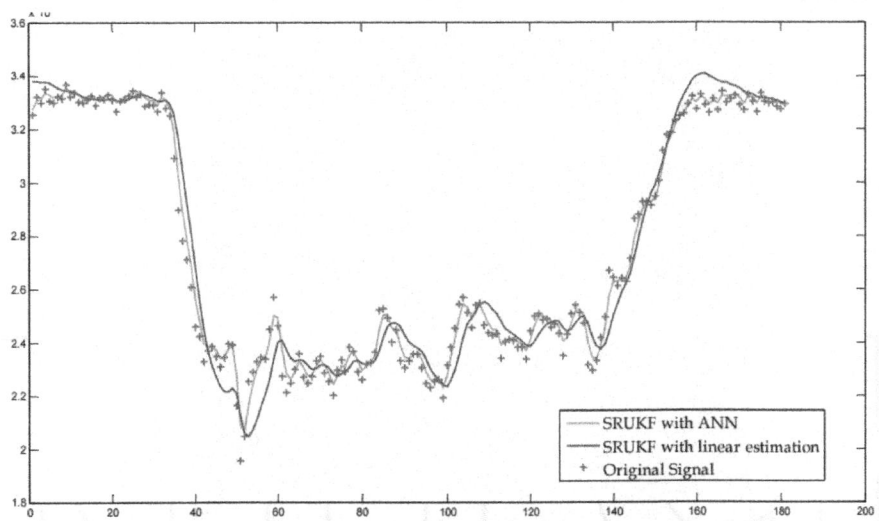

Figure 15: Comparison between the projections of cemented soil samples.

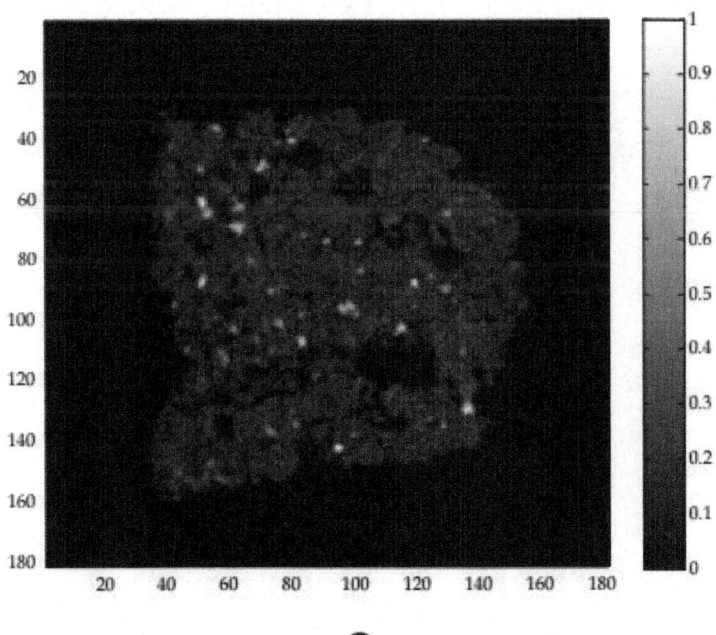

a

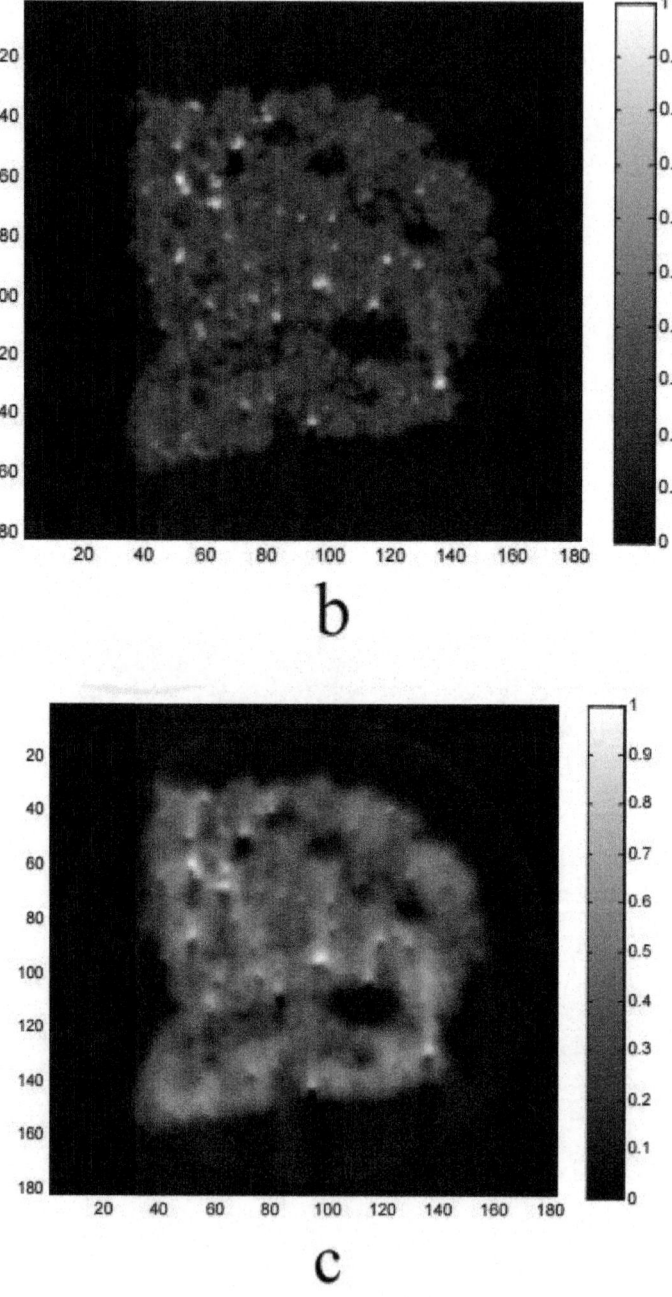

Figure 16: Comparison between the reconstructed images of cemented soil sample: a. Original projections, b. Projections filtered with the SRUKF and the ANN and c. Projections filtered with linear estimation.

In the signal comparisons it is possible to see the estimation errors. These errors in linear estimation promote false artifacts and excessive anti-aliasing. Also, the excessive filtering process with linear estimation has promoted losses in the images details. The SRKUF with the ANN eliminates any noise in a dynamic way and, thus, there is a higher predominance of low values of photon counting. Every single detail is preserved in the signal filtered with the SRUKF.

The signal filtered with the ANN algorithm presents the details better, a precise correction and the interest object (the porosity of soil bulk) is preserved while the signal filtered with the linear estimation promotes excessive anti-aliasing and detail loss. The false porosity (due to granularity of Poisson noise) has been eliminated in the images as it can be observed in the outer parts of the sample. In the image obtained by the filtering with the ANN, the contrast has been preserved as well as small-pointed (critical) elements in the image.

The low-contrast in this image is changed due to the false estimation in the projections like a high value estimated by filter. The single elements are either preserved or changed a little due to the contrast. With the linear estimation, the excessive filtering provoked a decrease in the contrast, which lead to the elimination of pores and high-contrastive elements. The outlines of the objects in the image are also lost when there is an excessive smoothing.

Another important factor is the separation between porosity and granularity of the images. The granularity can generate false micro-pores or fake elements. On the other hand, excessive smoothing can hide the pores and the major elements in the soil composition

CONCLUSIONS

The Kalman filtering that uses linear estimation promotes a smoothing of values and do respect the nature of data distribution (linear attenuation coefficients), which present a uniform distribution, while the filter is limited to working with a Gaussian process. By analyzing the results presented in Figures 4, 5, 6, 7, 8 and 9, it was possible to observe smoothed projections and estimation errors, while Figures 10, 11, 12, 13, 14 and 15 presented losses of important details, such as micro pores and other important elements of the soil, which enables to characterize it.

Nowadays, with the use of artificial neural networks, the results already show a better transition among the variables for the estimation process. Besides mapping the behavior of the sample data, the nonlinear function also makes the necessary transformation of the process uncertainties, which helps to preserve a greater number of details on an original image.

Besides, the regular presence of artifacts or distortions on the image were due to three factors: the limitation of the reconstruction algorithm because of the ramp filter, which is necessary to observe the contrast of different materials and soil porosity; the inaccurate choice of the noise variances because of the different resolutions of the samples obtained and due the equipment noise during data collection.

A better measurement of the noise variations in the detector should be made in real time. The closer the variance value is to the real value, the less it is prone to reconstruction errors or smoothing. Tests conducted earlier have proved that the increase of neurons, layers or number of states has not affected the filter efficiency but the processing time was longer. More recent researches involve the implementation of the filters shown in this study in an embedded system in order to ensure better results when filter variables are fed directly according to the state of the CT scanners.

ACKNOWLEDGMENT

This work was supported in part by the National Council for Research and Development (CNPq) under Grant 306988/2007-0 and Brazilian Enterprise for Agricultural Research (Embrapa) under Grant 03.10.05.011.00.01. Also we acknowledge the Coordination for the Improvement of Higher Education (CAPES, process 573963/2008-9 and 08/57870-9).

REFERENCES

1. Anscombe, F. J. (1948). The Transformation of Poisson, Binomial, Negative-Binomia Data. Vol. 15, in Biometrika. 1948.
2. Crestana, S. (1985). A Tomografia Computadorizada com um novo método para estudos da física da água no solo, São Carlos, USP, 140 p., 1985.
3. Crestana, S. & Nielsen, D. R. (1990). Investigações não-destrutivas de sistemas porosos multifásicos através de microtomografia de raios-X, gama e ressonância magnética nuclear (RMN). In: Encontro nacional sobre escoamento em meios porosos-enemp, 18., 1990.
4. Cruvinel, P. E. (1987). Minitomógrafo de raios X e raios gama computadorizado para aplicações multidisciplinares. Campinas, UNICAMP, 329 p., 1987.
5. Cruvinel, P. E.; Cesareo, R.;Crestana, S.;Mascarenhas, S. (1990). X and ⊣-ray computerized minitomograph scanner for soil science, IEEE Transactions on Instrumentation and Measurement, V.39, N.5, p.745-750, October, 1990.

6. Cruvinel, P. E. & Balogun, Fatai A. (2006). Tomógrafo de espalhamento Compton para medidas agrícolas. In Engenharia Agrícola; vol. 26 n.1 p. 151-160, April 2006.
7. Cruvinel, P. E. ; Pereira , M. F. L. ; Saito, J H ; Costa, L F . Performance Improvement of Tomographic Image Reconstruction Based on DSP Processors. IEEE Transactions on Instrumentation and Measurement, v. 58, p. 3295-3304, 2009.
8. Deremack, E. & Crowe, D. G. (1984), Optical Radiation Detectors. John Wiley & Sons, Inc., 1984.
9. Duerinckx, A. J., & Macocski, A. (1978), Polycromatic Streak Artifacts in Computed Tomography Images. In J. Comput. Assist. Tomogr., 1978.
10. Granato, L. F. (1998). Algoritmo adaptativo para a melhoria em imagens tomográficas obtidas em múltiplas energias, São Carlos, UFSCar, 135 p., 1998.
11. Hainsworth, J. M. & Aylmore, L. A. G. (1983). The use of the computed-assisted tomography to determine spatial distribution of soil water content, Aust. Journal Soil Res., N. 21, p.1435-1443,1983.
12. Haykin, S. (2001). Kalman Filtering and Neural Networks, John Wiley & Sons, Inc.,ISBN 978- 0471369981,USA
13. Hender, W. R. (1983). The physical principals of computed tomography, USA, 1983.
14. Ibbott, G. S. (1980),Radiation therapy treatment planning and the distortion of CT images. Med. Phys, 1980.
15. Joseph, P. M. & Spital, R. D. (1978). A method for correction bone-induced artifacts of CT scanners. 1978: 100–108.
16. Julier, S. J. & Uhlmann, J. K.. (1997). A new extension of Kalman filter to nonlinear systems. Symp. Aerospace/Defense Sensing, Simul. and Controls, 1997.
17. Kalman, R. E. (1960). A new approach to Linear Filtering and Prediction Problems. Transaction of the ASME - Journal of basic Engineering, 1960.
18. Laia, M A. M.; Cruvinel, P. E. & Levada, A. L. M. (2007). Filtragem de projeções tomográficas da ciência do solo utilizando transformada de Anscombe e Kalman, DINCON'07, São José do Rio Preto, 2007.
19. Laia, M. A. M. & Cruvinel, P. E. (2008a), Filtragem de projeções tomográficas utilizando Kalman Discreto e Redes Neurais, IEEE América Latina,vol. 6, ed. 1, march, 2008.
20. Laia, M. A. M.; Levada, A. L. M.; Botega, L. C.; Cruvinel, P. E.; Pereira, M. F. L. & Macedo, A. (2008b). A Novel Model for Combining Projection

and Image Filtering Using Kalman and Discrete Wavelet Transform in Computerized Tomography. In: 2008 11th IEEE International Conference on Computational Science and Engineering, 2008, São Paulo. 2008. 11th IEEE International Conference on Computational Science and Engineering, 2008. v. 11. p. 219-226.

21. Laia, M. A. M.& Cruvinel, P. E. (2009). Applying an Improved Square Root Unscented Kalman Filtering in Tomographic Projections of Agricultural Soil Samples. In: The 3rd Southern Conference on Computational Modeling (3MCSUL), 2009, Rio Grande - RS. The 3rd Southern Conference on Computational Modeling (3MCSUL),2009.v. 1.

22. Laia, M. A. M.& Cruvinel, P. E. (2011). Evaluation of an embedded unscented Kalman filter for soil tomography. In: 1a. Conferência de Sistemas embarcados Críticos. (CBSEC) São Carlos – SP. 2011.

23. Macedo, A. (1997). Construção e uso de um tomógrafo com resolução micrométrica para aplicações em ciências do solo e do ambiente. São Carlos: USP-EESC, 1997. 129p.

24. Mascarenhas, N. D. A.; Santos, C. A. N. & Cruvinel, P. E. (1999). Transmission Tomography Under Poisson Noise Using the Anscombe Transformation and Wiener Filtering of Projections. Nuclear Instruments and Methods in Physics Research Section A, 1999.

25. Mcculloch, W. S. & Pitts, W. (1947), A logical calculus of the ideas immanent in Nervous Activity. Bulletin of mathematical biophysics, 1947.

26. Minatel, E. R. (1997). Desenvolvimento de Algoritmo para Recontrução e Visualização Tridimensional de Imagens Tomográficas com uso de Técnicas Frequenciais e Wavelets. São Carlos-SP: UFSCar, 1997.

27. Pereira, G. A. S. (2000).Filtro de Kalman: Teoria e Aplicações. Centro de Pesquisa e Desenvovilmento em Engenharia Elétrica., 2000.

28. Petrovic, M.; Siebert, J. E. & Rieke, P. E. (1982). Soil bulk analysis in three dimensions by computed tomographic scanning, Soil Sci. Soc. Am. J., n.46, p.445-450, 1982.

29. Naime, J.M. (1994). Projeto e construção de um tomógrafo portátil para estudos de ciência do solo e planta, em campo. São Carlos: USP-EESC, 1994. 87p. il..

30. van der Merve, R. & Wan, E. A. (2001). The square-root unscented Kalman Filter for state and parameter-estimation. Acoustics, Speech, and Signal Processing, 2001. Proceedings. (ICASSP '01). 2001 IEEE International Conference on, 2001: 3461-3464.

31. Venturini, Y. R. (1995).Análise quantitativa da qualidade de imagens digitais com o uso de espectro de Wiener. São Carlos – SP, UFSCar, 1995.
32. Wan, E. A. & Merwe, R. (2000). The Unscented Kalman Filter for Signal Processing. Proceedings of Symposium 2000 on Adaptive Systems for Signal Processing, Communication and Control, 2000, october ed.
33. Welch, G. & Bishop, G. (2004). An Introduction to the Kalman Filter. Chapel Hill: Departament of Computer Science, University of North Carolina, 2004.
34. Ziel, A. D. (1976). Noise in measurements. John Wiley & Sons, inc, 1976.

CITATION

CHAPTER 1
Ali Ateş, "The Effect of Polymer-Cement Stabilization on the Unconfined Compressive Strength of Liquefiable Soils," International Journal of Polymer Science, vol. 2013, Article ID 356214, 8 pages, 2013. doi:10.1155/2013/356214.

CHAPTER 2
Shih-Hao Jien 1, Chung-Chi Wang, Chia-Hsing Lee and Tsung-Yu Lee, Stabilization of Organic Matter by Biochar Application in Compost-amended Soils with Contrasting pH Values and Textures, doi:10.3390/su71013317.

CHAPTER 3
B.Suneel Kumar & T.V.Preethi, Behavior of Clayey Soil Stabilized with Rice Husk Ash & Lime, ISSN: 2231-5381.

CHAPTER 4
H. Ismaiel, "Cement Kiln Dust Chemical Stabilization of Expansive Soil Exposed at El-Kawther Quarter, Sohag Region, Egypt," International Journal of Geosciences, Vol. 4 No. 10, 2013, pp. 1416-1424. doi: 10.4236/ijg.2013.410139.

CHAPTER 5
E. Saljnikov and D. Cakmak (2011). Phosphorus: Chemism and Interactions, Principles, Application and Assessment in Soil Science, Dr. Burcu E. Ozkaraova

Gungor (Ed.), ISBN: 978-953-307-740-6, InTech, DOI: 10.5772/29556.

CHAPTER 6

Michael A. Blazier, Hal O. Liechty, Lewis A. Gaston and Keith Ellum (2011). Poultry Litter Fertilization Impacts on Soil, Plant, and Water Characteristics in Loblolly Pine (Pinus taeda L.) Plantations and Silvopastures in the Mid-South USA, Principles, Application and Assessment in Soil Science, Dr. Burcu E. Ozkaraova Gungor (Ed.), ISBN: 978-953-307-740-6, InTech, DOI: 10.5772/29356.

CHAPTER 7

T. Watanabe, M. S. H. Khan, I. M. Rao, J. Wasaki, T. Shinano, M. Ishitani, H. Koyama, S. Ishikawa, K. Tawaraya, M. Nanamori, N. Ueki and T. Wagatsuma (2011). Physiological and Biochemical Mechanisms of Plant Adaptation to Low-Fertility Acid Soils of the Tropics: The Case of Brachiariagrasses, Principles, Application and Assessment in Soil Science, Dr. Burcu E. Ozkaraova Gungor (Ed.), ISBN: 978-953-307-740-6, InTech, DOI: 10.5772/30334.

CHAPTER 8

Johan van Tol, Pieter Le Roux and Malcolm Hensley (2011). Soil Indicators of Hillslope Hydrology, Principles, Application and Assessment in Soil Science, Dr. Burcu E. Ozkaraova Gungor (Ed.), ISBN: 978-953-307-740-6, InTech, DOI: 10.5772/28724.

CHAPTER 9

Marcelo Muniz Benedetti, Nilton Curi, Gerd Sparovek, Amaury de Carvalho Filho and Sérgio Henrique Godinho Silva (2011). Updated Brazilian's Georeferenced Soil Database – An Improvement for International Scientific Information Exchanging, Principles, Application and Assessment in Soil Science, Dr. Burcu E. Ozkaraova Gungor (Ed.), ISBN: 978-953-307-740-6, InTech, DOI: 10.5772/29627.

CHAPTER 10

Marcos A. M. Laia and Paulo E. Cruvinel (2011). An Application Approach to Kalman Filter and CT Scanners for Soil Science, Principles, Application and Assessment in Soil Science, Dr. Burcu E. Ozkaraova Gungor (Ed.), ISBN: 978-953-307-740-6, InTech, DOI: 10.5772/31030.

INDEX

A
Artificial neural network (ANN) 249

B
Best management practices (BMPs) 33

C
Calcium aluminosilicate hydrates (CASH) 65
Calcium aluminum hydrate (CAH) 61
Calcium silicate hydrate (CSH) 61
California bearing ratio (CBR) 41, 44
California Bearing Ratio (CBR) 41
Cation exchange capacity (CEC) 22
Cement kiln dust (CKD) 51
Computerized tomography (CT) 247

D
Diethylenetriaminepentaacetic acid (DTPA) 72, 90
Discrete Kalman filter (DKF) 256
Drained upper limit (DUL) 188, 190
Drained Upper Limit (DUL) 200

E
Electrical conductivity (EC) 22
Extended Kalman filter (EKF) 257

H
Hydrology of Soil Types (HOST) 187

L
Least significant difference (LSD) 24

M
Monoammonium phosphate (MAP) 71

N
Nuclear Magnetic Ressonance (NMR) 247

O
Optimum moisture content (OMC) 13, 41, 44

P
Permanent Wilting Point (PWP) 200
Phosphoenolpyruvate carboxylase (PEPC) 144
Phosphoenolpyruvate/phosphate translocator (PPT) 162
Plant Available Water (PAW) 200
Plasticity index (PI) 57, 62
Pyruvate kinase (PK) 144

R

Readily-oxidizable carbon (ROC) 22
Rice Husk Ash (RHA) 39, 41, 43, 48

S

Scanning electron microscope (SEM) 51, 57, 64, 65
Sodium dodecyl sulfate (SDS) 147
Soil organic matter (SOM) 20
Soil's Conservation and Survey National Service\" (SNLCS) 219
Square-Root Unscented Kalman Filter (SRUKF) 261
Sucrose phosphate synthase (SPS) 157
Sulfoquinovosyl diacylglycerol (SQDG) 160
Switchgrass silvopasture (SILVO) 119

T

Total Kjeldahl phosphorus (TKP) 114

Triose phosphate translocater (TPT) 162
Triose-phosphate translocator (TPT) 159

U

Ultrasonic pulse velocity (UPV) 7
Unconfined compressive strength (UCS) 41
Unsatured Soil Hydraulic Database (UNSODA) 219
Unscented Kalman filter (UKF) 248, 257

V

Volatile matter (VM) 31

X

X-ray fluorescence (XRF) 56